A History of Science in Society

A History of Science in Society

From Philosophy to Utility

Third Edition

Andrew Ede and Lesley B. Cormack

UNIVERSITY OF TORONTO PRESS

utppublishing.com

Library and Archives Canada Cataloguing in Publication

Ede, Andrew, author

A history of science in society : from philosophy to utility / by Andrew Ede and Lesley B. Cormack.—Third edition.

Includes bibliographical references and index.

Issued in print and electronic formats.

ISBN 978-1-4426-3500-5 (hardcover).—ISBN 978-1-4426-3499-2 (paperback).—ISBN 978-1-4426-3501-2 (html). —ISBN 978-1-4426-3502-9 (pdf) .

1. Science—History. 2. Science—Social aspects—History. I. Cormack, Lesley B., 1957–, author II. Title.

Q125.E33 2016   509   C2016-901619-6
                                            C2016-901620-X

We welcome comments and suggestions regarding any aspect of our publications—please feel free to contact us at news@utphighereducation.com or visit our Internet site at www.utppublishing.com.

North America

5201 Dufferin Street

North York, Ontario, Canada, M3H 5T8

2250 Military Road

Tonawanda, New York, USA, 14150

ORDERS PHONE: 1-800-565-9523

ORDERS FAX: 1-800-221-9985

ORDERS E-MAIL: utpbooks@utpress.utoronto.ca

UK, Ireland, and continental Europe

5201 Dufferin Street

Estover Road, Plymouth, PL6 7PY, UK

ORDERS PHONE: 44 (0) 1752 202301

ORDERS FAX: 44 (0) 1752 202333

ORDERS E-MAIL: enquiries@nbninternational.com

Every effort has been made to contact copyright holders; in the event of an error or omission, please notify the publisher.

The University of Toronto Press acknowledges the financial support for its publishing activities of the Government of Canada through the Canada Book Fund.

Printed in the United States of America.

CONTENTS

ILLUSTRATIONS

Figures

Plates

CONNECTIONS BOXES

ACKNOWLEDGMENTS

To Graham and Quin, who have grown up with this book—and who put up with two authors working in the house at the same time. We would also like to thank those people who helped make this book possible: our editor and publisher; friends and colleagues who read early drafts and gave advice; reviewers and users who have offered helpful criticism and forced us to defend our position; and all the amazing historians of science on whose shoulders (or toes) we stand.

INTRODUCTION

Science has transformed human history. It has changed how we see the universe, how we interact with nature and each other, and how we live our lives. It may, in the future, even change what it means to be human. The history of such a powerful force deserves a full and multifaceted examination. Yet a history of science is unlike a history of monarchs, generals, steam engines, or wars because science isn't a person, an object, or an event. It is an idea, the idea that humans can understand the physical world.

This is a history of what happens when a legion of thinkers, at different times and from different backgrounds, turned their minds and hands to the investigation of nature. In the process, they transformed the world.

The history of science is such a vast subject that no single book about it can really be comprehensive, and so the story we tell examines science from a particular point of view. Some histories of science have focused on the intellectual development of ideas, while others have traced the course of particular subjects such as astronomy or physics. In this book, we have chosen to look at science from two related perspectives that we believe offer a window onto the historical processes that shaped the study of nature. First, we have examined the link between the philosophical pursuit of knowledge and the desire of both the researchers and their supporters to make that knowledge useful. There has always been a tension between the intellectual aspects of science and the application of scientific knowledge. The ancient Greek philosophers struggled with this problem, and it is still being debated today. The call in every age by philosophers and scientists for more support for "research for its own sake" is indicative of the tension between the search for knowledge and the pressure to apply that knowledge. What counts as useful knowledge differed from patron to patron and society to society, so that the Grand Duke Cosimo de' Medici and the United States Department of Energy looked for quite different "products" to be created by their clients, but both traded support for the potential of utility.

The tension between intellectual pursuits and demands for some kind of product not only was felt by many natural philosophers and scientists but has also led to controversy among historians of science. Where does science end and technology begin? they have asked. Perhaps the most famous articulation of this is the "scholar and craftsman debate." Historians of science have tried to understand the relationship between those people primarily interested in the utility of knowledge (the craftsmen) and those interested in the intellectual understanding of the world (the scholars). Some historians have denied the connection, but we feel it is integral to the pursuit of natural knowledge. The geographers of the early modern period provide a good example of the necessity of this interconnection. They brought the skills of the navigator together with the abstract knowledge of the mathematician. Translating the spherical Earth onto flat maps was an intellectual challenge, while tramping to the four corners of the globe to take measurements was an extreme physical challenge. Getting theory and practice right could mean the difference between profit or loss, or even life and death.

Our second aim has been to trace the history of science by its social place. Science does not exist in disembodied minds, but is part of living, breathing society. It is embedded in institutions such as schools, princely courts, government departments, and even in the training of soldiers. As such, we have tried to relate scientific work to the society in which it took place, tracing the interplay of social interest with personal interest. This has guided our areas of emphasis so that, for example, we give alchemy a greater allocation of space than some other histories of science because it was more socially significant than topics such as astronomy or physics in the same period. There were far more alchemists than astronomers, and they came from all ranks and classes of people, from peasants to popes. In the longer term, the transformation of alchemy into chemistry had a very great impact on the quality of everyday life. This is not to say that we neglect astronomy or physics, but rather that we have tried to focus on what was important to the people of the era and to avoid projecting the importance of later work on earlier ages.

In each chapter, we have highlighted one aspect of this interaction of science and society, from politics and religion to economics and warfare, under the heading

"Connections." While each of these vignettes is part of the larger narrative of the book, they can also be read as individual case studies.

It is from the two perspectives of utility and social place that our subtitle comes. As we began to look at the work of natural philosophers and scientists over more than 2,000 years, we found ourselves more and more struck by the consistency of the issue of the utility of knowledge. Plato disdained the utility of knowledge, but he promoted an understanding of geometry. Eratosthenes used geometry to measure the diameter of the Earth, which had many practical applications. In the modern era, we have seen many cases of scientific work unexpectedly turned into consumer goods. The cathode ray tube, for instance, was a device created to study the nature of matter, but it ended up in the heart of the modern television. Philosophers and scientists have always walked a fine line between the role of intellectual and the role of technician. Too far to the technical side and a person will appear to be an artisan and lose their status as an intellectual. Too far to the intellectual side, a person will have trouble finding support because they have little to offer potential patrons.

Although the tension over philosophy and utility has always existed for the community of researchers, we did not subtitle our book "Philosophy *and* Utility." This is because the internal tension was not the only aspect of philosophy and utility that we saw over time. Natural philosophy started as an esoteric subject studied by a small, often very elite, group of people. Their work was intellectually important but had limited impact on the wider society. Over time, the number of people interested in natural philosophy grew, and as the community grew, so too did the claims of researchers that what they were doing would benefit society. Through the early modern and modern eras, scientists increasingly promoted their work on the basis of its potential utility, whether as a cure for cancer or as a better way to cook food. And, in large part, the utility of science has been graphically demonstrated in everything from the production of colour-fast dyes to the destruction of whole cities with a single bomb. We have come to expect science to produce things we can use, and, further, we need scientifically trained people to keep our complex systems working—everything from testing the purity of our drinking water to teaching science in school. Our subtitle reflects the changing social expectation of science.

We have also made some choices about material based on the need for brevity. This book could not include all historical aspects of all topics in science or even introduce all the disciplines in science. We picked examples that illustrate key events and ideas rather than give exhaustive detail. For instance, the limited amount of medical history we include looks primarily at examples from medicine that treated the body as an object of research and thus as part of a larger intellectual movement in natural philosophy. We also chose to focus primarily on Western developments in natural philosophy and science, although we tried to acknowledge that natural philosophy existed in other places as well and that Western science did not develop in isolation. Especially in the early periods, Western thinkers were absorbing ideas, materials, and information from a wide variety of sources. By the seventeenth and eighteenth centuries, Western scholars were interacting with other cultures and exchanging information, although not on an equal footing. In later periods, Western science became a powerful tool for modernization and internationalization of countries around the world. *A History of Science* tells a particular—and important—story about the development of this powerful part of human culture, which has and continues to transform all our lives. To study the history of science is to study one of the great threads in the cloth of human history.

CHAPTER TIMELINE

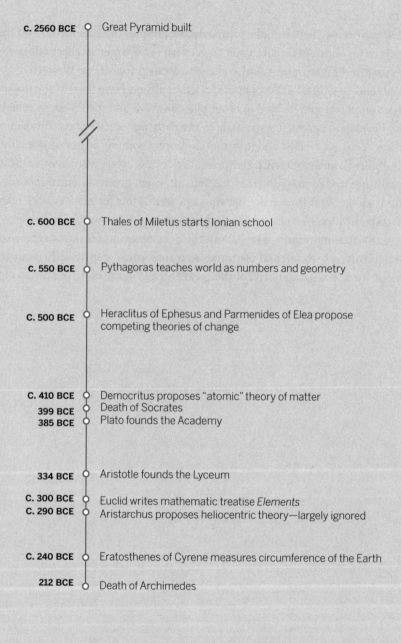

C. 2560 BCE	Great Pyramid built
C. 600 BCE	Thales of Miletus starts Ionian school
C. 550 BCE	Pythagoras teaches world as numbers and geometry
C. 500 BCE	Heraclitus of Ephesus and Parmenides of Elea propose competing theories of change
C. 410 BCE	Democritus proposes "atomic" theory of matter
399 BCE	Death of Socrates
385 BCE	Plato founds the Academy
334 BCE	Aristotle founds the Lyceum
C. 300 BCE	Euclid writes mathematic treatise *Elements*
C. 290 BCE	Aristarchus proposes heliocentric theory—largely ignored
C. 240 BCE	Eratosthenes of Cyrene measures circumference of the Earth
212 BCE	Death of Archimedes

THE ORIGINS OF NATURAL PHILOSOPHY

1

The roots of modern science are found in the heritage of natural philosophy created by a small group of ancient Greek philosophers. The voyage from the Greeks to the modern world was a convoluted one, and natural philosophy was transformed by the cultures that explored and re-explored the foundational ideas of those Greek thinkers. Despite intellectual and practical challenges, the Greek conceptions of how to think about the world and how the universe worked remained at the heart of any investigation of nature in Europe and the Middle East for almost 2,000 years. Even when natural philosophers began to reject the conclusions of the Greek philosophers, the rejection itself still carried with it the form and concerns of Greek philosophy. Today, when virtually nothing of Greek method or conclusions about the physical world remains, the philosophical concerns about how to understand what we think we know about the universe still echo in our modern version of natural philosophy.

To understand why Greek natural philosophy was such an astounding achievement, we must consider the conditions that led to the creation of a philosophy of nature. Since the earliest times of human activity, the observation of nature has been a key to human survival. Knowledge of everything—from which plants are edible to where babies come from—was part of the knowledge acquired and passed down through the generations. In addition to practical knowledge useful for daily life, humans worked to understand the nature of existence and encapsulated their knowledge and conclusions in a framework of mytho-poetic stories.

Humans have always wanted to know more than just what is in the world; they want to know why the world is the way it is.

Early Civilizations and the Development of Knowledge

With the rise of agriculture and the development of urban civilization, the types of knowledge about nature were diversified as new skills were created. There arose four great cradles of civilization along the river systems of the Nile, the Tigris-Euphrates, the Indus-Ganges, and the Yellow. They shared the common characteristic of a large river that was navigable over a long distance and that flooded the region on a periodic basis. The Nile in particular flooded so regularly that its rise and fall was part of the timekeeping of the Egyptians. These floods renewed the soil, and the lands in temperate to subtropical zones were (and are) agriculturally abundant, providing food to support large populations.

A growing group of people were freed from farm work by the surplus the land provided. These people were the artisans, soldiers, priests, nobles, and bureaucrats who could turn their efforts to the development and running of an empire. The mastery of these skills required increasingly longer periods of study and practice. Artisans required apprenticeships to acquire and master their arts, while the priest class took years to learn the doctrine and methods of correct observance. The military and ruling classes required training from childhood to grow proficient in their duties. Because the empires were long-lasting, especially the Egyptian empire, the rulers planned for the long term, thinking not just about the present season but about the years ahead and even generations into the future. Thus, these civilizations could take on major building projects such as the Great Wall of China or the Great Pyramid of Giza.

In addition to the obvious agricultural and economic advantage provided by the rivers, they had a number of subtle effects on the intellectual development of ancient civilizations. Dealing with large-scale agricultural production required counting and measurement of length, weight, area, and volume, and that led to accounting skills and record-keeping. Agriculture and religion were intertwined, and both depended on timekeeping to organize activities necessary for worship and production, which in turn led to astronomical observation and calendars. As these societies moved from villages to regional kingdoms and finally became empires, record-keeping exceeded what could be left to memory. Writing and accounting developed to deal with the problems of remembering and recording the myriad activities of complex religions, government bureaucracies, and the decisions of judges at courts of law.

Another aspect of intellectual development that came from the periodic flooding had to do with the loss of local landmarks, so skills of surveying were developed. Rather than setting the boundaries of land by objects such as trees or rocks, which changed with every inundation, the land was measured from objects unaffected by the flooding. In addition to the practical skills of land measurement, surveying also introduced concepts of geometry and the use of level and angle measuring devices. These were then used for building projects such as irrigation systems, canals, and large buildings. In turn, surveying tools were closely related to the tools used for navigation and astronomy.

These kinds of practical skills contributed to a conception of the world based on abstract models. In other words, counting cattle contributed to the concept of arithmetic as a subject that could be taught independent of any actual object to be counted. Similarly, getting from place to place by boat led to the development of navigation. The skill of navigation started as local knowledge of the place a pilot frequently travelled. While a local pilot was useful, and the world's major ports still employ harbour pilots today, general methods of navigation applicable to circumstances that could not be known in advance were needed as ships sailed into unknown waters. The skill of navigation was turned into abstract ideas about position in space and time.

The various ancient empires of the four river systems mastered all the skills of observation, record-keeping, measurement, and mathematics that would form the foundation of natural philosophy. While historians have increasingly acknowledged the intellectual debt we owe these civilizations, we do not trace our scientific heritage to the Egyptians, Babylonians, Indians, or Chinese. Part of the reason for this is simply chauvinism. Science was largely a European creation, so there was a preference for beginning the heritage of natural philosophy with European sources rather than African or Asian ones.

There is, however, a more profound reason to start natural philosophy with the Greeks rather than the older cultures, despite their many accomplishments. Although these older cultures had technical knowledge, keen observational skills, and vast resources of material and information, they failed to create natural philosophy because they did not separate the natural world from the supernatural world. The religions of the old empires were predicated on the belief that the material world was controlled and inhabited by supernatural beings and forces, and that the reason for the behaviour of these supernatural forces was largely unknowable. Although there were many technical developments in the societies of the four river cultures, the intellectual heritage was dominated by the priests, and their interest in the material world was an extension of their concepts of theology. Many ancient civilizations, such as the Egyptian, Babylonian, and Aztec empires, expended a large proportion of social capital (covering such things as the time, wealth, skill, and public space of the society) on religious activity. The Great Pyramid, built as the tomb for the Pharaoh Khufu (also known as Cheops), rises 148 metres above the plain of Giza and is the largest of the pyramids. It is an astonishing engineering feat and tells us a great deal about the power and technical skills of the people who built it. But the pyramids also tell us about a society that was so concerned about death and the afterlife that its whole focus could be on the building of a giant tomb.

The very power of the four river centres may have worked against a change in intellectual activity. Social stratification and rigid class structure kept people in narrowly defined occupations. Great wealth meant little need to explore the world or seek material goods from elsewhere since the regions beyond the empire contained little of interest or value compared to what was already there. Although it was less true of the civilizations along the Indus-Ganges and Tigris-Euphrates river systems, which were more affected by political instability and invasions, both the Egyptian and Chinese civilizations developed incredibly complex societies with highly trained bureaucracies that grew increasingly insular and inward-looking.

The Greek World

It is impossible to be certain why the Greeks took a different route, but aspects of their life and culture offer some insight. The Greeks were not particularly well-off, especially when compared to their neighbours the Egyptians. Although unified by language and shared heritage, Greek society was not a single political entity but a collection of city-states scattered around the Aegean Sea and the eastern end of the Mediterranean. These city-states were in constant competition with each other in a frequently changing array of partnerships, alliances, and antagonisms. This struggle extended to many facets of life, so that it included not just trade or military competition but also athletic rivalry (highlighted by the athletic and religious festival of the Olympics); the pursuit of cultural superiority by claiming the best poets, playwrights, musicians, artists, and architects; and even intellectual competition as various city-states attracted great thinkers. This pressure to be the best was one of the spurs to exploration that allowed the Greeks to bring home the intellectual and material wealth of the people they encountered.

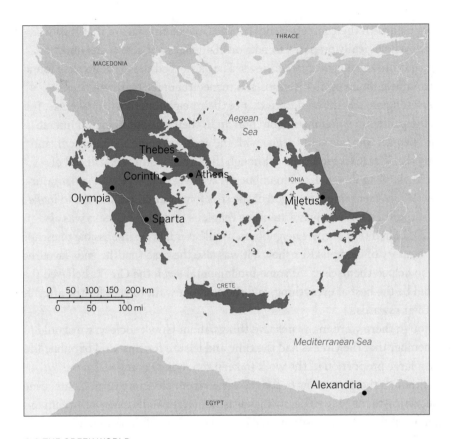

1.1 THE GREEK WORLD

Another factor was the degree to which Greek life was carried out in public. Much of Greek social structure revolved around the marketplace or *agora*. This was not just a place to shop but a constant public forum where political issues were discussed, various medical services were offered, philosophers debated and taught, and the news and material goods of the world was disseminated. The Greeks were a people who actively participated in the governance of the state and were accustomed to debate and discussion of matters of importance as part of the daily course of life. Greek law, while varying from state to state, was often based on the concept of proof rather than the exercise of authority. The public exchange of ideas and demand for individual say in the direction of their political and cultural life gave the Greeks a heritage of intellectual rigour and a tolerance for alternative philosophies. The vast range of governing styles that coexisted in the city-states, from tyranny to democracy, show us a willingness to try new methods of dealing with public issues.

Combined with the competitiveness of the Greeks, this meant that they were not only psychologically prepared to take on challenges but also accustomed to hearing and considering alternative views. They absorbed those things they found useful from neighbouring civilizations and turned them to their own needs.

Greek religion also differed from that of their neighbours. For the Greeks, the gods of the pantheon were much more human in their presentation and interaction with people. Mortals could argue with the gods, compete against them, and even defy them, at least for a time. Although the Greek world was still full of spirits, Greeks were less inclined to imbue every physical object with supernatural qualities. While there might be a god of the seas to whom sailors needed to make offerings, the sea itself was just water. The religious attitude of Greeks was also less fatalistic than that of their neighbours. While it might be impossible to escape fate, as the story of Oedipus Rex shows, it was also the case that the gods favoured those who helped themselves. At some fundamental level, the Greeks believed that they could be the best at everything, and they did not want to wait for the afterlife to gain their rewards.

Although there were many positive things about Greek society, we should also remember that the Greeks had the time and leisure for this kind of public life because a large proportion of the work to keep the society going was done by slaves. Although the conditions of slavery varied from city-state to city-state, even in democratic Athens (where democracy was limited to adult males of Athenian birth), most of the menial positions and even the artisan class were made up of slaves. Those who worked with their hands were at the bottom of the social hierarchy.

Thales to Parmenides: Theories of Matter, Number, and Change

Whether these elements of Greek society and social psychology are sufficient to explain why the Greeks began to separate the natural from the supernatural is difficult to prove. Yet this separation became a central tenet for a line of philosophers who began to appear in Ionia around the sixth century BCE. The most famous of these was Thales of Miletus (c. 624–c. 548 BCE). We know very little about Thales or his work. Most of what comes down to us is in the form of comments by later philosophers. He was thought to have been a merchant, or at least a traveller, who visited Egypt and Mesopotamia where he was supposed to have

learned geometry and astronomy. Thales argued that water was the prime constituent of nature and that all matter was made of water in one of three forms: water,
earth, and mist. He seems to be borrowing from the material conception of the
Egyptians, who also considered earth, water, and air to be the primary constituents
of the material world, but he took it one step further by starting with one element.
Thales pictured the world as a sphere (although it might have been drum-shaped)
that floated on a celestial sea.

Even in this fragmentary record of Thales' philosophy, two things stand out.
First, nature is completely material; there are no hints of supernatural constituent
elements. This does not mean that Thales discarded the gods but rather that he
thought that the universe had a material existence independent of supernatural
beings. The second point is that nature functions of its own accord, not by supernatural intervention. It follows that there are general or universal conditions
governing nature and that those conditions are open to human investigation and
understanding.

Following Thales was his student and disciple Anaximander (c. 610–c. 545 BCE).
Anaximander added fire to the initial three elements and produced a cosmology
based on the Earth at the centre of three rings of fire. These rings were hidden
from view by a perpetual mist, but apertures in the mist allowed their light to
shine through, producing the image of stars, the sun, and the moon. Like Thales,
Anaximander used a mechanical explanation to account for the effects observed
in nature. His system presented some problems since it placed the ring of fire for
the stars inside the rings of fire for the moon and the sun. He may have addressed
these issues elsewhere, but that information is lost to us.

Anaximander also tried to provide a unified and natural system to account for
animal life. He argued that animals were generated from wet earth that was acted
upon by the heat of the sun. This placed all four elements together as a prerequisite
for life. This conception of spontaneous generation was borrowed from earlier
thinkers and was likely based on the observation of events such as the appearance
of insects or even frogs from out of the ground. Anaximander took the theory a step
further by arguing that simpler creatures changed into more complex ones. Thus,
humans were created from some other creature, probably some kind of fish. This
linked the elements of nature with natural processes rather than supernatural
intervention to create the world that we see.

The Ionian concern with primary materials and natural processes would
become one of the central axioms of Greek natural philosophy, but by itself
it was insufficient for a complete philosophical system. At about the time

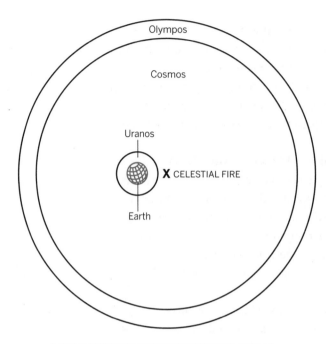

1.2 THE UNIVERSE ACCORDING TO PYTHAGORAS

Anaximander was working on his material philosophy, another group of Greeks was developing a conception of the world based not on matter but on number. This thread of philosophy comes down to us from Pythagoras (c. 582–500 BCE). It is unclear if there actually was a single historical figure named Pythagoras. Traditionally, he was thought to have been born on the island of Samos and to have studied Ionian philosophy, perhaps even as a student of Anaximander. He was supposed to have threatened the authority of the tyrant Polycrates on Samos and was forced to flee the island for Magna Graecia (Italy).

Because Pythagoras's followers became involved in conflicts with local governments, the Pythagoreans should not be regarded as simply a wandering band of mathematicians. Their lives were based, in fact, on a religion full of rituals. They believed in immortality and the transmigration of souls, but at the heart of Pythagoreanism was the conception of the universe based on number. All aspects of life could be expressed in the form of numbers, proportions, geometry, and ratios. Marriage, for example, was given the number five as the union of the number three representing man and the number two representing woman. Although there were mystical aspects of the number system, the Pythagoreans attempted to use mathematics to quantify nature. A good example can be seen in their demonstration of musical harmony. They showed that the length of a string determined the note produced, and that note was then related exactly to other notes by fixed ratios of string length.

The Pythagoreans developed a cosmology that divided the universe into three spheres. (See figure 1.2.) Uranos, the least perfect, was the sublunar realm or terrestrial sphere. The outer sphere was Olympos, a perfect realm and the home of the gods. Between these two was Cosmos, the sphere of moving bodies. Since it was governed by the perfection of spheres and circles, it followed that the planets and fixed stars moved with perfect circular motion. The word "planet" comes from the Greek for "wanderer," and it was used to identify these spots of light that constantly moved and changed position against the fixed stars and relative to each other. The planets were the Moon, Sun, Mercury, Venus, Mars, Jupiter, and Saturn. The fixed stars orbited without changing their position relative to each other, and it was from these that the constellations were formed.

While this arrangement was theologically satisfying, it led to one of the most perplexing problems of Greek astronomy. The philosophy of perfect circular motion did not match observation. If the planets were orbiting the Earth at the centre of the three-sphere universe, they should demonstrate uniform motion—and they did not. To resolve this problem, the Pythagoreans moved the Earth out of the centre of the sphere and created a point—home to a celestial fire—that was the centre of uniform motion. This kept the Earth motionless and resolved the issue of the observed variation in the velocity and motion of the planets. The desire to keep the Earth at the centre of the universe and preserve the perfection of circular motion led most of the later Greek philosophers to reject the Pythagorean solution. A radical solution to this problem was proposed by Aristarchus of Samos (c. 310–230 BCE), who argued for a heliocentric (sun-centred) model, but his ideas gained little support because they not only violated common experience but ran against religious and philosophical authority on the issue.

One of the most famous geometric relations comes down to us from the Pythagoreans, although they did not create it. This is the "Pythagorean theorem" that relates the length of the hypotenuse of a triangle to its sides. This relationship was well known to the Egyptians and the Babylonians and probably came from surveying and construction. The relationship can be used in a handy instrument by taking a rope loop marked in 12 equal divisions that when pulled tight at the 1, 4, and 8 marks produces a 3–4–5 triangle and a 90° corner. (See figure 1.3.) The Pythagoreans used geometric proof to demonstrate the underlying principle of this relationship.

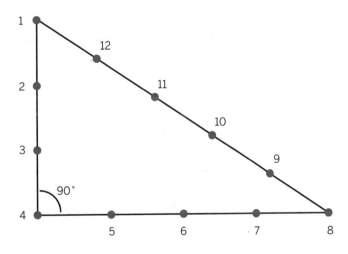

1.3 USING THE PYTHAGOREAN RELATION TO CREATE A RIGHT ANGLE

A rope with 12 evenly spaced knots when pulled at 1, 4, and 8 creates a right angle at 4. This simple device was known to the Egyptians and used for surveying and building.

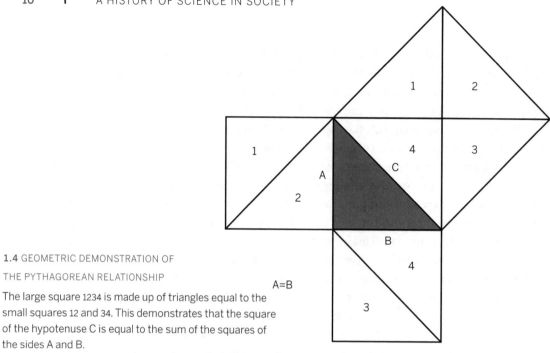

1.4 GEOMETRIC DEMONSTRATION OF
THE PYTHAGOREAN RELATIONSHIP
The large square 1234 is made up of triangles equal to the
small squares 12 and 34. This demonstrates that the square
of the hypotenuse C is equal to the sum of the squares of
the sides A and B.

Despite the mystical aspects of a world composed of number, the foundation
of Pythagorean thought places the essential aspects of natural phenomena within
the objects themselves. In other words, the world works the way it does because
of the intrinsic nature of the objects in the world and not through the intervention of unknowable supernatural agents. Ideal forms, especially geometric objects
such as circles and spheres, existed as the hidden superstructure of the universe,
but they could be revealed, and they were not capriciously created or changed by
the gods.

The degree to which the Pythagoreans desired a consistent and intrinsically
driven nature can be seen in the problem created by "incommensurability," referring
to things that had no common measure or could not be expressed as whole number
proportions such as 2:3 or 4:1. The Pythagoreans argued that all nature could be
represented by proportions and ratios that could be reduced to whole-number
relationships, but certain relationships cannot be expressed this way. In particular,
the relationship between the diagonal and the side of a square cannot be expressed
as a ratio of integers such as 1:2 or 3:7. As figure 1.4 demonstrates, the relationship
can be shown geometrically, but the arithmetic answer was not philosophically
acceptable since it required a ratio of 1:$\sqrt{2}$, which could not be expressed as an

integer relation. No squared number could be subdivided into two equal square numbers, nor in the case of √2 can the number be completely calculated.[1] According to legend, the Pythagorean Hippapus, who discovered the problem, was thrown off the side of a ship by Pythagoras to keep incommensurability secret.

The problems of Greek mathematics were compounded by two practical issues. The Greeks did not use a decimal or place-holder system of arithmetic but used letters to represent numbers. This made calculations and more complex forms of mathematics difficult. In addition, even though the Greeks and the Pythagoreans in particular were extremely powerful geometers, they did not have a system of algebra, and proofs were not based on "solving for unknowns." Geometric proofs were created to avoid unknown quantities. These two aspects of Greek mathematics put limits on the range of problems that could be addressed and probably encouraged their concentration on geometry.

While the Ionians investigated the material structure of the world and the Pythagoreans concentrated on the mathematical and geometric forms, another aspect of nature was being investigated by Greek thinkers. This was the issue of change. Motion, growth, decay, and even thought are aspects of nature that are neither matter nor form. No philosophy of nature could be complete without an explanation of the phenomena of change. At the two extremes of the issue were Heraclitus of Ephesus (c. 550–475 BCE) and Parmenides of Elea (fl. 480 BCE). Heraclitus argued that all was change and that nature was in a constant state of flux, while Parmenides asserted that change was an illusion.

Heraclitus based his philosophy on a world that contained a kind of dynamic equilibrium of forces that were constantly struggling against each other. Fire, at the heart of the system and the great image of change for Heraclitus, battled water and earth, each trying to destroy the others. In a land of islands, water, and volcanoes, this had a certain pragmatic foundation. Heraclitus's most famous argument for change was the declaration that you cannot step into the same river twice. Each moment, the river is different in composition as the water rushes past, but, in a more profound sense, you are as changed as the river and only the continuity of thought gives the illusion of constancy.

For Parmenides, change was an illusion. He argued that change was impossible since it would require something to arise from nothing or for being to become non-being. Since it was logically impossible for nothing to contain something

---

1. Like π, √2 is part of a collection of numbers that were later called "irrational," because they do not form proper ratios.

(otherwise it would not have been nothing in the first place), there could be no mechanism to change the state of the world.

Parmenides' best-known pupil, Zeno (fl. 450 BCE), presented a famous proof against the possibility of motion. His proof, called Zeno's paradox, comes in a number of forms but essentially argues that to reach a point, you must first cover half the distance to the point. To get to that halfway point, you would first need to cover half the distance (i.e., one-quarter of the full distance), and therefore one-eighth, one-sixteenth, and so on. Since there are an infinite number of halfway points between any two end points, it would take infinite time to cover the whole distance, making it impossible to move. (See figure 1.5.)

Our modern perception seems to favour Heraclitus over Parmenides, but they share a common concern. Each philosopher was attempting to establish a method for understanding the events in the world based on the intrinsic or natural action of the world. They were also attempting, as the Ionians and the Pythagoreans did, to establish a method for determining what certain knowledge was. Statements about the condition of the world had to be supported by a proof that could be examined by others and did not rely on special knowledge. They were asking epistemological questions, that is, questions about how someone could come to know something and just what that "something" could be. The Greek natural philosophers did not frame their questions as inquiries into the behaviour of gods or supernatural agents but rather asked such questions as: What in the world around us is fundamental and what is secondary? What system (outside revelation) can a thinker use to determine what is true and what is false? To what degree should the senses be trusted?

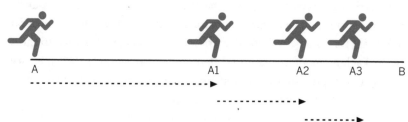

1.5 ZENO'S PARADOX
As the runner covers half the distance from "A" to "B," he must first cover the distance from "A1" to "B," then half the distance from "A2" to "B," and so on. Since there are an infinite number of halfway points, and it takes a finite amount of time to move from point to point (even though the time to cover the distance is very small), it will thus take an infinite amount of time to get from "A" to "B."

For Parmenides, the senses were completely untrustworthy and only logic could produce true or certain knowledge. Heraclitus at first seemed to have more faith in the senses, but in fact he reached a very similar conclusion. Any appearance of stasis, even something as simple as one rock resting on another, is an illusion, and only logic can be relied upon to make clear what is actually happening in nature.

Socrates, Plato, Aristotle, and the Epicureans: The Ideal and the Real

The philosophical threads of Thales, Pythagoras, Heraclitus, Parmenides, and many others came together in the work of the most powerful group of Greek thinkers, who were at the intellectual hub of Athens in the fifth century BCE. Socrates (470–399 BCE) established a context for natural philosophy by completely rejecting the study of nature as being largely unworthy of the philosopher's thought and by creating the image of selfless dedication to the truth that helped form the image of the "true" intellectual to this very day. Socrates' rejection of the study of nature mirrored the increasing disdain the intellectual elite felt for the merchant and craft class and their material concerns. Philosophy was supposed to be above the petty concerns of the day-to-day world, and philosophers were not, both literally and figuratively, to get their hands dirty.

For Socrates, the real world was the realm of the Ideal. Since nothing in the material world could be perfect, it followed that the material world must be secondary to the ideal. For example, while one could identify a beautiful person, the concept of beauty must have been present prior to the observation or we would be unable to recognize the person as beautiful. Further, while any particular beautiful material thing must necessarily fade and decay, the concept of beauty continues. It thus transcends the material world and is eternal.

This idealism also applied to the comprehension of the structure of the material world. Any actual tree was recognizable as a tree only because it reflected (imperfectly) the essence of "tree-ness," or the form of the ideal tree. These ideal forms were available to the human intellect because humans had a soul that connected them to the perfect realm. Socrates believed that, because of this, we actually had within ourselves the knowledge to understand how things worked. With skillful questions, this innate knowledge could be revealed, and from this process we get the Socratic method, a form of teaching based not on the instructor giving information to the student but asking a series of questions that guides the student's thoughts to the correct understanding of a topic.

Socrates' philosophy led him to question everything, including the government of Athens. He was convicted of corrupting the city's youth, but rather than asking for exile, he chose death. He drank a potion of the poison hemlock, with the firm belief that he was leaving the imperfect, corrupt material world for the perfection of the Ideal realm.

Socrates left no written material, so what we know of his teachings largely comes to us from his most famous pupil, Plato (427–347 BCE). The son of an aristocratic Athenian family, Plato wrote a series of dialogues based on Socrates' ideas and likely drawn from actual discussions. Although Plato's later work shifted away from its Socratic roots, he preserved the general premise of Ideal forms. One of Plato's other teachers was Theodorus of Cyrene, a Pythagorean, who taught him the importance of mathematical idealism. Although Plato accepted the primacy of the Ideal, he did not go as far as Socrates in his rejection of the material world.

Plato's primary interests were ethical and political. In his most famous work, *The Republic*, he explored what he considered ideal society and the problems of social organization. He did introduce natural philosophy, but it was in a lower realm of consideration and used mostly as a tool for consideration of the underlying structure of the cosmos. In the allegory of the cave, found in Book VII of *The Republic*, Plato argued that people are like prisoners in a dark cave who, from childhood, see only a strange kind of shadow play. Because the prisoners have no other reference, the shadows are taken to be reality. To see reality, the prisoners must free themselves and look upon the real world under the light of the sun. In this story, Plato argued that what we perceive through our senses is an illusion, but logic and philosophy can reveal the truth. The material world was explored in more detail in his *Timaeus*, where he presented a system of the four terrestrial elements of earth, water, air, and fire. The supralunar or celestial realm was made of a perfect substance, the ether, and was governed by a different set of physical rules. This system gained general acceptance among Greek philosophers and became one of the axioms of natural philosophy.

Plato, unlike his teacher Socrates, was not content to espouse his philosophy in the *agora*. The solution to the problems of society was education, which meant training students in a philosophy based on logic and a pursuit of knowledge of the Ideal. To this end, Plato founded a school in 385 BCE. Constructed on land once owed by the Athenian hero Academos, it became known as the Academy. It did

CONNECTIONS

Natural Philosophy
and Patronage:
Aristotle and
Alexander the Great

The relationship between patron and client has been an important part of the development of natural philosophy and science from the time of the Ancient Greeks. Aristotle was heavily influenced by the materials he received from Alexander the Great, and his fame spread even farther because of the king's patronage.

In 343 BCE, King Philip II of Macedon asked Aristotle to join his court as the tutor to his son Alexander. Aristotle's father had been Philip's personal physician, so there was already a connection between Aristotle and Philip's family. The call to go to Macedon came at a time when Aristotle was pursuing biological and philosophical research on his own because he had quit his teaching position at the Academy, the school established by Plato in Athens.

Aristotle remained at court for seven years, teaching the sons of Macedonian nobles. Aristotle found Alexander a good, if somewhat mercurial, student who wanted to be the best at whatever he did. When Philip was assassinated in 336 BCE, Alexander became the king and went on to conquer Greece and then most of the known world, including Asia Minor, Egypt, and Persia. He remained close friends with Aristotle, corresponding with his teacher throughout his life. He also sent Aristotle hundreds of samples of plants and animals, and over 10,000 scrolls from distant lands.

In 334 BCE Aristotle returned to Athens and established a new school called the Lyceum. Under the patronage of Alexander, the school thrived and Aristotle wrote a number of his most important works in this period, including *Physics*, *Parts of Animals*, and *De Anima*. The vast library created from Alexander's gifts helped Aristotle with his philosophical work, while the plant and animal samples helped him with his biological research. Aristotle described fish, for example, that were not noted again in Europe for hundreds of years, and developed a robust classification system because of this wide experience.

Alexander was a philosopher-king: literate, well-educated, and curious about more than just the necessities of warfare and politics. His relationship with Aristotle became a model of patronage that many later natural philosophers from Alcuin to Descartes hoped to find for themselves.

not have the formal structure of modern schools, but in many ways it was the foundation for the concept of higher education. Students who had already been tutored in the basic principles of subjects such as rhetoric and geometry travelled to the Academy to engage in discussion and debate under the auspices of a more senior philosopher in a kind of seminar atmosphere.

Plato's most famous student was Aristotle (384–322 BCE). A brilliant thinker, Aristotle had expected to become the head of the Academy when Plato died, but this position was denied him, going instead to Plato's cousin Speusippas, of whom little is known. Disappointed at having been passed over, Aristotle left Athens and travelled north. In 343 BCE he became the tutor to Alexander, son of Philip II, King of Macedon. When Philip died, Alexander became the leader of the Macedonians and proceeded to unify (that is, conquer) all of Greece. Once that was accomplished, he set out to conquer the rest of the world. With the patronage of Alexander the Great, Aristotle returned to Athens and founded a rival school, the Lyceum, in 334 BCE. It was sometimes called the peripatetic school because the instructors and scholars did their work while walking around the neighbourhood.

Aristotle did not reject all of Plato's philosophy, sharing a belief in the necessity of logic and some aspects of Platonic Idealism. He was, however, far more interested in the material world. Although he agreed with Plato that the world was impure and our senses fallible, he argued that they were actually all we had. Our intellect could be applied only to what we observed of the world around us. With this as a basis, Aristotle set out to create a complete system of natural philosophy. It was a powerful and extremely successful project.

At the heart of Aristotle's system were two fundamental ideas. The first was a system to provide a complete description of natural objects. The second was a system to verify knowledge that would satisfy the demands of proof necessary to convince people who lived in a competitive, even litigious, society. The combination of these two components produced the apex of Greek natural philosophy. No aspect of Aristotle's philosophy depended on supernatural intervention, and only one entity, the unmoved mover, existed outside the system of intrinsic or natural action.

The first step in the description of natural objects was identification and classification. Aristotle was a supreme classifier. Much of his work was on biology, and among other things he grouped what we call reptiles, amphibians, and mammals by their characteristics, even grouping dolphins with humans. He also observed the development of chicks in hen eggs and tried to make sense of sexual reproduction.

As astute as many of his observations were, Aristotle saw them as an examination of a level of superficial distinction; it was the job of the philosopher to look beyond these secondary characteristics and seek the underlying structure of nature. To do this, it was necessary to determine what aspects of nature could not be reduced to simpler components. The simplest material components were the four elements, and all material objects in the terrestrial realm were composed of these four substances. The superficial distinction between objects was the result of the different proportions and quantities of the elements that made up the objects in the world.

The elements by themselves were not sufficient to account for the organization and behaviour of matter. Matter also seemed to have four irreducible qualities, which Aristotle identified as hot/cool and wet/dry. These were always present as pairs (hot/wet, cool/wet, hot/dry, cool/dry) in all matter, but were separate from the material. A loose analogy would be to compare the bounce of a basketball and a bowling ball. The degree of bounce of a basketball and a bowling ball are very different and depend on the material that each is made of, but the "bounciness" of the two balls can be studied separately from the study of the materials that compose the two types of ball.

While the four elements and the four qualities could describe the matter and quality of composed things, they did not explain how a thing came to be. For this, Aristotle identified four causes: formal, material, efficient, and final. The formal cause of a thing was the plan or model, while the material cause was the "stuff" used to create the object. The efficient cause was the agent that caused the object to come into being, and the final cause was the purpose or necessary condition that led to the object's creation.

Consider a stone wall around a garden. The formal cause of the wall is its plans and drawings. Without a plan detailing dimensions, it is impossible to know how much stone will be required to build it. The material cause of the wall is the stones and mortar. These materials impose certain restrictions on the finished wall; it might be possible to draw a plan for a 30-metre high wall with a base only 20 centimetres wide, but such a wall cannot be constructed in reality. The efficient cause is the stonemason; again, certain restrictions will be imposed on the wall by the limits of the mason's abilities. The final cause is the reason to build the wall—to keep the neighbour's goat out of our garden, for example.

Although Aristotle and Plato's conception of the four elements could be reduced to a kind of particle model with a geometric structure (fire, for example, was composed of triangles), in general they treated the elements as a continuous substance. This view was challenged by the Epicureans, who proposed an even more materialistic model of nature. The philosopher Epicurus (342–271 BCE), like Plato, was from an aristocratic Athenian family. He founded a philosophical school known as the Garden and revived the work of an earlier philosopher, Democritus (c. 460–c. 370 BCE). Democritus had argued for a materialistic understanding of the universe, and the Epicureans pictured the world as constructed of an innumerable (but not infinite) number of atoms that were indestructible. The appearance and behaviour of matter were based on the varying size, shape, and position of the particles.

Epicurean natural philosophy was the most mechanistic Greek philosophy. In addition to challenging the material foundation of nature, the Epicureans also challenged the path to knowledge of nature, arguing that knowledge could only come from the senses. Because knowledge of nature did not require the intellectual refinement of logic or mathematics, it was knowledge open to all, not just learned men. This belief in knowledge from the senses contributed to the reputation of the Epicureans as sensualists, which did not help the philosophy when it was attacked as atheistic and decadent by Jewish, Islamic, and Christian scholars in later years. Although there was suspicion of all Greek philosophy by later theological thinkers, Aristotle's system was more easily revised than the Epicurean because it ultimately depended on axioms that could be ascribed to God. Thus, Epicurean thought was largely condemned or ignored until the seventeenth century when it gained a titular place as the foundation of modern studies of matter because of its proto-atomic model. Thus, it is seen as the ancient precursor to modern chemistry.

Aristotelian Theories of Change and Motion

The three fundamental aspects of matter (elements, qualities, and causes) in the Aristotelian system cannot assemble themselves into the universe; to bring everything together there must be change and motion. There are two kinds of motion. The first, natural motion, is an inherent property of matter. In the terrestrial realm all elements have a natural sphere, and they attempt to return to their natural place by moving in a straight line. However, because many objects in the world are mixtures of the four elements, natural motion is restrained in various ways. A tree, for example, contains all four elements in some proportion, but it grows a certain way with the roots going down because the earth element wants to go down while the crown grows up as its air and fire elements try to go up.

Plato and Aristotle accepted the Pythagorean idea that the matter in the celestial realm was perfect and that its inherent natural motion was also perfect, travelling in a uniform and immutable circle, which was the perfect geometric figure. Aristotelian astronomy thus required the objects in space to move according to this dictum. While this was a reasonable assumption for most of the objects that could be observed, such as the sun, moon, and stars, it created problems for later astronomers. (See figure 1.6.)

Other forms of motion, particularly locomotion, required motion to be introduced to the universe. For this, Aristotle traced a chain of motion back from observation to origin. Anything moving had a mover, but that mover had to have something moving it, and so on. Take as an example an archer shooting an arrow. We see an arrow fly through the air, and we can observe that it was the bow moving that moved the arrow. The archer

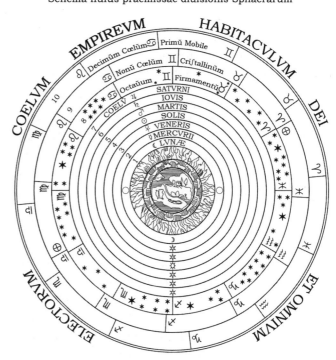

Schema huius praemissae diuisionis Sphaerarum

1.6 THE ARISTOTELIAN COSMOS

Wikipedia user Cipozy. Licensed under the terms of CC-BY-SA.

makes the bow move by the motion of muscles, and the muscles are made to move by the will of the archer. The mind thinks (which is a kind of motion as well) because of a soul, and the body exists because it was the product of the athlete's parents. Birth and growth are also forms of motion. The archer's parents were created by the grandparents, and so on. To prevent this from becoming a completely infinite regress, there has to be some point at which a thing was moved without being moved itself by some prior thing. This is the unmoved mover. In a sense, the unmoved mover kick-started motion in the universe by starting the great chain of action by a single act of will.

Let us return to the arrow as it flies along. As long as the bow is in contact with it, we can see that it is the bow and the muscles that are making it move, but what keeps it moving after it has left the bowstring? The aspect of its motion toward the ground is covered by its natural place as the heavy earth element of the arrow attempts to return to its proper sphere. The continuation of motion,

1.7 THE ARROW'S MOTION
ACCORDING TO ARISTOTLE

The arrow interacts with the air as it moves to continue its "unnatural" motion. This system may seem awkward, but it was likely based on observation of motion through water. An oar pulled through water seems to compress the water (it clearly mounds up) on the front surface, while eddies and voids seem to form around the back surface of the oar. The water in front then rushes around the oar to fill in the space at the back.

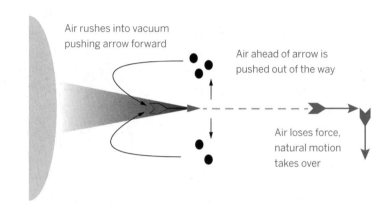

Air rushes into vacuum pushing arrow forward

Air ahead of arrow is pushed out of the way

Air loses force, natural motion takes over

Aristotle reasoned, had to have something to do with motion being added to the object as it moves. He concluded that the arrow was being bumped along by its very passage through the air. The arrow was pushing the air out of its natural place, in effect compressing it at the front and creating a rarefied or empty area at the back. The air rushed around the arrow to restore the natural balance and, in doing so, bumped the arrow ahead. Since the air resisted being moved from its natural place, it would eventually stop the forward flight of the arrow. (See figure 1.7.)

It also followed from Aristotle's system that the amount of element in an object governed its rate of motion. An arrow, constructed of wood and thus not containing a large amount of earth element, would stay in motion over the ground longer than a rock composed almost completely of earth element. This led Aristotelians to argue that if a small rock and a large rock weighing ten times as much were dropped together, the large rock would fall ten times faster than the small rock.

Aristotelian Logic

While understanding the structure of matter and motion was important, such knowledge was not by itself sufficient to understand the world. This was, in part, because the senses could be fooled and were not entirely accurate, but it was also because observation was confined to the exterior world and could not by itself reveal the underlying rules or structure that governed nature. That could be discovered only by the application of the intellect, and that meant logic. While Aristotle returned to the subject of logic repeatedly, his logic was most clearly presented in his two works on the subject, the *Posterior Analytics* and the *Prior Analytics*. At the heart of his logical system was the syllogism, which offered a method to prove a relationship and thereby produce reliable or certain knowledge. We continue to use syllogistic logic today as a method of verifying the reliability of statements. One of the most famous syllogisms says:

1. All men are mortal.   *Major premise, derived from axioms or previously established true statements.*
2. Socrates is a man.   *Minor premise. This is the condition being investigated.*
3. Therefore, Socrates is mortal.   *Conclusion, which is deduced from the premises.*

The syllogism was a powerful tool to determine logical continuity, but it could not by itself reveal whether a statement is true, since false but logical syllogisms can be constructed.

1. All dogs have three legs.
2. Lassie has four legs.
3. Therefore, Lassie is not a dog.

The second syllogism is as consistent as the first, but because the major premise is false, the conclusion is false. The axiom "dogs have three legs" does not stand the test of observation or definition, and so the syllogism fails. Thus, it is not surprising that Greek philosophers expended a great deal of effort on the discovery and establishment of axioms. Axioms were irreducible, self-evident truths. They represented conditions that must exist if the world was to function, but recognizing them was difficult. Aristotle concluded that axioms could be recognized only by the agreement of all learned men, which echoed Greek political discourse. An example of an axiom is the operation of addition, which must be accepted as a necessary mathematical operation or all of arithmetic collapses. The property of addition cannot be broken down into simpler operations; multiplication, on the other hand, can be broken down into repeated addition and is thus not axiomatic.

The problem of what was axiomatic and how to be sure of axiomatic statements was at the centre of debates over natural philosophy and science, in part because the axioms of previous generations often became the target of investigation and reduction for new thinkers. The philosophical and practical attacks on axioms at times made some scholars unsure whether any knowledge was reliable, while it set others, such as René Descartes (1596–1650), on a search for a new foundation of certainty.

The power of Aristotle's system was its breadth and completeness. It integrated the ideas that had been developed and philosophically tested, in some cases for several hundred years, with his own observations and work on logic. It presented a system for understanding the world that was almost completely intrinsically derived. With the exception of the unmoved mover, no part of his system required supernatural intervention to function, and further, it was based on the belief that all of nature could be understood. The comprehensibility of nature became one of the characteristics of natural philosophy that separated it from other studies such as theology or metaphysics.

Aristotle's system was a masterful use of observation and logic, but it did not include experimentation. Aristotle understood the concept of testing things, but he rejected or viewed with distrust knowledge gained by testing nature, because such tests only showed how the thing being tested acted in the test rather than in nature. Since testing was an unnatural condition, it was not part of the method of natural philosophy, which was to understand things in their natural state. It is tempting to find fault with Aristotle because of his rejection of experimentation, but this would be to argue that Aristotle's objectives must have been the same as those of modern science. The object of study for Aristotle and modern science was nature and how nature functions, but the forms of the questions asked about nature were very different. One of the central questions for Aristotle and other natural philosophers was teleological, asking "To what end does nature work?" They assumed that only through observation and logic could this question be answered.

Euclid and the Alexandrians

After the death of Aristotle, both the Academy and the Lyceum continued to be major centres for philosophical education, but the heart of Greek scholarship began to shift to Alexandria. This movement was spurred after 307 BCE when the ruler of Egypt, Ptolemy I (who had been one of Alexander's generals) invited Demetrius Phaleron, the deposed dictator of Athens, to move to his capital at Alexandria. Alexandria was an ideal location as a trade hub that linked Africa, Europe, the Middle East, and Asia. Demetrius was credited with advising Ptolemy to establish a collection of texts and establish a temple to the Muses, who were the patrons of the arts and sciences. Although its exact founding and early history are unclear, the temple to the Muses became the Museum, from which our modern use of the term descends. Part of the Museum was the library, which became increasingly important and eventually overshadowed the Museum in historical recollection. The Great Library of Alexandria eventually housed the greatest collection of Greek texts and was the chief repository and education centre for Aristotelian studies after the decline of Athens.

One of the great figures to be associated with the Museum was Euclid (c. 325–c. 265 BCE).[2] His most enduring work was the *Elements*, a monumental compilation of mathematical knowledge that filled 13 volumes. While the majority of the material in the *Elements* was a recapitulation of earlier works by other scholars, two factors raised it above a kind of mathematical encyclopedia. The first was the systematic presentation of proofs, so that each statement was based on a logical demonstration of what came before. This not only gave the mathematical proofs reliability but also influenced the method of presenting mathematical and philosophical ideas to the present day. These proofs were based on a set of axioms such as the statement that parallel lines cannot intersect or that the four angles created by the intersection of two lines are two pairs of equal angles and always equal 360° in total.

The second factor was the scope of the work. By bringing together the foundation of all mathematics known to the Greeks, the *Elements* was a valuable resource for scholars and became an important educational text. It covered geometric definitions and construction of two- and three-dimensional geometric figures, arithmetic operations, proportions, number theory including irrational numbers, and solid geometry including conic sections. In a time when all manuscripts had to be copied by hand, the *Elements* became one of the most widely distributed and widely known texts.

Greek natural philosophy was most notable for its philosophical systems, but those systems should not be seen as being removed from the real world or as some kind of irrelevant intellectual pastime. One of the purposes of Aristotelian natural philosophy was to make the world known, and a known world was a classified and measured world. Eratosthenes of Cyrene (c. 273–c. 192 BCE) set out to measure the world. Eratosthenes was a famous polymath who worked in many fields, especially

2. Like Pythagoras, there is some dispute as to whether Euclid was a real person or a name applied to a collective of scholars. From later commentators and internal evidence, Euclid may have been educated in Athens, perhaps at Plato's Academy, and then moved to Alexandria.

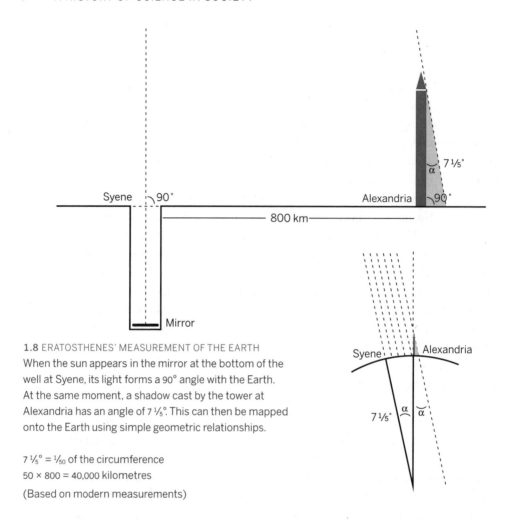

1.8 ERATOSTHENES' MEASUREMENT OF THE EARTH
When the sun appears in the mirror at the bottom of the
well at Syene, its light forms a 90° angle with the Earth.
At the same moment, a shadow cast by the tower at
Alexandria has an angle of 7 ⅕°. This can then be mapped
onto the Earth using simple geometric relationships.

7 ⅕° = ⅟₅₀ of the circumference
50 × 800 = 40,000 kilometres
(Based on modern measurements)

mathematics, and who became the chief librarian of the Museum in Alexandria
about 240 BCE. He applied his concepts of mathematics to geography and came up
with a method to measure the circumference of the Earth. That the Earth was a
sphere was long understood by the Greeks and was taken as axiomatic in
Aristotle's philosophy, but an accurate measurement was a challenge. Eratosthenes
reasoned that by measuring the difference in the angle of a shadow cast at two
different latitudes at the same moment, he could calculate the circumference. By
knowing the angle formed by the two lines radiating from the centre of the Earth
to the measuring points and the distance between the two points at the surface,
he was able to determine the proportion of the globe that distance represented.

(See figure 1.8.) From this, it was a simple matter to work out the circumference of the whole globe. His answer was 250,000 *stadia*. There has long been an argument about just how accurate this measurement was, since it is not clear what length of stadia Eratosthenes was using, but it works out to about 46,250 kilometres, which is close to the current measurement of 40,075 kilometres at the equator.

Archimedes, the Image of the Philosopher

The intellectual heritage of the Greeks, particularly that of Aristotle and Plato, was profound, but it was not solely their thought that they contributed. The Greeks also helped to create the image of the philosopher, an image that persists in various forms to the present day. Long before students have learned enough to comprehend the complex ideas of the philosophers, they have been exposed to the image. Even more famous than Socrates accepting death, the story of Archimedes (c. 287–212 BCE) has shaped the cultural view of philosophers.

Archimedes lived most of his life in Syracuse. He may have travelled to Alexandria and studied with Euclidean teachers at the Museum; it is clear that later in his career he knew and corresponded with mathematicians there. Among his accomplishments Archimedes determined a number for *pi*—relating the circumference, diameter, and area of a circle—and then extended this work to spheres. He established the study of hydrostatics, investigating the displacement of fluids, asking why things float, and the relationship between displaced fluids and weight. This has come down to us as Archimedes' principle that a body immersed in a fluid is buoyed up by a force equal to the weight of the fluid displaced by the body. Archimedes also determined the laws of levers through geometric proof.

As powerful as Archimedes' mathematics and philosophical work might have been, it was the legends that grew up around him that made him a memorable figure. His work was not confined to intellectual research, since he also created mechanical devices. Chief among these were the war machines he built to help defend Syracuse from the Romans during the Second Punic War. These included various ballistic weapons and machines to repel ships from docking. Although Archimedes did not invent Archimedes' screw (which consists of a rotating spiral tube used to lift water), his name was attached to it as the kind of thing he would have invented.

The famous story about Archimedes inventing burning mirrors or using polished shields to set fire to Roman ships using the reflected light of the sun was a myth created long after his death. Although theoretically possible, most modern recreations of the burning mirrors have shown that it would have been at best impractical, requiring the Roman ships to remain still for a significant period, and having no Roman notice the fire until it was large enough to have done significant damage.

Archimedes in the bath is the best-known tale from the philosopher's life. Hiero, the king of Syracuse, was concerned that the gold he had given craftsmen to make a crown had been adulterated with less valuable metal, but once the crown was made, how could the fraud be detected? Archimedes was supposed to have solved the problem while in the public baths when he realized that it was a hydrostatic problem. The gold would displace less water than a similar weight of silver because the gold was denser. He leapt from the bath and ran naked through the city, exclaiming "Eureka!" meaning "I have found it." No historical record exists that this happened, and it would have been difficult to use the displacement method with the tools available to Archimedes, but he could easily have solved this problem using a hydrostatic balance, a device that he wrote about and used.

Archimedes' death also became the stuff of legend. Plutarch (45–120 CE) tells the story in *Plutarch's Lives*:

> Archimedes, who was then, as fate would have it, intent upon working out some problem by a diagram, and having fixed in his mind alike and his eyes upon the subject of his speculation, he never noticed the incursion of the Romans, nor that the city had been taken. In this transport of study and contemplation, a soldier, unexpectedly coming up to him, commanded him to follow to Marchellus; which he declining to do before he had worked out his problem to a demonstration, the soldier, enraged, drew his sword and ran him through.[3]

Whether the legends are based on actual events is less important than the image of the ideal scholar they have come to represent. While the historical image of Archimedes has ranged from absent-minded philosopher to man of action to the "Divine Archimedes" as Galileo called him, the image of the true philosopher is that of a person above mundane concerns or personal self-interest. He is selfless, absorbed in study to the exclusion of all else, and perhaps a touch socially unaware. While Archimedes made mechanical devices and thus has also been associated with engineers, he was far more interested in philosophy than such contrivances. He

....

3. Plutarch, *Plutarch's Lives*, trans. John Dryden (New York: Random House, 1932) 380.

became the exemplar of a good scientist who can turn his hand to both theoretical and practical projects. While Aristotle and Plato can be revered as great intellects, they seem a bit distant and dry, always theorists looking at the big picture, while Archimedes is a much more comfortable role model for the modern experimentalist.

Conclusion

By the time the Greek world came under the control of Rome, a powerful group of Greek thinkers had completed the creation of the study of nature as a discipline and removed all but the most tangential connection to supernatural beings or forces. They made the universe measurable, and thus it could be known. They set the framework for intellectual inquiry that would be used in the Mediterranean world for over 1,000 years, and a number of ideas from Aristotle and Plato still provoke debate to this day. Under Roman control, Alexandria became even more important as a centre of learning, and the basis of Aristotelian philosophy was exported to the far-flung reaches of the Empire, from Roman Britain to the Fertile Crescent in the Middle East. Along with the philosophy went a new image of the sage, the scholar, the intellectual, whose job was not to interpret the mysteries of a world full of spirits but to read and reveal the text of the book of nature.

Essay Questions

1. Why did natural philosophy develop in the Greek world rather than in Egypt or the Fertile Crescent?

2. What were the principle concerns of Greek natural philosophers?

3. Comparing Plato's and Aristotle's systems, what were similar concerns and how did they differ?

4. What was Aristotelian logic and why was it so important for natural philosophy?

CHAPTER TIMELINE

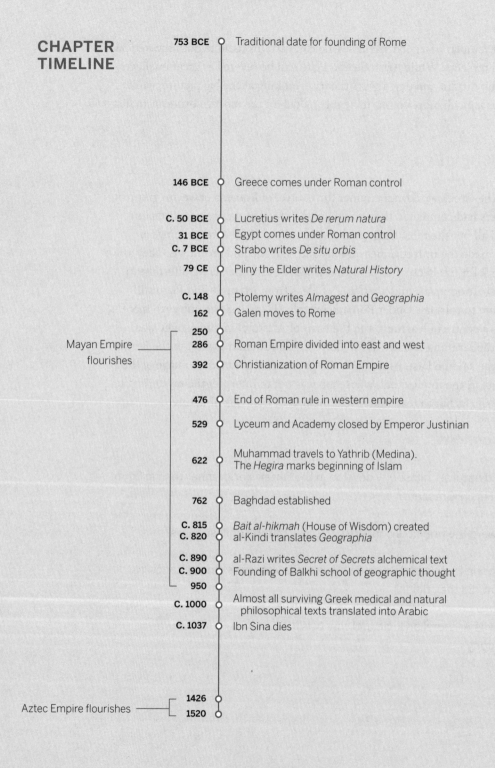

753 BCE	Traditional date for founding of Rome
146 BCE	Greece comes under Roman control
C. 50 BCE	Lucretius writes *De rerum natura*
31 BCE	Egypt comes under Roman control
C. 7 BCE	Strabo writes *De situ orbis*
79 CE	Pliny the Elder writes *Natural History*
C. 148	Ptolemy writes *Almagest* and *Geographia*
162	Galen moves to Rome
250	
286	Roman Empire divided into east and west
392	Christianization of Roman Empire
476	End of Roman rule in western empire
529	Lyceum and Academy closed by Emperor Justinian
622	Muhammad travels to Yathrib (Medina). The *Hegira* marks beginning of Islam
762	Baghdad established
C. 815	*Bait al-hikmah* (House of Wisdom) created
C. 820	al-Kindi translates *Geographia*
C. 890	al-Razi writes *Secret of Secrets* alchemical text
C. 900	Founding of Balkhi school of geographic thought
950	
C. 1000	Almost all surviving Greek medical and natural philosophical texts translated into Arabic
C. 1037	Ibn Sina dies
1426	
1520	

Mayan Empire flourishes (250–950)

Aztec Empire flourishes (1426–1520)

THE ROMAN ERA AND
THE RISE OF ISLAM

2

While the Greek philosophers were struggling with the structure of the cosmos, across the Adriatic Sea a small group of people living on the east bank of the Tiber River were in the process of creating a powerful military state. Traditional legends claim Romulus and Remus founded Rome in 753 BCE, but the origins of the city were probably Etruscan. Around 500 BCE Etruscan rule ended and Roman rule began. Rome expanded its area of control through the fourth and third centuries BCE, conquering or absorbing its neighbours. When Rome fought the Punic Wars against Carthage between 264 and 146 BCE, it established its military prowess and began its rise to empire.

As Rome expanded, it came into contact with Greek culture both through Greek colonies on the Italian peninsula and later by conquest of Greece itself. Roman dominance of Greece was completed by 146 BCE, and with the occupation the intellectual heritage of Greece came largely under the control of the Roman Empire. Greek scholarship was not destroyed by Rome, and in fact the Roman elite adopted Greek education and studied Greek philosophy, holding many Greek philosophers in high regard. This regard was not generally for the sake of philosophy but for a more practical purpose. Mastering Greek philosophy was seen as a good method to discipline the mind just as the legionnaires disciplined the body; both prepared the elite of Rome for their role as masters of the world. The Romans were at heart a people interested in practical knowledge. Their engineers created buildings, roads, aqueducts, and many other magnificent structures that have

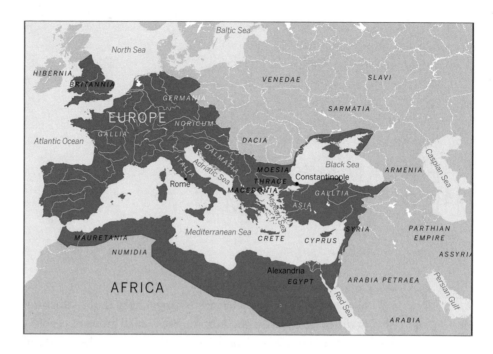

2.1 THE ROMAN EMPIRE

survived into the modern world. As impressive as the end products of Roman industry were, even more important was the power of the organizational system that could conceive, manage, and expand the enormous empire. In the Roman Empire nature was to be bent to useful ends.

The study of nature for the Romans was, therefore, oriented more toward practicality than philosophical speculation. Roman intellectuals were more concerned that a thing worked than about demonstrating the truth of the knowledge of that thing. Thus, they were more concerned with machines, studies of plants and animals, medicine, and astronomy than epistemology or philosophy. The Roman Empire was not based, as the Greek city-states had been, on public discourse and democracy but on public demonstrations of power. Making nature do your bidding was more essential than right reasoning. The Romans took the Greek heritage, in natural philosophy as in much else, and transformed it to aid their own objectives.

For the Roman elite, learning Greek philosophy might not be an end in itself but a way of training the mind. Intellectual acuity, even if the ends were material, still required a sound foundation. This heritage led a number of Roman intellectuals to preserve and propound Greek thought. For example, around 75 BCE the famous orator and politician Marcus Tullius Cicero (106–43 BCE) located and restored the tomb of Archimedes, who despite fighting against the Romans was well liked for his facility with machines. In 50 BCE the poet Titus Lucretius Carus (c. 95–55 BCE) wrote *De rerum natura*, a defence of Epicurean philosophy, and expounded the theory of Democritean atomism. In 40 BCE Marc Antony gave some 200,000 scrolls (primarily from the library at Pergamum) to Cleopatra (a descendent of the Greek rulers established by Alexander), who added them to the library of the Museum at Alexandria, making it the largest collection in the world. The gift was not completely altruistic, as Antony hoped to extend Rome's influence in Egypt, but it certainly confirmed the library's value. When Rome subjugated Egypt in 31 BCE, the conquerors, well aware of the Museum's worth, preserved the greatest centre of learning in the Mediterranean world both as an ornament in their empire and for the practical value of its materials.

The Romans developed a taste for large-scale projects. One of the keys to their success was the widespread use in their architecture of the arch, which allowed them to create much larger and much more open structures than the Egyptians or Greeks had been able to build using the column and lintel system. An arch rotated in three dimensions produces a dome, which was another innovation in Roman architecture. They also introduced the use of hydraulic cement as a mortar; because it set even under water, it was a very useful tool for building bridges, piers, and docks.

The greatest engineering accomplishment of the Roman era was the road system. While the majority of Roman roads did not represent the most complex engineering problems that had to be mastered, they were the key to the centralized control of the empire. Roman power functioned because the roads not only provided a communications system and a safe trade route but also allowed the rapid deployment of military forces.

Natural Philosophy in the Roman Era: Ptolemy and Galen

While Roman engineers were inventing and developing solutions to the problems of empire, natural philosophers in the Roman era were not as innovative. They did not create a new system of natural philosophy but turned their energy to continuing and extending the philosophical systems that came from Greece, particularly the Aristotelian (which dominated at Alexandria) and the Platonic systems. One way this extension took shape was in the commentaries and encyclopedic work of a number of scholars such as Posidonius (c. 135–51 BCE), who wrote commentaries on Plato and Aristotle, and Pliny the Elder (23–79 CE), whose massive work *Natural History* was nothing less than a complete survey of all that was known about the natural world presented to an educated but general audience. It was reported that Pliny and his assistants reviewed more than 2,000 volumes to compile their information. Some of this material was fantastic and mythical, such as descriptions of strange beasts and people with no heads, but Pliny also reiterated Eratosthenes' measurement of the size of the globe.

Two exceptions to the largely derivative natural philosophy in the Roman era were advances in astronomy and medicine. In both cases, the philosophical foundation came primarily from Aristotle but was extended well beyond any work of the earlier Greek period. In addition to the importance of the work itself, both the astronomy of Ptolemy (c. 87–c. 150 CE) and the medical discoveries of Galen (129–c. 210 CE) were significant conduits for the transmission of Greek philosophy to scholars after the fall of Rome.

Ptolemy's Astronomy

Although we know almost nothing about Ptolemy's life, his work is recognized as the cornerstone of natural philosophy to this day. His full name was Claudius Ptolemaeus, which suggests both Greek and Roman roots. Living in Alexandria, he produced material on astrology, astronomy, and geography, using complex mathematics and a large body of observations. His methods of astronomical calculation in particular shaped the Western view of the heavens for more than 1,000 years. In terms of accuracy, his observations were not surpassed until the beginning of the seventeenth century in the era of Tycho Brahe and with Galileo's introduction of the telescope.

Ptolemy's work on astronomy, collected in the *Mathematical Syntaxis*, commonly known as the *Almagest* (from the Arabic *al-majisti* meaning "the best"), accomplished two things. First, he created a mathematical model that reconciled Aristotelian cosmology with observation. Second, he provided a comprehensive tool, including tables and directions, to make accurate observations. His work extended both that of Hipparchus of Rhodes (fl. second century BCE), who had made numerous precise observations of the stars and planets, worked out the precession of the equinoxes, and measured the length of the year and the lunar month; and of Eudoxus (c. 390–c. 337 BCE), whose system of nested spheres each with a slightly different axis of rotation was a creative solution to the problem of retrograde motion.

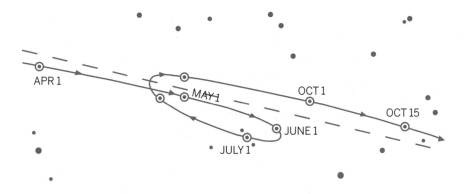

APR 1
MAY 1
OCT 1
OCT 15
JUNE 1
JULY 1

2.2 RETROGRADE MOTION

Aristotle's geocentric or Earth-centred system seemed to be obvious from general experience and is philosophically consistent, but there were several problems with it when it came to detailed observation. One of the most difficult observations to reconcile was retrograde motion. (See figure 2.2.) If an observer traced the course of the planets Venus, Mercury, Mars, Jupiter, and Saturn over an extended time relative to the stars (which move eastward in a yearly cycle), each planet gradually moved eastward, and then seemed to slow down and loop back westward for a time before continuing their west to east movement. This was most noticeable in the orbit of Mars.

In addition to the problem of retrograde motion, a number of the planets seemed to move at different speeds in different parts of their orbits, while the fixed stars moved in a very regular pattern. The combined problems of motion and time seemed to contradict the axiom of the perfect circular and spherical nature of the heavens. Retrograde motion also presented practical problems, since precise knowledge of the objects of the skies was necessary for casting horoscopes, aiding navigation, and telling time. Ptolemy created a working model of the heavens that resolved all these problems. It is important to understand that he regarded his model not as a true description of the universe but rather as a mathematical device that allowed observers to track the movement of the celestial bodies. Because of the utility of his system and its fit with the philosophy and theology of later scholars, the Ptolemaic system became synonymous with the actual structure of the heavens.

Ptolemy based his deductions on a large body of observations that came from the resources of the Museum's library and from his own work and that of assistants. To reconcile the necessity of circular motion (as required by Aristotelian cosmology) with the observed motion of the planets, he introduced geometric

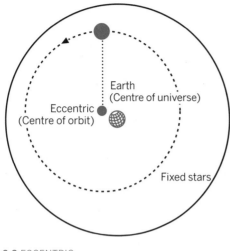

2.3 ECCENTRIC

"fixes" that allowed for a mechanization of the process of mapping the movement of the celestial bodies. These fixes were the eccentric, the epicycle, and the equant.

If a planet moved uniformly around the Earth, there was no problem describing its orbit as circular, but most planets seemed to move differently at different points in their orbit. Ptolemy reasoned that the problem could not be the planet actually going faster and slower (what mechanism could cause such a change?), but our perception of the motion. By moving the centre of the planet's orbit away from the Earth (at the centre of the universe), the eccentric replicated the observed non-uniform motion while allowing the planet to follow a perfect circular orbit in uniform motion. (See figure 2.3.)

The eccentric did not solve all the problems, so Ptolemy also introduced the epicycle, a small circle centred on a larger circle or deferent. (See figure 2.4.) This fix neatly accounted for retrograde motion. Later astronomers realized that epicycles could be added to solve observational problems, as we see particularly in late medieval and Renaissance mechanical models.

The equant was the most complex of Ptolemy's devices. (See figure 2.5.) The equant is not at the centre of the orbit but is displaced from it. However, the motion of the planet on the deferent is uniform around the equant. This means that the planet's apparent motion will be faster and slower in different parts of the orbit because the region swept out by the planet will not be equal.

Using these three geometric devices, Ptolemy was able to account for all the varied motions of the heavens and to predict future celestial activities. The *Almagest* was a brilliant achievement, and his system was so powerful that it became the basis for Western and Middle Eastern astronomy for over 1,300 years; a version of it survives to this day for small craft navigation at sea. Although much of the *Almagest* was complicated, part of its power was that it was not mathematically complex. All the elements of Ptolemy's models were based on the geometry of the circle, which was well understood. While there could be many epicycles employed to establish the orbit of a planet, they were all constructed the same way. The *Almagest* provided a complete account of celestial motion of all the objects that could be seen with the naked eye. The observations were so accurate and the method of calculation so complete that from a practical point of view Ptolemy had resolved the issue of astronomy. There was some tinkering with the distribution of epicycles and the exact location of the eccentrics, but the model worked so well that it could be made into a mechanical device. This celestial clock was perfected by Giovanni de Dondi of Padua around 1350 CE, and a working copy of his masterpiece of clockwork and Ptolemaic astronomy can be found at the Smithsonian Institution in Washington, DC. (See Plate 7.)

Ptolemy's other great work, the *Geographia*, applied his powerful mathematical tools and the resources of the Museum to the terrestrial realm. In a sense, the *Almagest* and the *Geographia* represent two parts of the same system, the first representing the supralunar realm and the second the terrestrial or sublunar realm. To achieve good astronomical results it was necessary to know where you were on the globe; to know that, the globe had to be treated mathematically. Ptolemy summarized the work of other geographers and examined aspects of cartography including various methods of projection, longitude, and latitude; he then provided lists of some 8,000 places and their coordinates. He treated the celestial and terrestrial globes as equivalent, applying the same grid system to each, and using the same spherical geometry to plot points. He divided the globe into a series of parallel belts or "climates" and developed a grid of longitude and latitude coordinates. In doing so, he created a map projection that has never been completely superseded and that was of immense importance to later European exploration and contact with other parts of the world.

Ptolemy's mathematical geography contrasts with the earlier descriptive geography of Greek scholars such as Strabo (c. 63 BCE–c. 21 CE). Strabo wrote an eight-volume geography, *De situ orbis*, around 7 BCE, in which he set out to describe every detail of the known world, based both on his own extensive travels and on the accounts he gathered from other travellers. This was an enterprise closely tied to history and politics. Ptolemy made a distinction between his mathematical rendering of the globe, which he called "geography," and Strabo's type of terrestrial research, labelled "chorography."

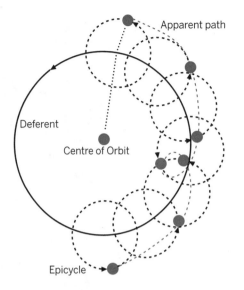

2.4 EPICYCLE

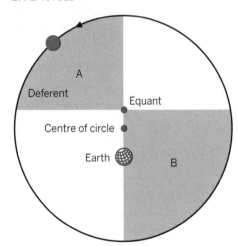

2.5 EQUANT
The orbiting planet takes equal time to travel around quadrant "A" as around quadrant "B."

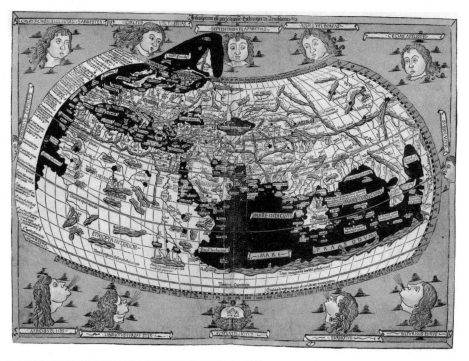

2.6 PTOLEMY'S WORLD MAP FROM *GEOGRAPHIA* (1482)

The most useful parts of Ptolemy's *Almagest*, as well as his "Table of Important Cities" from the *Geographia*, were compiled as the *Handy Tables*. This reference allowed a quick way of doing celestial calculations and was easier than the methods outlined in the *Almagest*. The *Handy Tables* became a standard tool for astronomy. The geographical material was not as well known, nor as widely circulated, as the astronomical, and it faded from sight after the fall of Rome. Its rediscovery in the fifteenth century had a major impact on geographical thought and exploration in Renaissance Europe. Because Ptolemy's works were so useful, they were widely disseminated, which in turn helped them to survive the turmoil of the end of a number of empires including the Roman and the Byzantine. Where Ptolemy's work survived, the Aristotelian foundation of his work also persisted.

Galen's Medicine

In contrast to Ptolemy, we know much more about Galen's life. Born in 129 CE at Pergamum, second only to Alexandria as a centre of learning in the period, Galen studied mathematics and philosophy before beginning his medical training at the age of 16. In 157 CE he became surgeon to the gladiators at Pergamum. In many ways, it was this first professional work that allowed him to begin creating his own system of medical knowledge, particularly of anatomy. At a time when human dissection was forbidden, he got first-hand experience of human anatomy by tending to wounded and dead gladiators. He saw the structure of muscle and bone, sinew and intestine laid bare by violent injury and was responsible for trying to set the parts back in place when possible. In 162 CE he travelled to Rome, remaining for four years before returning to his home town. When a plague struck, Emperor Marcus Aurelius called him back to Rome, where he settled permanently as the personal physician to four emperors: Marcus Aurelius, Lucius Verus, Commodus, and Septimius Severus.

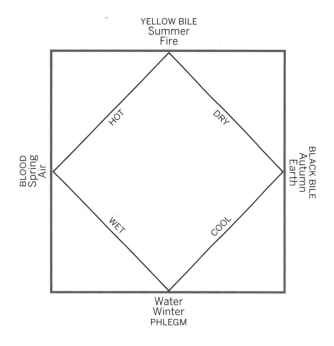

2.7 THE FOUR GALENIC HUMOURS

Medical philosophy of the time was dominated by Hippocratic theory. Hippocrates of Cos (c. 460–c. 370 BCE) may have been a single individual, a mythic figure, or a name given to a collective. The Hippocratic system of medicine was based on the concept of regimen and balance. Regimen covered not only the physical aspects of a patient's health but also social, mental, and spiritual aspects. Hippocratic doctors conducted long interviews with patients, asking about their diet, work, home life, sex life, and spiritual health. Horoscopes were cast, and even geography was considered, since living near a swamp or exposure to certain winds was considered harmful to well-being. Although Hippocratic doctors considered spiritual aspects and used horoscopes, they regarded illness as primarily natural rather than supernatural and thus treated disease with material solutions such as drugs, diet, and exercise. Health was considered to be the correct balance of physical action, diet, and lifestyle; illness represented an unbalancing of the elements.

The basis of the Hippocratic theory of balance was the four humours, which covered the four bodily fluids: phlegm from the head, blood from the liver, yellow bile from the stomach, and black bile from the stomach or intestine. (See figure 2.7.) Each of these had a paired quality of hot/cold and wet/dry as well as one of the four elements that fit nicely with the Aristotelian system. The objective of medical intervention was to balance the four humours. Too much or too little of any humour resulted in illness. For example, a person with too much blood was sanguine and the treatment was bleeding, while a bilious person had too much bile in their system and needed a purgative. These physical conditions were, as the modern use of the terms sanguine and bilious suggest, also associated with temperament.

While modern medicine often traces its heritage back to the Hippocratic doctors of Greece, especially through the Hippocratic Oath, there were in fact few unified theories of medicine in Greek or Roman times. Although some Hippocratic doctors had experience with medical trauma such as wounds, fractures, and other injuries, there were other medical practitioners, such as surgeons, who dealt with the body directly. Hippocratic doctors did not deal with women, who were treated by another group of practitioners including midwives. To a large extent, medical treatment for men and women was separate. If a man could afford a physician, he sought a philosophically trained physician like Galen. For women and the poor, there were a range of practitioners and a range of treatments from the most practical to the most spiritual. Aid was often provided to the sick at temples, where prayer and supplication were part of treatment.

At least four general groups of medical philosophy have been identified by historians at the time Galen began to practise medicine: the rationalist, empiricist, methodist, and pneumatist. Even within those groups there was no united form of practice. Each school and each doctor who took on apprentices taught a different version of medical theory. Further, each doctor was also his own salesman seeking clients and patrons, often literally in the marketplace. Because of this competition for clients, a doctor had to be able to persuade potential clients that his brand of medicine was the best, so medical education also included training in philosophy, rhetoric, and disputation.

When Galen became physician to the gladiators of Pergamum, he was taking a lucrative job but one with relatively low status. Gladiatorial combat was a big money enterprise, but Galen's work, primarily dealing with wounds, was considered very practical and thus of a lower status than the intellectual diagnosis of disease. Despite the issue of status, treating the gladiators gave Galen something he could not get elsewhere—detailed exposure to human anatomy. Religious and cultural taboos prevented the dissection of humans, so in the Roman world anatomical training was theoretical or based on animal dissection, particularly of monkeys.

Galen brought to his work his philosophical training, which covered Plato, Aristotle, and the Stoics, who believed in a physics based on the material world, as well as many other elements of classical thought. He accepted the Hippocratic humours but wanted to make clear the functions of the human organs, so he applied Aristotelian categories, particularly the four causes, to his anatomical work. His close observation demonstrated, for example, that arteries carried blood rather than the older theory that they carried "pnuema," or air. Each organ and structure in the body had a purpose, and dissection and vivisection were the key tools to establishing what that purpose was. His anatomical work became a powerful tool not only for physiology but also for persuasion. His demonstrations put him ahead of other physicians who depended on rhetoric to sell their brand of medicine, because he could show people his system through actual dissections of animals. Galen's success was so great that he received the patronage of four emperors and was physician to the elite of Roman society. The support afforded by such patronage allowed him the time to write, and he may have authored as many as 500 treatises, of which more than 80 survive to the present.

Galen's productivity and patrons would have been enough to ensure him a place in medical history, but a further element helped preserve Galenic medicine when other aspects of Greco-Roman thought were repudiated or lost. Galen adopted a strongly teleological philosophy in which nothing existed without a purpose and all of nature was constructed in the best possible way and for the best possible good. This demonstrated the existence and perfection of the Demiurge, the fashioner of the world. While extending Platonic idealism to the body, this teleological philosophy fit well with that of Islamic, Jewish, and Christian thinkers. His practical medicine was considered one of the few things worth keeping from the pagan, decadent, and materialistic world of the Roman Empire by these religions, which had strong precepts about care for the sick as a religious duty. When Galenic texts survived, the philosophical foundations of Plato and Aristotle were also preserved.

The Decline of Rome

By the time of Galen's death around 210 CE, cracks were beginning to appear in the Roman Empire. In 269 CE, the library of Alexandria was partly burned when Septima Zenobia captured Egypt. In 286 CE, Diocletian divided the empire into eastern and western administrative units. Alexandria's great library was burned to the ground in 389 during a riot between pagans and Christians, which closed the Museum. Emperor Theodosius I ordered the destruction of pagan temples in 392 CE as the last vestiges of Roman religion were officially replaced by Christianity.

The Roman Empire might have continued to weather these problems had it not been for pressure from the outside. In the east the Persians resisted Roman control and were a constant threat. In the northwest a whole host of peoples were either pushing against the frontier at the Rhine or struggling against Roman control. Attacks across the frontier increased in the fifth century because the Germanic tribes were being pushed from behind by the Huns while at the same time being attracted to the wealth of the faltering empire. The Roman roads, which had once been the means of controlling the empire, guided the invaders into the heart of the Roman world.

In the eastern part of the empire Constantinople struggled to continue the traditions of empire and learning after Rome itself had fallen to barbarian invaders. Established in 330 CE by Emperor Constantine I, it became the capital of a separate eastern empire in 395 CE after the death of Theodosius I. With its Greek heritage and strategic location, Constantinople preserved elements of Greek learning and culture during the years of decline caused by the end of central rule by Rome.

Rome, once the greatest city of an empire that spanned much of Europe, northern Africa, and the Middle East, was sacked by the Visigoths in 410 CE, then again by the Vandals in 455 CE. The fall was partly the result of internal problems such as increased levels of taxation, the disparity of rich and poor, the grievances of the conquered peoples, a decline in Roman participation in the military, and constant political infighting and civil war. Romans' ability to deal with these problems may have been significantly diminished by the effects of lead poisoning, as the heavy metal leached not only from alloy plates, bowls, and cups but also from the lead pipes extensively used for plumbing (*plumbum* being the Latin for lead). Moreover, acetate of lead was used to sweeten wine.

The western empire finally ceased to function in 476 CE when German invaders established themselves as the emperors of Rome. The new rulers were not particularly interested in philosophy, natural or otherwise. The last great natural philosopher in Rome was Anicius Manlius Severinus, better known as Boethius (480–524 CE). His best remembered work was *On the Consolation of Philosophy*, but his most lasting impact came from his role as a translator. He translated a number of Aristotle's logical works into Latin, as well as Porphyry's *Introduction to Aristotle's Logic* and Euclid's *Elements*. Because of his work, Aristotle was almost the only available source for Greek natural philosophy until the twelfth century. Boethius was jailed for treason and finally executed by Theoderic the Great in 524 CE, ending Greek-oriented natural philosophy in Western Europe for almost seven centuries.

Early Christianity and Natural Philosophy

The other element that resulted in a decline in interest in Greek philosophy was the rise of Christianity. Whether it also contributed to the decline of the Roman Empire is a matter of historical debate. On the one hand, Christians were frequently associated with dissident elements in Roman society, and efforts by the government to fight the spread of the religion at home took resources and attention away from other problems such as the barbarian pressures on the borders. On the other hand, the Christians did not create the external problems, and with the Christianization of the Roman Empire under Theodosius I starting in 392 CE, Christianity offered the possibility of a unifying force in a badly fractured society. In the short term, the Roman Christians suffered during the collapse of the western empire, but in the longer term the Church was able to preserve and rekindle the intellectual aspects of philosophy.

Christianity had (and continues to have) an uneasy relationship with natural philosophy. The messianic and evangelical aspects of the religion pointed people away from the study of nature and toward the contemplation of God. The Greek philosophers were pagans and, therefore, to be rejected, but they were also part of the extraordinary Roman Empire and closely linked with the intellectual power and managerial skills, particularly literacy and bookkeeping, that the Church needed to survive. In addition, many of the most important leaders of the early Church were trained in Greek philosophy. Augustine (354–430 CE) in particular was well trained in Greek philosophy, and he became the voice of the intellectual Church. Even so, the last vestiges of Greek philosophical education, the Academy and the Lyceum, were closed by Emperor Justinian in 529 CE. Although they had been little more than tattered remnants of their former glory for years, the closing marked the end of Greco-Roman philosophical power.

Christianity was fraught with internal controversy, and the problem of heresy was frequent. The Donatists' and Arians' challenge to the theology that emanated from Rome as early as the third century led to the Council of Nicaea in 325 CE and the promulgation of the Niceaen Creed to establish orthodoxy. Even ownership of

Bibles became an issue, and in many places only the clergy were legally allowed to own them. With literacy largely confined to members of the Church, this was not a difficult restriction to enforce, and the lack of literacy meant that Greek material was equally inaccessible.

Through deliberate and accidental destruction, loss, or rejection, the intellectual heritage of natural philosophy largely disappeared in the Latin West (the western lands of the Roman Empire where Latin was the language of the Church and the small educated class). Some texts and ideas survived in scattered pockets in the West, while in the eastern empire (where Greek was still used) Byzantium held on to more material. What survived were parts of the work of Hippocrates and Galen on medicine, because of the duty of the Church to care for the sick; parts of Euclid; fragments of Aristotle's logic; parts of Plato, particularly *Timaeus*; and some of the ideas of astronomy from Ptolemy, which were used to help keep up the calendars. Many Christians thought they were living in the end days of biblical prophecy, so there was little impetus to preserve or study the old knowledge even if it were available. While the light of philosophy never went out completely in the West, it was dimmed considerably, and what remained was folded back into theology.

The Rise of Islam and Its Effect on the Development of Natural Philosophy

The void created by the collapse of Rome also had an effect on the southeast side of the Mediterranean. The people who lived on the Arabian Peninsula and in the Middle East were conveniently placed to trade with Asia, Africa, and Europe, and a number of important centres developed that had access to both Persian and Byzantine markets. Territorial conflicts as well as struggles based on cultural and religious differences resulted in the emergence of a number of independent states in the region. In this period of turmoil, Muhammad began his efforts to convert the people of Arabia to his new religion. Unable to gain significant inroads in Mecca, he travelled to Yathrib (later Medina) in 622. This trip, called the *Hegira*, marked the first step toward the foundation of the Islamic world.

Muhammad's return to Mecca in 630 was followed by a wave of conversions both through peaceful exhortations by preachers and traders and by the sword and the *jihad* or holy war. By the time of his death in 632, most of the Arabian Peninsula had been converted to Islam. Syria (previously a Byzantine province),

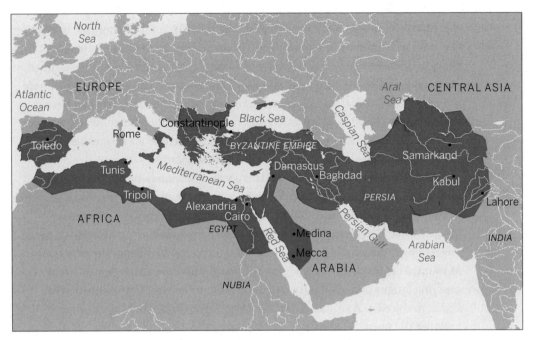

2.8 ISLAMIC AND BYZANTINE EMPIRES 750–1000

Palestine, and Persia followed by 641; a year later, Egypt came under the control of the first caliphs, who were both political and religious rulers. Under the Umayyad caliphs the conquests continued, and by 750 the Islamic empire ran from Spain in the west to the Indus River in the east.

The capture of many of the most important centres of learning in the Middle East, particularly Alexandria, gave Islamic scholars control of vital intellectual resources. As the strength of the Arabic world grew, so did Islamic scholarship, first translating and integrating the philosophies of the Greco-Roman world and then establishing a very high level of competence in research and critical analysis. Yet, until recently, Western historians of science have regarded Islamic natural philosophers as little more than imitators of Greek work and a conduit through which it passed to European scholars. So clear was this prejudice that many older history texts used only the Latin version of Arabic names, thereby suggesting that the only significant Islamic works were those used by Western European scholars. The reason for this dismissal or denial of authentic and innovative study was the idea that Islamic natural philosophers, despite access to the Greek material, failed to expand upon it, whereas the same material in the hands of European scholars led to a revolution in natural philosophy and the creation of modern science.

More recent scholarship awards Islamic thinkers a far greater role in shaping the work of later natural philosophers. Islamic scholars did not accept Greek thought unchallenged and added not only their critical thinking to the body of material available but also their own original research. They were also far more willing to test ideas than the Aristotelian or Platonic philosophers had been, and although this should not be confused with experimentalism (which uses a different philosophic conception of certainty), it became an acceptable tool for natural philosophy because of its use by these scholars.

As in Christianity and Judaism, there was a tension between the intellectual and spiritual aspects of Islamic theology, but certain of its tenets made it amenable to the study of nature, particularly if the elements of faith were interpreted broadly. One of the five pillars of the faith was *Shahadah*, the profession of the creed, which essentially called all the faithful to read the Q'ran, the holy book of Islam. This resulted in a push toward literacy and the promotion of Arabic as a unifying language from the Iberian Peninsula in the west to the border with China in the east. A second pillar, the *Hajj*, or pilgrimage to Mecca, brought together the people of the far-flung empire, even when different regions were not politically unified. This created ties of personal contact, trade, and intellectual exchange. Indian mathematics, Chinese astronomy and inventions, and Hellenized Persian culture flowed up and down the pilgrimage and trade routes along with silk, ivory, and spices. The most famous of these arteries was the Silk Road, the lengthy trade route that connected China to the Arabic world. While the Silk Road is best remembered for the exotic products that moved from east to west, it also brought ideas including Hindu mathematics and Chinese alchemy.

The Q'ran itself also contributed to a more positive attitude toward the study of nature than the Bible did for Europeans. It was more precise about creed and liturgy (reducing the potential for schism), but was also more worldly, calling on the faithful to study nature as part of God's creation. Many of its passages present knowledge and the acquisition of knowledge as sacred. One of Muhammad's most famous sayings was "Seek knowledge from the cradle to the grave." The centre of Islamic religious life was the mosque, which, particularly outside the Arabic regions, served as a school of Arabic literacy. Many mosque schools, or *maktab*, developed into more extensive educational institutions and became essentially the first universities, offering advanced studies for students and research facilities, such as libraries, for scholars.

Another aspect that should not be overlooked was the sheer wealth of the Islamic empires. The ability of caliphs to order the creation of schools, libraries, hospitals, and even whole cities demonstrates their economic power. With those resources available, even a low level of interest in natural philosophy could produce significant results. The Islamic world received large collections of Greek and Roman material along with their conquests, and its proximity to the Byzantine Empire meant, at least in times of peace, a potential for intellectual exchange. Educated Persians and Syrians, with their knowledge of Greek culture running back to the time of Alexander the Great, became bureaucrats within the empires and brought with them their intellectual heritage.

The Islamic Renaissance

When a new dynasty started under the Abbasids, there was increased interest in the intellectual heritage of the Greeks. The early Abbasids were intellectually tolerant and had a strong interest in practical skills, employing educated Persians and even Christians in government. In particular, the Nestorians (a Christian sect from Persia) served as court physicians. They practised Galenic medicine, preserving not only the practical aspects of Galen's work but its Aristotelian and Platonic foundation as well.

In 762 the Abbasid Caliph al-Mansur established a new capital, Baghdad, on the Tigris River. He also began a tradition of translation of Greek and Syriac texts into Arabic. His grandson, Harun ar-Rashid, continued this work and even sent people to Byzantium to look for manuscripts. However, the greatest intellectual developments came under Harun's son, al-Mamun, who around 815 created the *Bait al-hikmah* or House of Wisdom. This was part research centre, containing an extensive library and an observatory, and part school, attracting many of the most important scholars of the day. This state-supported enterprise was also responsible for the majority of the translation of Greek, Persian, and Indian material into Arabic.

The head of al-Mamun's research centre was Hunayn ibn Ishaq (808–873), a Nestorian Christian and physician, who grew up bilingual (Arabic and Syriac) and later learned Greek, perhaps in Alexandria. He translated over 100 works, many of them medical. His son and other relatives continued the translation work, in particular Euclid's *Elements* and Ptolemy's *Almagest*, both of which became important foundational texts for Islamic scholars. By 1000 almost every surviving work of Greek medicine, natural philosophy, logic, and mathematics had been translated into Arabic.

The interest in education in Islam fostered the appearance and high status of the *hakim*, a sage or wise man, and philosophers such as Aristotle were revered as sages. The educational system included philosophy and natural philosophy as components of a well-rounded education. There was a great flowering of culture, known as the Islamic Renaissance, starting in the ninth century and running until about the twelfth century. During this period Islamic scholars continued the intellectual traditions of the Greeks, but there were important differences. Islamic scholars had to conduct their work within the framework of their religion. While there were liberal and conservative periods, often varying with a change in rulers, Greek material could not simply be adopted outright. Some aspects were accepted with little change, such as Ptolemaic astronomy; some were modified, such as the introduction of God rather than an indefinite "unmoved mover" in Aristotelian physics; and some elements were rejected outright, such as various cosmological creation stories that came from Greek and Roman sources.

In addition to the questioning inherent in the ratification of pagan material, Islamic scholars pursued new ideas in natural philosophy. This was partly a result of circumstances, since scholars often lacked access to the complete corpus of Greek thought and so might have only a fragmentary idea of, for example, Aristotelian optics. It was then necessary to do independent work on the topic. Islamic scholars were also more interested in testing observations than Aristotle or Plato had been, in part because they had a less intellectualized concept of the acquisition of natural philosophic knowledge. In other words, they had a more hands-on approach. This attitude toward knowledge acquisition coincided with expectations for the educated class in Islamic society, since the educated and affluent were supposed to be able to turn their hand to poetry and music, history and philosophy, and martial arts such as riding and swordplay, as well as understanding practical matters such as commerce and trade. Scholarship and courtly behaviour were intimately linked in the lives of many of Islam's greatest natural philosophers, and many of the traits associated with chivalry for the European knights were in fact adopted from the Islamic world.

Another reason Islamic scholars were more willing to test nature was because they lived in a more materially oriented and technically advanced society. The craft skills of the Arabic world were extremely accomplished, surpassed in this period only by China, which was a trade partner. Artisans made a wide range of tools and instruments, and there was both an appreciation for fine work and the money to support it. Two examples of this high level of skill can be seen in glass-making and metallurgy. Glass-making was a large-scale industry that produced many of the tools used by Islamic scholars to investigate optics and alchemy, while metalworkers produced instruments such as astrolabes and armillary spheres. Another development in metalwork that intrigued (and terrified) Europeans was Damascus steel, which in the form of swords gave the armies of Islam an edge (literally and figuratively) over the weapons used by the Crusaders.

Many of the greatest Islamic natural philosophers were educated as physicians, which perfectly combined practical and theoretical training. This meant that they were first introduced to Greek philosophy through Galenic material. While a distinction existed between the intellectual understanding of health and disease and the practical matters of surgery and bone-setting, the technical skills of Islamic practitioners surpassed those of the Greco-Roman world and far outstripped their European neighbours. Technical abilities and tools extended to abdominal surgery and cataract removal. Eye surgery was linked to theories of vision and the more theoretical study of optics. Thus, medicine was a perfect conduit for natural philosophy in the Islamic world. It was theologically sound, since care for the ill was part of the charity requirements of the faith, and it was both practical and intellectual without being a craft, and thus acceptable for the upper class. With these characteristics, physicians frequently held high posts in government and at court.

Agriculture was another area of expertise for Islamic scholars and practitioners. The coming of Islam freed many farmers from their previous overlords; this freedom combined with increased literacy encouraged a burgeoning of practical and theoretical work on agriculture and botany. In part because of the lines of communication that were established within the Islamic world, and in part because of the freedom that farmers enjoyed (in comparison to the peasants of Latin Europe), interest in useful plants led to one of the largest transfers of biological material in history, as crop plants and their particular farming needs were transferred from China in the east throughout the Islamic world to the Iberian Peninsula in the west. A partial list of transplanted crops includes bananas, cotton, coconut palms, hard wheat, citrus fruit, plantain, rice, sorghum, watermelons, and sugar cane. A somewhat less practical biological exchange occurred in 801 when Caliph Harun al-Rushid of Baghdad sent an elephant as a present to Charlemagne. The collecting of plants, both useful and decorative (roses, tulips, and irises were also part of the great plant transfer) led Islamic scholars to create encyclopedias of plants such as al-Dinawari's (828–896) *The Book of Plants* and Ibn al-Baitar's (c. 1188–1248) *Kitab al-jami' li-mufradat al-adwiya wa al-aghdhiya*, a pharmacopoeia listing over 1,400 plants and their medicinal uses. One of the world's largest botanical gardens was established in Toledo in the eleventh century.

One of the most powerful minds of the Islamic Renaissance was Abu 'Ali al-Husain ibn Abdallah ibn Sina (980–1037), whose life was chronicled in his autobiography and the memoirs of his students. He was a child prodigy who had memorized the Q'ran by the age of ten and had begun training as a physician when he was 13, although he also studied widely in philosophy. After curing the Samanid ruler Nuh ibn Mansur of an illness, he was allowed to use the Royal Library. It was then that ibn Sina began to explore a vast range of material from mathematics to poetics. Because of his skills as a physician, he found employment at the courts of various rulers, but it was a turbulent time, and he was involved in a number of political struggles that saw him made a vizier by Prince Shams ad-Dawlah in Hamadan, a region in west-central Iran, only to be forced from office and jailed for a time.

Ibn Sina left Hamadan in 1022, on the death of the Buyid prince he had been serving, and moved to Isfahan. He entered the court of the local prince and spent the last years of his life in relative calm, completing the major works he began in Hamadan. He was prolific, producing over 250 works covering medicine, physics, geology, mathematics, theology, and philosophy. He wrote so much that he had a special pannier made so he could write while on horseback. His two most famous books were the *Kitab al-Shifa'* (*The Book of Healing*) and *Al Qanun fi al-Tibb* (*The Canon of Medicine*). Despite its title, *Kitab al-Shifa'* is actually a scientific encyclopedia covering logic, natural philosophy, psychology, geometry, astronomy, arithmetic, and music. Although including many aspects of Greek thought, particularly Aristotle and Euclid, it does not simply recount those works. The *Al Qanun fi al-Tibb* became one of the most important sources of medical knowledge. It was both a translation of and a commentary on Galenic medicine and contains what is perhaps the first discussion of mental illness as a form of disease. When ibn Sina's work was discovered by Latin scholars, his name was translated as Avicenna, and his books helped fuel a drive to rediscover Aristotle.

A contemporary of ibn Sina was Abu Ali al-Hasan ibn al-Haytham (c. 965– c. 1039). Although ibn al-Haytham was not trained as a physician, he worked on vision, diseases of vision, and the theory of optics. In his *Kitab al-Manazir* (*Book of Optics*) he presented the first detailed descriptions and illustrations of the parts of the eye in optical terms and challenged the Aristotelian optics of Ptolemy. Where Ptolemy had supported the extramission theory of vision that was based on a kind of ray coming out of the eye and intersecting objects to produce sight, ibn al-Haytham supported the intromission theory that posited that light struck objects and that rays then travelled from the object into the eye. He also described refraction mathematically and undertook a series of experiments to demonstrate optical behaviour. Like ibn Sina, ibn al-Haytham was prolific, writing over 200 treatises and through which he became known to European scholars as Alhazen.

In addition to the social and philosophical space created for natural philosophy by the physicians, Islamic scholars also gained a powerful new tool in the form of an improved mathematical system. The tool was Hindu-Arabic numerals and placeholder mathematics. Originally an import from India, it profoundly changed Islamic scholarship, opening up new classes of problems and methods of calculation. It was pioneered by Muhammad ibn Musa al-Khwarizmi (c. 780–c. 850) in a work called *Concerning the Hindu Art of Reckoning*, which, in addition to the symbol set that was the precursor to the modern notation system, introduced zero as a mathematical object. While the Greeks had understood the concept of nothing, they had explicitly rejected "nothing" as a mathematical term, and it was not a necessary concept for geometry.

Al-Khwarizmi went on to produce *Al-jabr wa'l muqabalah*, which became known in the West as *Algebra*; we get the term "algorithm" from his name. It was from this work that solving for unknowns was developed. Al-Khwarizmi also demonstrated solutions for various quadratic equations including the use of square roots. Historians have argued over whether he was an original thinker or was only a compiler of earlier work such as parts of Euclid's *Elements* and Ptolemy's *Almagest*. Although the answer cannot be definitive unless new material is found, clues such as the fact that al-Khwarizmi's calculation of coordinates for locations were more accurate than those of Ptolemy suggest an intellect capable of difficult and exacting work.

The greatest Islamic natural philosopher of the era was Abu Arrayhan Muhammad ibn Ahmad al-Biruni (973–1048). A polymath by any standard, al-Biruni's studies covered astronomy, physics, geography and cartography, history, law, languages (he mastered Greek, Syriac, and Sanskrit and translated Indian manuscripts into Arabic), medicine, astrology, mathematics, grammar, and philosophy. He was taken by the ruler Mahmud (whether as a guest or prisoner is unclear) to India, where he composed *India*, a massive work that covered social, geographic, and intellectual aspects of Indian culture. He corresponded with ibn Sina and became known as al-Ustadh, meaning "Master" or "Professor." Among his accomplishments were calculating the radius of the Earth, finding that it was 6,339.6 kilometres (extremely close to the modern value); making detailed observations of a solar and lunar eclipse; and writing about the use of mathematics and instruments in his work *Shadows*.

CONNECTIONS

Intercultural Exchange: The Development of Islamic Cartography

Geography and cartography are probably the branches of science most likely to draw on knowledge and traditions of a number of different cultures and societies, since they require extensive travel or knowledge of other parts of the world. The development of Islamic cartography and geography demonstrate the complex interconnections among different knowledge communities, as information was communicated, appropriated, and adapted for use by the growing Islamic empires.

The earliest cartographic traditions in this area were an amalgam of pre-Islamic Arabian, Persian, and Indian influences. The earliest mapping was done during the Abbasid rule in Baghdad after 762, where the rulers encouraged science and literature and recognized that the conquered nations such as the Sassanids and Byzantines had much to offer. Indian knowledge was also seen as important and transmitted to the court through traders and scholars. Early geographical work owed much to the Indian traditions, especially placing the prime meridian at Ujjain (Arin), and seeing Lanka (Sri Lanka) as the "Cupola of the Earth" (the central point of the inhabitable world). From the Persians, geographers took up the concept that the inhabited world was divided into seven *kishrars* or regions, with six regions encircling the central one of the Iranian area, in a sort of lotus flower image.

Under the Caliph al-Ma'mun (r. 813–833), mapping began to develop. Al-Khwarizmi produced tables of longitude and latitude coordinates for places, influenced by Ptolemy's *Geographia*, which was translated during the

Alchemy

Islamic scholars were not content to confine their understanding of the world to a philosophical system. They wanted to utilize their knowledge, and the greatest exploration of the application of philosophy to the material world was in the study of alchemy. The etymology of the name encapsulates the intellectual heritage of the study. The origin of the term probably came from the ancient Egyptian *khem*, meaning black. The Greek *khēmia* meant "art of transmutations practised by the Egyptians," since Egypt was the Black Land. In Arabic, the Greek root was transformed into *al-kimiyā*, meaning "the art of transmutation" and hence from Arabic into Latin and English.

caliphate by Abu Yusuf Ya'qub ibn 'Ishaq aṣ-Ṣabbaḥ al-Kindi (c. 801–c. 873). Thus, Greek ideas started to interact with the earlier Persian and Indian views on the placement of landmasses and inhabited places on the earth. While Muslim geographers rejected Ptolemy's map projections (based on a grid of latitude and longitude coordinates), they were interested in establishing the coordinates for particular places. They were also able to correct Ptolemy's length of the Mediterranean. A deeply scholarly debate about the location of the prime meridian developed since Ptolemy had placed it at the Fortunate Isles in the west, in contrast to the Indian placement of the prime meridian in the east at Ujjain.

The earliest Islamic maps that still exist today came from a separate tradition, that of the Balkhi school of geographers of the tenth and eleventh centuries. These maps were based on knowledge from travel, trade routes, and postal routes of the far-flung empire. The cartographers were themselves extensive travellers, many from the western caliphates of Egypt and Palestine. For example, Abu al-Qasim Muhammad ibn Hawqal (travelled 943–969) was born in Upper Mesopotamia and spent his life travelling through Islamic Africa, Persia, Turkestan, and Sicily. In this way, practical experience was as important to geographical knowledge as the scholarly traditions of the earlier geographers.

By the mid-eleventh century, Abu Rayan al-Biruni was creating another tradition of Islamic geography and cartography. A prolific translator and mathematician, Abu Rayan al-Biruni became interested in the mathematical aspect of geography and cartography. He combined a strong knowledge and understanding of both Greek and Indian sources, as well as the work of the Balkhi school, to develop some new theories about the Earth. He remeasured a degree of latitude and tried (somewhat unsuccessfully) to measure the difference in longitude between locations. He developed a method to determine the direction of Mecca from any location, indicating the interconnection of geographical investigation and social and religious life. Mapping the Muslim world was only possible with these sorts of interactions among knowledge communities both east and west.

Alchemy was in some ways the precursor to the modern material sciences of pharmacology (iatrochemistry), chemistry, mining and smelting, and parts of physics and engineering, as well as aspects of biology such as the study of fermentation, decay, and reproduction. At a basic level, alchemists were trying to identify, classify, and systematically produce useful or interesting substances. Yet this aspect of alchemy, which may seem eminently useful and complete to us, was regarded as mere craft and not the objective of the study at all. The true study of alchemy was the manipulation of the material world, particularly the transformation of substances from one kind to another. It was in this study that alchemists ventured into a mystical realm that had spiritual and religious implications.

The transformation of material is in many ways an everyday occurrence. Wood turns into flame, ice turns into water, and seeds turn into plants. Some transformations seem more magical than others; smelting, for example, takes hunks of rock and transforms them into metal. All societies that manipulated materials developed systems of explanation that covered both the process and the reason materials could be transformed. These explanations were often kept secret not only for trade and safety reasons but also because powerful supernatural forces were involved and so involved religious concerns as well. Thus, alchemy developed both an exoteric, or public, aspect and an esoteric, or secret, element.

Islamic alchemy was founded on Egyptian and Greek ideas about the material world. Through the Egyptian connections came Hermeticism, from Hermes, the Greek name for the Egyptian deity Thoth, the father of book learning and creator of writing. Hermeticism was a blend of Egyptian religion, Babylonian astrology, Platonism, and Stoic thought. The Hermetic documents were likely compiled in the second century BCE and had strong occult aspects. To round out the spiritual side, alchemy was also affected by Gnosticism, which started in Babylon and influenced early Christianity. The Gnostics were strong dualists who saw the world in terms of antagonistic pairs such as good and evil, light and dark. Knowledge of certain things could be gained only by "gnosis," or enlightenment that came from inner awareness rather than reason or study. Both Hermetic and Gnostic studies acquired a heritage of secrecy because of potential conflicts with more powerful religions and the desire of adherents to guard their esoteric knowledge.

From the Greeks came the Aristotelian description of matter combined with neo-Platonic concepts of the Ideal. In addition to Aristotle's division of matter, in his *Meteorologica* (which discusses the condition of the terrestrial realm) the earth is described as a kind of womb in which metals grow. Less perfect or base metals such as lead have a natural inclination to become noble metals, seeking ultimately the perfection of gold if the conditions were right. This was linked to both Aristotelian and Platonic ideas about the original source of differentiated matter (the four elements) from a single undifferentiated prime matter. Prime matter had no "pattern," so the alchemists thought it could be made to take on the pattern of terrestrial matter. The key to this transmutation process was often thought to be a kind of catalyst. This agent was known by a number of names, but the most common was the "Philosopher's Stone," which was mentioned as early as 300 CE in the alchemical collection *Cheirokmeta* attributed to Zosimos (fl. 300 CE). Whether the Philosopher's Stone was an actual object, the product of alchemical processes, or a spiritual state depended on the theory of the alchemist.

It is also from Zosimos that we learn of one of the earliest women to practise alchemical research: Miriam, sister of Moses (c. third century BCE), later known as Maria the Jewess, although it is not certain that she was Jewish and she was not the sister of the biblical Moses. Miriam lived in Alexandria and was interested in chemical processes. Zosimos attributes to her the creation of a high-temperature double-boiler for experiments using sulphur and other equipment, and her name survives to the modern age in the French term *bain-marie*, referring to a double-boiler in cooking.

The intertwining of practical skills, religious and mystical thought, and philosophy plus the secrecy of the practitioners makes alchemy a difficult practice to trace or understand. Greek material from the earlier period is not extensive and mostly practical, dealing with dyeing, smelting, and pharmacology. One of the interesting connections that does survive is the association of the planets with various metals.

As bits and pieces of Greek natural philosophy were disseminated through the Arabic-speaking world, the texts on the nature and structure of the material world hinted at the ability to manipulate it. The beauty of secret knowledge is that it makes all things possible, so the lack of clear antecedents, rather than hindering the interest in alchemy, actually spurred its creation among Islamic thinkers. One of the greatest sources for both Islamic and later European alchemy was the work attributed to Jābir ibn Hayyān. His dates are uncertain, but most likely c. 722 to c. 815. While there may have been a real person with that name, it is clear that the majority of work ascribed to him was compiled by the Ism'iliya, a tenth-century Muslim sect; it is not certain which, if any, texts were written by him.

Over 2,000 pieces of text have been attributed to Jabir ibn Hayyan, most of them of much later production, but the *Books of Balances* and the *Summa Perfectionis* (in its Latin form) cover the central aspects of his alchemy. Jabir starts from an Aristotelian foundation, accepting the four elements and the four qualities, but extends Aristotle's idea of *minima naturalia*, or smallest natural particles, as the basis for the difference between metals. The more densely packed the particles, the denser and heavier the metal. The objective of the alchemist was to transform less noble metals into gold by manipulating the structure and packing of the particles by a process of grinding, purification, and sublimation. The process was also governed by mercurial agents that were either catalysts or active components in the process of change. These agents were referred to by Jābir as medicines, elixirs, or tinctures, which reinforced the biological model of metals—the purification of metal was seen as akin to curing disease or purification of the body.

○ or ☉	Gold [Sun]
☽	Silver [Moon]
♀	Copper [Venus]
♂	Iron [Mars]
☿	Mercury
♄	Lead [Saturn]
♃	Tin [Jupiter]

♈	01. Calcination	*Aries, the Ram*
♉	02. Congelation	*Taurus, the Bull*
♊	03. Fixation	*Gemini, the Twins*
♋	04. Solution	*Cancer, the Crab*
♌	05. Digestion	*Leo, the Lion*
♍	06. Distillation	*Virgo, the Virgin*
♎	07. Sublimation	*Libra, the Scales*
♏	08. Separation	*Scorpio, the Scorpion*
♐	09. Ceration	*Sagittarius, the Archer*
♑	10. Fermentation	*Capricornus, the Goat*
♒	11. Multiplication	*Aquarius, the Water-carrier*
♓	12. Projection	*Pisces, the Fishes*

2.9 ALCHEMICAL SYMBOLS The alchemical symbols linked the material world with the universe by assigning each element an astrological symbol and each operation a sign from the zodiac.

While Jabir's work (or that attributed to him) was very influential, especially in the Latin West where he was known as Geber, he was something of an anomaly among Islamic scholars because of his concentration on alchemy. More typical of those scholars engaging in alchemical work was Abu Bakr Mohammad Ibn Zakariya al-Razi (c. 841–925). Trained in music, mathematics, and philosophy and likely able to read Greek, al-Razi became a famous and sought-after physician. He was head of the Royal Hospital at Ray (near modern Tehran) and then moved to Baghdad where he was in charge of its famous Muqtadari Hospital. As a physician, he wrote *Kitab al-Hawi fi al-tibb* (*The Comprehensive Book on Medicine*), a massive 20-volume work that covered all of Greco-Roman and Islamic medicine, and *al-Judari wal Hasabah* (*Treatise on Smallpox and Measles*) that contained the first known description of chicken pox and smallpox.

For al-Razi, alchemy was less esoteric than it was for Jabir, and certain aspects of his work, such as the development of drugs and the use of opium as an anaesthetic, can be seen as an extension of his medical work. His most important alchemical text, *Secret of Secrets* or the *Book of Secrets*, does not, despite its title, reveal the secret of transmutation of base metals into gold. Rather, it is one of the first laboratory manuals. Divided into three sections, *Secret of Secrets* covers substances (chemicals, minerals, and other substances), apparatus, and recipes.

The list of equipment was extensive, covering beakers, flasks and jugs, lamps and furnaces, hammers, tongs, mortars and pestles, alembics (stills), sand and water baths, filters, measuring vessels, and funnels. This list remained quite literally identical to the standard equipment found in alchemical, chemistry, pharmaceutical, and metallurgical laboratories until the middle of the nineteenth century, and most of it is still familiar to chemists even today.

Although the *Secret of Secrets* did not offer a specific method of transmutation, it suggested strongly that it could be done. Al-Razi believed in transmutation and subscribed to the same general alchemical theory proposed by Jabir. What separates the two is al-Razi's concentration on practical issues and systematic approach. (See figure 2.10.) For al-Razi, alchemy developed from experience working with materials, rather than from a body of theory that presupposed chemical behaviour. Because of the practical advice he offered, his work became extremely popular in the Latin West, where he was known as Rhazes.

Astronomy in Islam

The stars had guided trade caravans from before the time of Muhammad, and astrology (developed by the Persian Zoroastrians) was important to Abbasid leadership, ensuring that astrologers had a high status in the courts of the early Islamic rulers. In addition to these uses for stellar observation, the injunction that the faithful should pray toward the Ka'bah in Mecca added a particular requirement that engaged astronomers and geographers (often, as in the case of Ptolemy, the same person) in a long and detailed program of

2.10 TABLE OF SUBSTANCES ACCORDING TO AL-RAZI IN *SECRET OF SECRETS*

MINERAL	VEGETABLE	ANIMAL	DERIVATIVE
(see chart below)	Little used	Hair	Litharge (yellow lead)
		Bone	Red lead
		Bile	Burnt copper
		Blood	Cinnabar
		Milk	White arsenic
		Urine	Caustic soda
		Egg	...
		Mother of pearl	Etc.
		Horn	
		...	
		Etc.	

TABLE OF MINERALS					
SPIRITS	BODIES	STONES	VITRIOLS	BORACES	SALTS
Mercury	Gold	Pyrite	Black	Bread borax	Sweet
Sal	Silver	Tutia	White	Natron	Bitter
Ammoniac	Copper	Azurite	Yellow	Goldsmith's boax	Soda
Orpiment	Iron	Malachite	Green		Salt of urine
Realgar	Tin	Turquoise	Red	...	Slaked lime
Sulphur	Lead	Haematite		Etc.	Salt of oak
	Chinese lead	White arsenic			(Potash)
		Kohl			...
		Mica			Etc.
		Gypsum			
		Glass			

1. These tables come from al-Razi, *Secret of Secrets*, in *Alchemy*, ed. E.J. Holmyard (New York: Dover, 1990) 90.

observation. Islamic astronomy was also necessary for timekeeping because the Q'ran mandated the use of the lunar calendar for all religious activities.

One of the first Islamic leaders to support astronomical work was the Abbasid Caliph al-Ma'mun in the ninth century. This helped to give astronomy a level of prestige that continued through the Golden Age. The first significant Arabic work on astronomy was *Zij al-Sindh* by al-Khwarizmi in 830. It was based primarily on Ptolemaic ideas, setting the theoretical framework for later astronomers, but it also marks the beginning of independent work in the Islamic world.

In 850 Abu ibn Kathir al-Farghani wrote *Kitab fi Jawani* (*A Compendium of the Science of the Stars*) that extended the Ptolemaic system introduced by al-Khwarizmi, corrected some of the material, and included calculations for the precession of the Sun and the Moon as well as a measurement of the circumference of the Earth.

The widespread interest in astronomy also led to the development of astronomical instruments. Although the astrolabe was well known to Greek astronomers, the technical skills of Islamic craftsmen led to the creation of very good astrolabes. One of the earliest surviving examples was made by Mohammad al-Fazari around 928. Using a variety of sundials, quadrants, and armillary spheres, Islamic astronomers compiled extensive star catalogues.

Although there were observatories in many of the major cities in the Islamic world, the most influential was established by Hulagu Khan in the thirteenth century at the city of Maragha. Its construction was overseen by the great Persian polymath Nasir al-Din al-Tusi (1201–74). In addition to his many scientific works, he identified the Milky Way as a collection of stars, an observation not confirmed in the West until the work of Galileo. Tusi is also famous for creating what is called the Tusi-couple, which places a small circle within a larger one so that a point on the small circle will oscillate in a regular fashion as the two circles rotate around their common centre. This mathematical device allowed Tusi to remove Ptolemy's awkward equant from astronomical calculations.

During the thirteenth and fourteenth centuries, astronomers following Tusi's lead were able to eliminate most of the extra motions associated with Ptolemy's schema, with the exception of the epicycles. Tusi's student, Qutb al-Din al-Shirazi (1236–1311), worked on the problem of Mercury's motion. Later, Ala al-Din ibn al-Shatir (1304–75), who worked as the religious timekeeper at the Great Mosque of Damascus, found a way to represent the motion of the moon. When Copernicus began his work, which would transform the model of the heavens by placing the sun at its centre, he appears to have had access to both Tusi's and al-Shatir's work, showing how instrumental Islamic astronomers were to the development of astronomy worldwide.

On the Heavens and Number around the Globe

The desire to understand nature through mathematics and astronomy has been a human impulse seen in almost every civilization. We can trace the development of these skills in the celestial observations left by people in the Americas, Australia, and the Pacific Islands. In particular, we have records from the Maya and the Aztec showing that they recorded the motion of the heavens and developed the mathematics needed to create calendars. These observations were primarily done for religious reasons, but they were also part of the practical planning of activities such as planting and harvesting.

The Mayan civilization built across the Yucatán Peninsula was ancient, beginning around 8000 BCE, but the age of greatest intellectual activity was during the Classical period from 250 CE to 950 CE. The Maya had good astonomers and mathematicians. Much of their work was done for religious purposes, but they left a record of significant mathematical and astronomical insight. The Maya recorded the motion of the sun, moon, Mercury, Venus, Mars, Jupiter, and many stars. Their system was geocentric and their observations allowed them to predict eclipses and chart the future position of stars, in many cases with greater precision than European astronomers of the period. They had a special interest in Venus, calculating its 584-day cycle with great accuracy. This may have been because Venus was astrologically associated with war and change.

The Maya created two calendars. The first was a solar count of 20-day periods known as *winal*. A year consisted of 18 *winal* plus a five-day period called the *wayeb*. This period was considered dangerous, when the division between natural and supernatural realms was opened. The Maya projected their calendar far into the future, calculating a span of 63 million years, although in practical terms the longest unit was the *ba'k'tun*, which recorded a period of 394 years. The second *tzolk'in* calendar was a 260-day cycle and was used for religious rituals. It is a subject of much debate about why the *tzolk'in* had 260 days, since this does not match up with any astronomical period. It may have been a numerological construct (13 × 20 for example) or even a measure of the human gestation period. Whatever the reason, the two calendars nonetheless spread throughout Mesoamerica.

According to one correlation between the Mayan *bak'tun* and the European Gregorian calendar, a *ba'k'tun* ended on December 20, 2012. This occurrence was taken by some to be a prophecy of apocalypse and was used as a plot element in the Hollywood movie *2012*. A number of pseudo-scientific documentaries such as

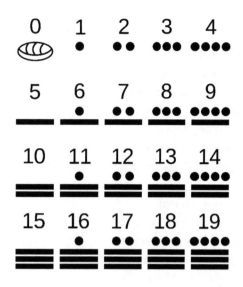

2.11 MAYAN NUMERALS

2012 Apocalypse presented the end of the *ba'k'tun* as a Mayan prophecy of doom. These shows had no foundation in history or science and are part of the problem with the misuse of science discussed in Chapter 13. In reality, the end of the *ba'k'tun* was simply followed by the start of the next.

Mayan calendars were in part so accurate because the Mayan civilization had a good mathematical system. The Maya used a placeholder vigesimal system (base 20) and had a symbol for zero. They used a series of dots and bars to write their numbers, making basic calculations easy. (See figure 2.11.) In many ways, the Maya were more advanced than their European or Asian counterparts when it came to number theory. What has not been discovered is any systematic use of geometry. Although Mayan architecture makes it clear that they could create a variety of angles, including a consistent 90° angle (probably using knotted ropes) and could align structures to the cardinal directions, we have not discovered any indication that they had developed theoretical principles of geometry.

The Aztec empire began around 1426 and centred on the city of Tenochtitlan (now Mexico City), north of the region controlled by the Maya. The Aztec empire lasted until the Spanish conquest in 1520. Given their proximity to the Mayan world, it is not surprising that the Aztecs had a similar affinity for mathematics and astronomy. Astronomy was so important that it was part of formal education at the *calmecac* schools, along with writing and theology. The Aztecs tracked the motions of the stars and planets and used the observations in the construction of temples and buildings. The best-known example of this is the Templo Mayor, oriented so that on the March 21 equinox, the sun was observed between the Tlaloc and Huitzilopochtli shrines. The Aztec astronomer-priests used the same two calendar systems as the Maya. One of the most important Aztec artefacts is the Calendar Stone (or Sun Stone). (See figure 2.12.) Carved c. 1479, it shows both calendars and a host of religious symbols.

Aztec mathematics borrowed ideas from the Maya and Olmec people. Like the Maya, the Aztecs had a vigesimal system and used dots and bars to represent numbers, adding other symbols for larger numbers. Unlike the Maya, the Aztecs do not appear to have used a symbol for zero, but understood the concept. Mathematics was used for calendars, but also for taxes based on land area of farms and surveying. The Aztec language includes words for tools such as a plumb line and a level.

There was no contact between the Americas and the rest of the world until the Europeans arrived in the late fifteenth century. Their ideas do not appear to have influenced other civilizations, but the history of this region tells us about the widespread interest in mathematics and astronomy. Anywhere people could look up at the sky over a period of time, they recorded and exploited their observations.

In the European context, Western historians have become increasingly aware that international contacts in the first centuries CE were greater than previously imagined. While contacts between the West and the Islamic world were direct, the vast distances and often rugged terrain that separated Europe and the Middle East from Asia were thought to have precluded or seriously limited contact between the two regions until the spread of Islam. It is now clear that this was not the case. For example, when Alexander the Great took his army east into India in 326 BCE, he already knew that great cities and empires existed on the subcontinent and beyond. This raises a complex question about the origins of Western science, since the ideas that have been associated with European natural philosophy may have been influenced by ideas from other parts of the world and in turn may have influenced other cultures. We can no longer claim that the origin of science was an independent product of the West; rather, it involved complex and long-standing trade in ideas and technologies with parts of the world far away from Greece and Rome.

We know that a host of inventions including gunpowder, paper, and the spinning wheel made their way from China to Europe, while from India came Hindu numbers (which we also call Arabic numbers in recognition of their path through the Muslim world), placeholder mathematics, and wootz steel, better known in the West as

2.12 AZTEC CALENDAR STONE

Wikipedia user Keepcases. Licensed under the terms of CC-BY-SA.

Damascus steel (again because the contact for Europeans was the Islamic world). What is less clear is whether particular explanations and discoveries were transmitted in a similar way, or whether they were arrived at independently in each culture. For example, every society that recorded the motion of the sun and moon independently discovered the solar and lunar calendars, and all societies interested in mathematics developed some version of the Pythagorean theorem.

The rise of the empires of the East was predicated on the same developments as those of the Mediterranean basin: agriculture, job specialization, bureaucracy, and urbanization. In India and China, natural philosophy became a part of the intellectual heritage of those regions, with Indian scholars more closely linking their observations of nature to the Vedic texts that were the foundational religious and social texts for many of the people of the region, while Chinese and Far Eastern scholars tended to be more pragmatic, in keeping with the less supernatural ideas of Taoism. Asian scholars were also influenced by Buddhism as it spread after the fifth century BCE.

The Vedic tradition in India came from a series of texts written between about 1500 and 900 BCE. These texts, while primarily religious, also include material about mathematics, geometry, biology, and medicine. Vedic mathematics was developed as part of the methods for correctly performing rituals and was studied as part of the six disciplines of the *Vedangās* (*Ancillaries of the Vedas*) starting around the sixth century BCE, especially *kalpa* (rituals) and *jyotiṣa* (astrology). Mantras from this period also showed an interest in large numbers, with some mantras naming units up to a trillion.

The *Baudhayana Sulba Sutra* by Baudhayana (fl. eighth century BCE) was an early mathematical text that identified what we would call the Pythagorean relationship and gave some of the common whole-number triplets (3, 4, 5 and 7, 12, 13). It also gave a formula for the square root of two, indicating that the mathematicians of the time understood that the most basic solution to the relationship of the sides of a right triangle to the hypotenuse could not be expressed as an integer.

One of the greatest gifts to mathematics within the Vedic tradition was the invention of positional notation using numerals. The origin of this system is not clear, but by 499 CE the astronomer and mathematician Aryabhata was using placeholders, although he used letters rather than numerals. Numerals much closer to modern symbols came into use around 600 CE. This system of notation was known to Syrian scholars by 662 CE and was later made much more common when they were adopted by Arabic scholars such as Al-Khwarizmi (which is why we now call them Arabic numerals). The first European mention of the new number system was in the second *Codex Vigilanus*, a compilation of different writings, completed in 976 CE.

The earliest known Indian astronomical text was the *Vedānga Jyotiṣa*, composed sometime between the sixth and fourth centuries BCE. Although it was religious in nature and created in part to regulate religious observance, it contained practical information on solar and lunar cycles, a list of planets, and guidance for celestial observation. Aryabhata argued that the Earth rotated, while still placing it in the centre of the universe in a geocentric model.

One of the most intriguing natural philosophical ideas came from the Hindu scholar Kanada (fl. second century BCE, although dates as early as sixth century BCE have been proposed). Kanada was interested in matter theory, studying a form of alchemy known as *rasavādam*. In part of his work, he argued for the existence of atoms, which he described as indivisible, indestructible, and eternal. Some modern scholars have suggested that Kanada's atomism, while more abstract than the Greco-Roman atomism of Democritus, was more complete.

Although Indian natural philosophers undertook a wide range of investigations and achieved many notable insights, they did not separate natural and spiritual studies of nature, as the Greeks did. Political, religious, and military turmoil such as the Arab conquest of the Sindh c. 712 may have disrupted the development of Indian natural philosophy in the sub-continent. Once India was part of the Muslim world, Islamic natural philosophy (borrowing useful concepts from the Indians) was supported by the rulers.

An important bridge between China, India, and the Mediterranean world was created around 263 BCE when Ashoka the Great (304–232 BCE), who controlled most of modern-day India, Afghanistan, Pakistan, and Bangladesh, converted to Buddhism. Ashoka sent emissaries to neighbouring regions as far away as Alexandria in the west and Burma in the east. Buddhism likely reached China around 70 CE, and opened a wider exchange of ideas between East and West. Some modern scholars have argued that Indian mathematics in particular was influenced by Chinese work, while Greek natural philosophy flowed into India through the remnants of the Greco-Bactrian kingdom established by Alexander the Great.

Natural philosophy in China represents a challenge to modern historians and philosophers of science. A massive 24-volume history of Chinese science and technology entitled *Science and Civilization in China* (1954–2004) was undertaken by the scholars Joseph Needham (1900–95) and Wang Ling (1917–94), so the vast range and depth of knowledge and invention are well known. China was, for most of its early history, the richest and most technologically advanced empire in the world. Its scholars were highly educated and commanded vast resources compared to European or even Islamic scholars. Several specific areas of research, such as alchemy and astronomy, were extremely well developed, but despite these advantages, a unified natural philosophical system did not develop.

Alchemy in China was closely tied to Taoist ideas and medicine, and there was no clear distinction between alchemical work and what we would call pharmacology. The study of alchemy extends back to at least the second century BCE, since concern about alchemists existed from at least 144 BCE, when the emperor issued an edict outlawing the making of "counterfeit" gold on pain of death. Although there was much work done on transmutation, the primary focus was on immortality. One of the oldest alchemical texts was *Tsan-tung-chi* (authorship unknown), that appeared around the third or late second century BCE. It described the way to make a golden pill that would make a person immortal. The noted Taoist scholar and high government official Ko-Hung (also Ge Hong, 283–343 CE) wrote extensively about alchemy and immortality. His general theory was based on the purification and transmutation of metals as a way to remove those negative aspects of biology that caused aging. Chinese alchemists, like Western alchemists, were in a difficult position, having to reveal enough to establish their skills, while needing to keep specific details of their work secret, both as trade secrets and to protect people from the dangers (spiritual and physical) of alchemy. While a strong *materia medica* developed, alchemy never became a comprehensive study of matter in China.

Chinese cosmology was a more unified study than alchemy, but it was also tied very closely to Taoist ideas about existence and the place of things in the universe. Early Chinese astronomers identified the solar and lunar calendars and plotted the paths of the visible planets. They were good observers and noted the passage of comets as well as the appearance of a supernova in 1054. There was a surge of astronomical work during the Han dynasty (25–220 CE), and then again during the Tang dynasty (618–907 CE) when an influx of Indian astronomers arrived in China with the spread of Buddhism.

Chinese observers were particularly adept at creating star catalogues, and more than 2,000 stars were listed by Zhang Heng (78–139 CE), who also calculated solar and lunar eclipses. The information from these catalogues was also used to create some very detailed armillary spheres, and the astronomer Su Song (1020–1101) and his colleagues created a massive mechanical clock tower in 1092 that included a moving armillary sphere and a celestial globe. (See figure 2.13.)

After the establishment of the Mongol Empire, Chinese astronomers worked more closely with Islamic scholars. Kublai Khan (1215–94) brought Iranian astronomers to

Beijing to build an observatory and started a school of astronomy around 1227. A series of observatories were built in Beijing, and the one completed in 1442 has been preserved and is one of the oldest pre-telescopic observatories still in existence. Through these contacts with Islamic astronomers, the Chinese learned about the Ptolemaic system.

By the sixteenth century, the Chinese invited a number of Europeans, particularly Jesuits, to teach them European natural philosophy and Ptolemaic astronomy. Although there was debate among Chinese astronomers about the Ptolemaic model, most of them rejected it because it required a material substance to occupy space, while their most widely held view was of celestial objects in an infinite void. As was the case with Indian science, Chinese science was powerful in particular subjects but never developed an overarching explanatory model or method that was not religious in nature.

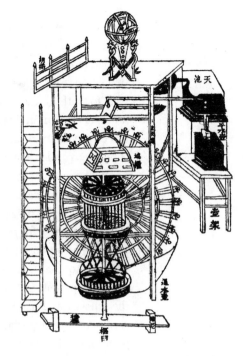

2.13 CHINESE MECHANICAL CLOCK (1092)

Chinese mechanical clock. From *Hsin I Hsiang Fa Yao*, ch. 3, p. 4a.

Conclusion: The End of the Islamic Renaissance

One of the last great thinkers of the Islamic Renaissance was Abul-Waleed Muhammad ibn Rushd (1126–98), who was known in both the Arabic and the Latin worlds as the Great Commentator or simply the Commentator. Rushd was educated in philosophy and trained as a physician, but he worked primarily as a judge and expert in jurisprudence and lived most of his life in Spain at Córdoba. In many ways Rushd represents both the power and the waning of power in Islamic natural philosophy. His commentaries on Aristotle were not based on primary sources but rather on Arabic translations, so he was not attempting to return to the original material. He wrote three sets of commentaries: the *Jami*, a simplified overview; the *Talkhis*, an intermediate commentary with more critical material; and the *Tafsir*, which represented an advanced study of Aristotelian thought in a Muslim context. These were fashioned as educational steps to take the novice from an introduction to an advanced understanding of Aristotle and, in effect, created an Islamic Aristotle.

What Rushd added to natural philosophy was not original work on nature, but a powerful synthesis that represented a well-established intellectual heritage. He presented the most dedicated version of Aristotelianism, essentially arguing for the perfection of Aristotle's system of logic and philosophy. From this position, Rushd argued that there were two kinds of knowledge of truth. The knowledge of the truth that came from religion was based on faith and thus could not be tested. It was the path to truth for the masses, since it did not require training to understand because it taught by signs and symbols. Philosophy, on the other hand, presented the truth directly to the mind and was reserved for an elite few who had the intellectual capacity to undertake its study. This did not mean that religion and philosophy conflicted but that they could *not* conflict. Philosophy might be an intellectually superior way of understanding truth, but it did not make truth. It followed that any truth revealed by religion would be the same as the truth arrived at by philosophy.

This support for philosophy and the declaration of the relation of philosophy and religion had a profound influence on medieval European scholars, particularly Thomas Aquinas (1225–74). Known as Averroes to the Latins, Rushd's work became a key component for much of their work on Aristotle, especially for a group called the scholastics. Rushd was both a source for medieval Aristotelianism through the commentaries (especially before Aristotle's work itself was widely available) and a philosophical anchor. His support for the intellectual superiority of philosophy also made his work the target of criticism, and supporters were accused of heresy or even atheism. The position Rushd argued has been echoed in considerations of the relationship between science and religion down to the modern day. For many natural philosophers and scientists the idea that the study of nature (philosophy) could reveal the truth of God's creation was not merely a justification for the reconciliation of reason and faith but a call to pursue the study of nature.

The interest in natural philosophy that grew during the Islamic Renaissance faded as the Islamic world became fractured and in general more conservative. The *hakim* were often brilliant individual thinkers, but for the most part they failed to reach a kind of critical mass that would make research in natural philosophy a desirable commodity. The natural philosophers were also victims of their own success, for, having created a model of ideal Arabic natural philosophy (particularly in areas such as medicine and astronomy), the schools slowly shifted from active research to perpetuation of established work. The failure to create an enduring research ethic was the result of a number of factors such as the political turmoil that disrupted all aspects of society and the swings from a high level of religious tolerance to strict fundamentalism that occurred almost overnight when leadership changed hands, making it potentially dangerous to engage in work that suddenly might be deemed unacceptable. It may also have reflected the level of respect accorded to the work of the great thinkers, which made new work more difficult to disseminate as the old work increasingly came to be seen as orthodoxy.

Cultural and religious changes also affected the place of natural philosophy. Mysticism on the one hand and more doctrinaire Islam on the other rose as the dominant religious forces in the thirteenth and fourteenth centuries. The Islamic Empire was increasingly under threat militarily, with incursions by the Mongols in the east, the reconquest of Spain in the west, and infighting among the kingdoms within the empire itself. Islamic law increasingly defined the proper sphere for human creativity; this did not include skepticism or any real place for personal opinion or secular corporate identity. In parts of the Muslim world pictures of people or nature were banned because it was thought they were idolatrous. This placed rather severe limits on certain kinds of investigations such as botany and hampered the communication of observations through texts. The rulers of religious states were also concerned that philosophy of any kind would conflict with theology, and so they were less willing to support work by scholars interested in those topics. There may also have been an element of psychological superiority that came from the power and glory of the richest Islamic states. In the early days of Islam, Greek knowledge and Roman power were still part of common knowledge, but 500 years after the fall of Rome and the end of the Byzantine Empire, the old world had clearly been surpassed by the new. Why then waste time and effort studying the remnants of a failed (and pagan) society?

Even the appearance of the barbarous and ill-educated knights of Western Europe seemed of little threat to the power of the Islamic world.

Essay Questions

1. What problems did Ptolemy's system solve and what problems were left unsolved?

2. Why do we consider Galen an Aristotelian?

3. How did natural philosophy develop in the Islamic world? Were Islamic scholars important innovators?

4. What topics were most important to Chinese natural philosophers and why did their study of nature develop as it did?

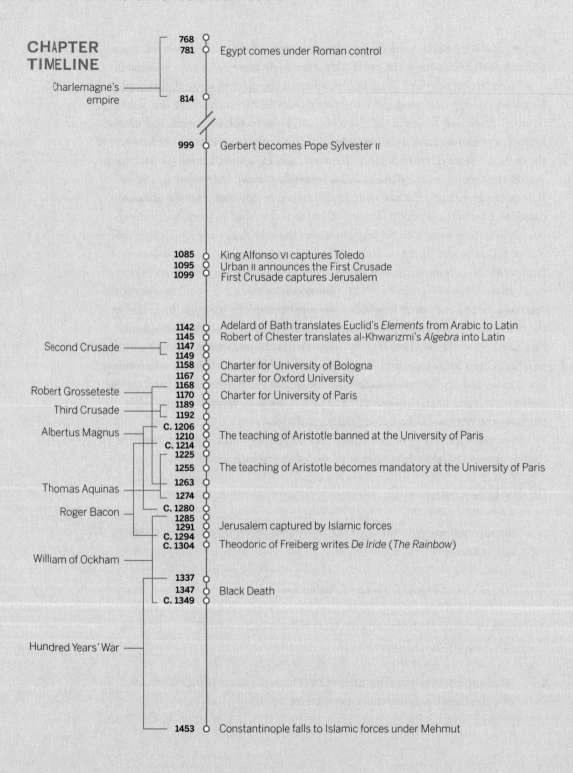

CHAPTER TIMELINE

Charlemagne's empire

768	
781	Egypt comes under Roman control
814	
999	Gerbert becomes Pope Sylvester II
1085	King Alfonso VI captures Toledo
1095	Urban II announces the First Crusade
1099	First Crusade captures Jerusalem
1142	Adelard of Bath translates Euclid's *Elements* from Arabic to Latin
1145	Robert of Chester translates al-Khwarizmi's *Algebra* into Latin

Second Crusade — 1147 / 1149

1158	Charter for University of Bologna
1167	Charter for Oxford University

Robert Grosseteste — 1168 / 1170 — Charter for University of Paris

Third Crusade — 1189 / 1192

Albertus Magnus — C. 1206 / 1210 — The teaching of Aristotle banned at the University of Paris

C. 1214 / 1225

1255	The teaching of Aristotle becomes mandatory at the University of Paris

Thomas Aquinas — 1263 / 1274

Roger Bacon — C. 1280 / 1285

1291	Jerusalem captured by Islamic forces

C. 1294 / C. 1304 — Theodoric of Freiberg writes *De Iride* (*The Rainbow*)

William of Ockham —

1337

1347	Black Death

C. 1349

Hundred Years' War —

1453	Constantinople falls to Islamic forces under Mehmut

THE REVIVAL OF NATURAL PHILOSOPHY IN WESTERN EUROPE

3

S

uccessive waves of invasions following the fall of Rome disrupted all aspects of life in Europe. The physical destruction of war and economic collapse destroyed many collections of texts, educational institutions fell into ruins, and society turned from the pursuit of knowledge and empire to basic survival. Despite the dire conditions, not all ancient knowledge was lost. Greek works survived in the Byzantine Empire, and certain texts remained known in the West, including most of Plato's *Timaeus*, parts of Galen's medical treatises, elements of Ptolemy's astronomy, and some of Boethius's studies in mathematics and astronomy, as well as Aristotle's logic. These resources were valuable but scattered and fragmentary. The best and brightest minds were gathered by the Church and turned their thoughts to questions of theology. Since certain aspects of Christian theology had to deal with issues of the physical world, there continued to be a need for information about the material realm, whether it was astronomy for calendars to keep track of feast days and observances or medicine to meet the Church's duty to care for the ill. In the early days of the Church, there was a struggle between those inclined to intellectual activities and those who favoured a more mystical approach. In the long run, the greater managerial skills of the intellectual wing came to dominate the administration of the Church, and the study of nature was included in Western intellectual practice.

In the Latin West during the course of the Middle Ages the Christian Church succeeded in establishing itself as the authority over intellectual as well as spiritual concerns. Therefore, just as had been the case in Islamic countries, supernatural and spiritual issues became intertwined with natural philosophical ones. Thus, despite the influence of Greek natural philosophy in the medieval West, the study of nature became a battleground for the primacy of natural or supernatural explanations once again. At stake were the questions of who controlled knowledge and who had the ultimate authority over truth claims. The answer was a reinventing and reordering of the intellectual universe, with a separation made between the spiritual and natural (or mystical and rational) which was different than that of the Greeks but similarly powerful. As long as that separation was controlled by the Catholic Church, the result was a well-ordered, carefully moderated series of disputations about nature and humanity's place within it. When the Church began to lose its authority in the sixteenth century, that very separation exploded into a cacophony of multiple voices.

The universities became the dominant and necessary space in the creation of both these careful rules of knowledge and the later tensions. The universities, founded in the twelfth century and beyond, provided space sanctioned by the Church and yet were not completely under the Church's control, since scholars were taught not only the prevailing system of scholasticism, which was focused on understanding the revealed truths of Christianity through rigorous syllogistic logic, but also to contest Greek philosophical ideas and methods of questioning while incorporating them into the powerful system of scholasticism. Competition was built into the system; ironically, those very places created to determine and preserve orthodoxy became the site for alternative natural philosophies in later centuries.

Those who studied nature in the medieval period were as concerned with the method of acquiring knowledge as with its application, and so there developed a complex dialogue concerning utility and practicality. Unlike Muslim scholars, who were interested first in applicable sciences such as medicine and astronomy, Latin scholars were first concerned with the use of natural knowledge as a path to salvation. While some of them explicitly experimented with nature and looked for applications in military, alchemical, and cartographic contexts, others were concerned about the implications of such action. Thus, the utility of natural knowledge, relatively unproblematic for Islamic scholars, was a hard-fought question for Europeans.

During the sixth and seventh centuries Europeans had only limited access to Greek, Roman, and Islamic natural philosophy. By the ninth century growing intellectual activity in Western Europe, particularly in parts of France and the rich Italian city-states, began to support new inquiries. This interest was further spurred as Europeans came into contact with the material and cultural wealth of the Islamic world. Material that Islamic scholars had preserved, commented on, and expanded, especially in the areas of logic and mathematics, medicine, alchemy, astronomy, and optics, increasingly came to the attention of Latin scholars. The "People of the Book," those who shared the Old Testament as a foundational religious document, were officially tolerated by Islamic rulers, and, as a result, Christian and Jewish, Muslim scholars were often able to visit and use the resources held in Islamic territory. Jewish scholars, who had ties with both Europe and the Middle East and were often multilingual, acted as a bridge between cultures. The libraries in Moorish Spain, particularly the one in Córdoba which contained over 400,000 volumes, became centres of education and recovery of the Greek texts that had been lost to the Latin West.

Charlemagne and Education

During the short-lived Carolingian Empire, which lasted only from 768 to 814 during the reign of Charlemagne (742–814), there was both a renewed interest in intellectual activity and the rebirth of the concept of an empire capable of matching the achievements of ancient Rome. Charlemagne claimed the title of Holy Roman Emperor and thereby established a new Roman era, if not exactly a new Roman Empire. His drive to create a European empire had more than a political effect, because it also shaped people's attitude toward the future. The early Middle Ages were tinged with a certain pessimism and a somewhat backward-looking view of society. This came in part from the belief held by many that the world was entering the end days as described in the Bible. Throughout Europe there were, literally, concrete examples that the past was better than the present, as the remains of the power of Rome dotted the landscape. Ruins of aqueducts, roads, and coliseums were a continual reminder of lost power and lost knowledge. Charlemagne's success started people thinking about the possibility of reclaiming the glory of Rome and about a future that might be better than the present. Matching the wonders of Rome required knowing what the Romans had achieved, and so attention turned to the Greco-Roman heritage.

Charlemagne was a superb general, but even more he was an astute politician who recognized that winning an empire was not the same as holding it together. Citizens must be persuaded to believe that they were better off in the empire than on their own, so Charlemagne worked to establish a uniform system of law, organized the military, improved the churches, and created public works. He placed education at the heart of his reforms, attracting Europe's foremost scholars to his court at Aachen (Aix-la-Chapelle) to manage the empire and help create this new culture.

CONNECTIONS

Natural Philosophy
and Education:
Alcuin and the
Rise of Cathedral
Schools

The growth of natural philosophy or science has always required an education system, since the principles necessary for the systematic study of nature are not self-evident and must be taught. Without such an education, knowledge was easily lost, and worse, the methods to acquire knowledge were also lost. The greatest example of this was the period after the fall of Rome, often called the "Dark Ages," when the light of learning nearly disappeared from Western Europe.

In the Ancient world, most formal education was provided by tutors and then only for the wealthiest families. Advanced education at schools like Plato's Academy or Aristotle's Lyceum trained the elite of the elite. The only other sources of education were the temples, some of which taught basic skills in literacy and mathematics to disciples. After the end of the Roman Empire, only a few churches, mosques, and synagogues offered basic education so that a handful of people could read the holy documents.

When Charlemagne became the Holy Roman Emperor in 800 CE, he had a big problem. Many priests were illiterate, so they could not read the Bible and perform the liturgy. The lack of literacy also meant that running an

The most prominent of these scholars was Alcuin (735–804), who had been educated in Ireland and was head of the cathedral school of York. There the monks had developed a curriculum based on a combination of classical training and Christian theology.

In 781 Alcuin met Charlemagne, who asked him to join his court and be his minister of education. Alcuin accepted and, in addition to developing a school system, educated the royal family and acted as a private tutor to the emperor.

Alcuin helped Charlemagne establish cathedral and monastery schools by imperial edict, and in turn these schools produced clerics with increasing levels of literacy and scholarship. Priests were to be literate, and Charlemagne charged the bishops with the responsibility of ensuring literacy and the delivery of proper religious observance, particularly the reading of the liturgy. While in Charlemagne's service, Alcuin was also instrumental in collecting manuscripts and establishing *scriptoria* for the copying and dissemination of the texts.

Alcuin's curriculum provided the foundation for education in Europe for over 1,000 years. His system was based on the study of the seven liberal arts, divided into two sections called the *trivium* and the *quadrivium*. From the Latin *liber* meaning empire was difficult, since everything from long-distance communications to government reports and tax collecting all required literary and mathematical skills that few people possessed.

Charlemagne gathered many scholars at his court at Aachen. He invited Alcuin of York, one of the most learned men in Europe, to join his court as a member of the Palace School and as his personal tutor. Alcuin influenced Charlemagne to enact educational laws for the Church and to require bishops to establish schools. These schools, called cathedral schools because they were housed in the home churches of the bishops, taught the clergy to read and write. The utility of literacy prompted the expansion of schools to monasteries and even lay (non-religious) schools in towns and cities. Monasteries began to copy and preserve texts in *scriptoria*. A number of

cathedral schools went on to become universities, with the University of Paris being the most famous. By the Third Lateran Council in 1179, the spread of literacy and education led to the Scholastic Movement to recapture the knowledge of antiquity.

Without education, natural philosophy would have disappeared completely from Western Europe. At first, the Church focused on practical aspects of natural philosophy, especially medicine (Galen) and astronomy (Ptolemy), but literacy opened the door for a much broader investigation of natural philosophy. Despite Alcuin's best efforts, however, Charlemagne never learned to read. It was said that he slept with books under his pillow in hopes that the knowledge might transfer by proximity (a practice some students are suspected of continuing to the modern day!).

free, the liberal arts served the purpose of educating the free man to be a good citizen, in contrast with the *artes illiberales*, which were studied for economic gain.

Trivium means place where three roads meet, but it also implies a public space. The three subjects of the trivium were logic, grammar, and rhetoric, and mastering these was the essential first step of education. Through clear thinking, clear writing, and correct speech in Latin (the *lingua franca* or universal language of European scholars), a person was prepared to participate in civilization. The *quadrivium* (or four roads) consisted of geometry, arithmetic, astronomy, and music. Music was the branch of mathematics that investigated proportions and harmony, which might include studying singing or playing instruments but was really concerned with the underlying mathematical theory. The two halves of the liberal arts curriculum represented the two ways of understanding the world, first through language and, once that was mastered, through the patterns of the world discernible only through mathematics.

One of the most gifted students to come out of the reformed schools was Gerbert (c. 945–1003). He studied in France and Spain before becoming headmaster of the cathedral school at Reims. He later became the archbishop of Reims, then of Ravenna in Italy. With the patronage of Otto III of Saxony, he was elected Pope Sylvester II in 999. Gerbert was deeply interested in logic and mathematics and was involved in efforts to find, translate into Latin, and copy Greek and Arabic texts on natural philosophy. When he became pope, he set the tone for the whole Church, raising the profile of natural philosophy and reinforcing the intellectual side of theology.

The Crusades and the Founding of the Universities

Although Alcuin and Gerbert established an intellectual tradition in the Church and began to prepare an audience for Greek and Islamic scholarship, they represented a tiny group interested in the still arcane study of philosophy. Churchmen of this period had a complex reaction to natural philosophy. Augustine, one of the most influential Christian thinkers, felt that natural philosophy could be an aid to theology, but revealed knowledge was always superior to discovered knowledge if there was any apparent conflict. Many theologians argued that the study of the natural world at best was irrelevant and at worst impeded one's hope of salvation. To place Greek natural philosophy at the heart of European scholarship required more than a slow acquisition of the ancient works and their Arabic commentaries and additions. What spurred the Europeans to the greatest action was the military struggle first against Islamic expansion and then for control of Jerusalem and the Holy Land. This both changed the culture of Europe and dramatically increased interest in the Greco-Roman world.

The Mediterranean Sea was almost completely under the control of Islamic forces who held Spain, North Africa, the Middle East, and Asia Minor. In 734 CE their western push into Europe stopped when Charles Martel defeated an Islamic army at Poitiers, ending further challenges to Frankish lands beyond the Pyrenees. Eventually Christian forces pushed the Moors out of Spain, starting with the capture of Toledo in 1085 by King Alfonso VI, although the last of the Islamic territory there was not captured until the late fifteenth century.

The expansion of Islam in the east was resisted by the Byzantine Empire, but under successive waves of Islamic forces starting with Suleman, the eastern European region was slowly conquered. Finally in 1453 Mehmut's army defeated the last holdouts in Constantinople, ending the Byzantine Empire. Refugees from the fall of Constantinople brought manuscripts and a knowledge of Greek to Western Europe, adding a second wave of interest in ancient philosophy. Constantinople was renamed Istanbul by the invaders, and from this base on the western side of the Bosphorus Islamic incursions into the west did not end until the second defeat of Ottoman Empire forces at the gates of Vienna in 1683. Many of the modern problems of the Balkans stem from the historical flux and mix of people and religions that long years of warfare brought to the region.

These external threats to Latin Christendom as well as domestic conditions led Pope Urban II to call Christians together for the First Crusade in 1095. Europe had entered a period of stability that left many of the nobility with little to do but fight among themselves. The knights of Europe were more Spartan than Athenian, mostly illiterate and trained from an early age to withstand the rigours of combat and not much else. With little in the way of new land available, the ruling class was under pressure to provide for second and later sons, since frequently little inheritance was left after the rules of primogenitor placed all the family lands in the hands of the eldest son. When Alexius I Comnenus, the Byzantine emperor, called for help against the Seljuk Turks, a crusade seemed a good way of dealing with many issues at once. Emboldened by the success of Alfonso in Spain, Urban believed that the Latin West could come to the aid of the Greek East against a much-feared enemy, while at the same time the largely idle knights could practise their profession far from home. For the nobility there was pious warfare, adventure, and the potential for land and wealth, while for the Church there was the possibility of controlling the Holy Land, conversions, and striking a blow against a competing faith. And for those supplying the Crusaders, there were significant profits.

The first three crusades—1096–99, 1147–49, and 1189–92—had some success from the Crusaders' point of view, with Jerusalem falling to Christian forces in 1099. Although the capture of Jerusalem was symbolic, the territorial gains were never great, and the European hold on the Holy Land was short-lived. What the Europeans really gained was renewed contact with a wider world. In a sense, natural philosophy returned to the Latin West because its people discovered a craving for spices, silk, fine china, ivory, perfume, and a host of exotic luxury items, many of which came from Asia along the Silk Road and through the Middle East. With these goods, they also traded ideas. Although east-west trade had never completely been cut off, the expansion of the trade in luxury items made cities such as Venice and Florence extremely wealthy. That wealth in turn financed the intellectual and artistic boom of the Renaissance, while adding to the wealth of the Arabic world that controlled the trade between Asia, Africa, and Europe. The desire of Europeans, especially those unable to participate in the Mediterranean trade, to avoid the middlemen and trade directly with the East was also the spur to global explorations in later years.

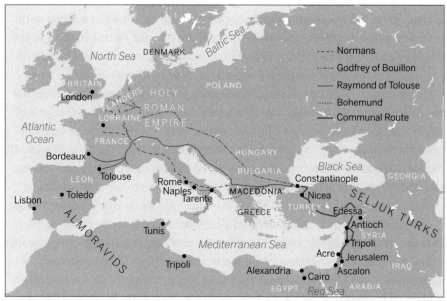

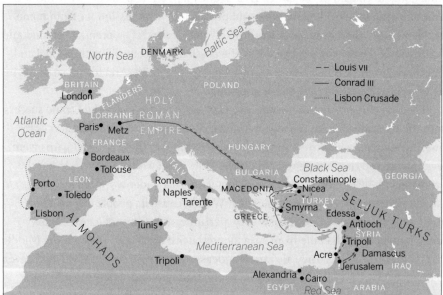

3.1 THE FIRST TWO CRUSADES

This expansion of commerce promoted urbanization, and, in turn, increasing urban populations could support education, including higher education in theology as well as secular topics such as law, the liberal arts, and medicine. Developing in large part out of the cathedral school system established by Charlemagne and in part from models copied from Islamic schools, the first European universities were founded in this period. The University of Paris claims it began in the early 1100s, making it the oldest institution of higher learning in Europe, but by charter the 1158 founding of the University of Bologna is probably the earliest officially organized university. Oxford University was founded in 1167, and the University of Paris was formally established by 1170.

The creation of the universities legitimized the study of natural philosophy and provided a place for scholars to live and work. They became the centres of intellectual debate and the repositories of manuscripts both old and new. As teaching organizations they produced more intellectually rigorous theologians and helped raise the level of literacy among the clergy. They also performed a vital role in training the growing secular managerial class. As well as holding positions of power in Church and government bureaucracies, these literate and university-trained students became essential members of the noble and princely courts.

Just as Islamic scholars had first gathered Greek philosophy and then produced Arabic translations, Western scholars eagerly sought out Arabic manuscripts and set them in Latin. During this period of rapid translation a number of scholars were key in introducing natural philosophy to the Latin audience. Adelard of Bath (c. 1080–c. 1152) undertook a number of translations, concentrating on mathematical texts such as al-Khwarizmi's *Astronomical Tables* and *Liber Ysagogarum Alchorismi* around 1126. In 1142 he translated Euclid's *Elements* from Arabic, opening the door to Greek geometry and mathematics. He also attempted to put together much of the new knowledge of natural philosophy in *Questiones Naturales*, written in 1111. Stephen of Antioch (fl. 1120) translated Haly Abbas's *Liber Regalis*, a medical encyclopedia, in 1127. Robert of Chester (fl. 1140) followed Adelard's mathematical work with a translation of al-Khwarizmi's *Algebra* in 1145. Eugenius of Palermo (fl. 1150) translated Ptolemy's *Optics* in 1154, and Henricus Aristippus (fl. 1150) finished Aristotle's *Meteorologica* in 1156. Galen was translated into Latin by Burgundio of Pisa around 1180, introducing another aspect of natural philosophy through medicine.

It was an exciting time for scholars as new knowledge was uncovered one manuscript at a time from the treasure trove of Arabic sources. One of the most important conduits for Greek and Arabic natural philosophy was the school of translation established by Archbishop Raymond at Toledo after its fall to Christian forces. Toledo was an ideal location, since it had long been a meeting place for Christian, Jewish, and Islamic scholars. It was there that Gerard of Cremona (1114–87) discovered the astronomical work of Ptolemy and translated the *Almagest* in 1175, placing the best of Greek astronomical knowledge in European hands. Gerard translated over 80 other works during his lifetime, including the works of al-Kindi, Thabit ibn Qurra, al-Razi, al-Farabi, Avicenna, Hippocrates, Aristotle, Euclid, Archimedes, and Alexander of Aphrodisias.

Natural philosophy represented only a small portion of the rediscovered texts, but it gave the intellectual class of Europe a greater taste for the ancients, whose work they were eager to adopt and adapt. Cicero and Seneca were popular, while Aristotle's system of logic was significant in a wide range of applications. The natural philosophers of the twelfth century privileged Plato's *Timaeus* over Aristotle's works, because Plato's idealism accorded well with Christian theology. Among Jewish scholars of this period Moses Maimonides (1135–1204) was the best known; his *Dalalat al-Hairin* (*Guide to the Perplexed*) attempted to place Jewish philosophy on a firm Aristotelian foundation. Written in Arabic (Maimonides was a physician at Saladin's court), it was translated into Hebrew and later into Latin.

Not all medieval scholars restricted their research to intellectual material and the rediscovery of the works of the ancient philosophers. There was enormous interest in the promise of the manipulation of nature offered by the alchemical texts. When Robert of Chester translated *Liber de Compositione Alchemie* (*Book of the Composition of Alchemy*) in 1144, he introduced alchemy to Europe. A compendium of Arabic chemistry, it was followed by a flurry of research as people hunted for more detailed work by Jābir (Geber) and al-Razi (Rhazes).

Early in the thirteenth century there was another burst of university founding that took advantage of the new knowledge and the growing market for education. The University of Padua was founded in 1222 and became a leading medical school. The University of Naples followed in 1224, with the University of Toulouse close behind in 1229. Starting in 1231 Cambridge became Oxford's chief rival. The University of Rome was founded in 1244, and the Sorbonne University in 1253.

Christian Theology vs. Aristotle's Natural Philosophy

The universities soon established themselves as *the* site of intellectual activity in Europe. While autodidacts (those who were self-taught) and those from earlier cathedral schools might once have claimed equal footing as scholars, by the end of the thirteenth century the Professor of Theology had much higher status. In this way the universities became both the protectors and creators of knowledge. However, they were essentially conservative institutions, so once something was made required reading, it became an unchallengeable authority. At the same time the universities stood in a complex relationship to the larger structure of the Catholic Church. They were seldom under the complete control of any one bishop, and thus they provided space that was sanctioned by the Church and yet not controlled by it. This allowed the debate about the primacy of faith or reason to be played out within their walls and cities. While several scholars were imprisoned for their impious views, the fact that these debates could take place at all speaks to the power and independence of these institutions.

Christian theologians were not universally pleased that the work of Arabic and Greek philosophers was being introduced to the Latin West. Aristotle was particularly subject to theological objections since he contradicted the Bible on many issues of natural philosophy, such as the infinite life of the universe, and as a pagan he offered an implicit challenge to Christian authority. Because of Aristotle's popularity among students, authorities at the University of Paris grew concerned over the effect of pagan philosophy on the future theologians and secular leaders being trained there; so, in 1210 they banned the reading and teaching of his works on natural philosophy. This was also a battle over authority, since the conservative Faculty of Theology effectively imposed the banning of Aristotle on the more progressive Faculty of Arts. The ban was renewed in 1215 by Robert de Courçon, a papal legate, and again by Pope Gregory IX in 1231. However, the general interest in Aristotle prompted Gregory to establish a commission to review Aristotle's work and clean up any theologically problematic elements.

Ironically, the banning of Aristotle's natural philosophy actually promoted the study of it, making it a kind of philosophical forbidden fruit. The ban applied only to the University of Paris, so other universities were free to offer Aristotelian instruction, and this was used as a selling feature to attract students. Further, the ban covered only natural philosophy, so Aristotle's work on logic, despite being intimately bound to the system of natural philosophy, was still available for study. Demand for Aristotle continued to grow, and the supply of texts and scholars also multiplied. Finally, in 1255 pressure to learn Aristotle and the wide availability of texts led the Faculty of Arts at Paris to pass new statutes that made instruction in Aristotle not just acceptable but a mandatory element of an arts education. Aristotle's works had gone from being outlawed to required knowledge in just 45 years.

The work of Aristotle became so fundamentally important to the intellectual life of the Latin West that he was referred to simply as "the Philosopher." Although translation efforts continued, the difficulty of his arguments and the often fragmentary nature of the available texts led to a heavy dependence on Arabic commentators.

In the early period of reintroduction the most popular commentator was ibn Sina (Avicenna). By the middle of the thirteenth century Rushd (Averroes) had become the chief commentator used by Latin scholars. Like Aristotle, Rushd was so important that he was referred to as "the Commentator."

This exalted treatment of Aristotle and his commentators gives us a false picture of medieval scholarship if it is taken to suggest a slavish or doctrinaire dedication to the Greek material. Historians for many years argued that medieval scholarship was largely derivative and thus an uninteresting but necessary path to later work that challenged Greek ideas. More recently, historians have realized that, while dedication to the texts was an important element of medieval scholarship, from the earliest times there was a constant debate about every aspect of Greek natural philosophy. One of the major concerns to plague medieval scholars was that the Greeks were not Christian, so every aspect of ancient philosophy had to be debated in the light of Christian orthodoxy. Since the majority of Latin scholars were members of the clergy, the pagan origin of Greek thought was seen by some as reason enough to reject it; this was part of the motivation for the banning of the study of Aristotle.

A more moderate group as typified by Pope Gregory IX was prepared to include aspects of Greek thought as long as they were not overtly theological or directly contradicted biblical authority. Indeed, one of the first challenges for medieval philosophers was to find a way by which Greek natural philosophy could coexist with revealed religion. The latter was necessary for salvation, but the former offered a path to understanding God's creation as well as a wealth of practical knowledge. Among those who made a formal attempt to align Aristotelian philosophy with Christian theology was Robert Grosseteste (c. 1168–1263). Grosseteste was the first chancellor of Oxford University and a man of enormous intellectual breadth. He worked on Aristotle's logic in the *Posterior Analytics* and on his physics and mechanics from *Physics, Metaphysics*, and *Meteorology*. Grosseteste reconciled Aristotelian ideas with biblical thought in commentaries on logic and natural philosophy. For example, he argued that while creation by God took precedence over the cosmology used by Aristotle, it did not follow that Aristotle was wrong about the composition of matter in the universe.

Grosseteste was also deeply interested in optics, working with Euclid's *Optica* and *Catoptica* as well as al-Kindi's *De aspectibus*. This fascination with light came in part from a belief that light in the material world was analogous to the spiritual light by which the mind gained certain knowledge about true forms or the essence of things. Light was the fundamental corporeal substance, and so the study of optics was the fundamental study in natural philosophy. Since understanding optics required mathematics, Grosseteste linked mathematics, natural philosophy, and religion together. His teaching, particularly to members of the Franciscan Order, led many scholars to the study of mathematics and natural philosophy.

Following Grosseteste was the great medieval thinker Albertus Magnus (c. 1206–80). Albertus held one of the two Dominican professorships at the University of Paris and was keenly interested in finding a place for Greek philosophy within the context of the Church and in challenging the intellectual place of the Franciscans. He wrote extensively on philosophy and theology and is remembered for many works on natural topics, ranging from geology to falconry and the powers of plants and magical beasts. Albertus was an energetic scholar who wrote commentaries on all available Aristotelian texts. Because of the range of his work, he became known as "Doctor Universalis," and he was not afraid to amend or correct "the Philosopher" on either natural or philosophical issues. Albertus did not propose a new orthodoxy based on Greek philosophy, but he argued that a corrected natural philosophy had great utility and could be exploited by the existing orthodoxy. As such, he expected the intellect to glorify the creation of God and the utility of natural philosophy to aid in making Christianity supreme.

Magic and Philosophy

Albertus Magnus was also the supposed author of one of the most popular medieval texts, *Liber Aggregationis*, or, by its English title, the *Book of Secrets*. The text was written by an unknown author or authors, perhaps even students of Albertus, and attributed to him. It is a compendium of treatises on "herbs, stones, and certain beasts," and from a modern perspective it seems to be mostly about magic, astrology, and mythical beasts such as the cockatrice and the griffin. Yet the work tries to set the world into a loosely Aristotelian framework, and there are aspects of both Pliny's encyclopedic descriptions of the world and the material inquiries of the Islamic alchemists in it as well. While most serious scholars of the age disdained this kind of mysticism, such compendiums were enormously popular. A large section is set in a kind of problem/solution format, offering formulas and methods of procedure to deal with specific problems, such as this defence against drunkenness:

If thou wilt have good understanding of things that may be felt, and thou may not be made drunken.

Take the stone which is called Amethystus, and it is of purple colour, and the best is found in India. And it is good against drunkenness, and giveth good understanding in things that may be understood.[1]

The *Book of Secrets* is a book of medieval magic and contains a powerful, if ill-defined, link between magic and natural philosophy. At the simplest level both were studies of an unknown world, and both offered the possibility of controlling the unknown through naming and describing it. Yet the magic of the *Book of Secrets* is instrumental rather than spiritual, and this distinction was important for practitioners interested in the unseen forces and powers of nature. The *Book of Secrets* carefully avoids the issue of witchcraft, supernatural powers of either good or ill, or calling on the powers of supernatural beings. As fantastic as were the properties of the items listed and described, they existed in the objects themselves, they were hidden except to the knowledgeable, and they were natural.

One of the most notable elements of the text is that it uses the terms *experimentari* and *experiri*, referring to experiments rather than just the experience of nature. While it is difficult to assess if or how much these descriptions and recipes were believed, it was certainly the case that many people took them seriously enough to try them. Even if Albertus was not the author of the *Book of Secrets*, he did favour a form of Aristotelian natural philosophy that was shaped by Arabic tradition and that included a more hands-on approach to the study of nature. This was a departure from the approach of earlier Latin scholars who were more interested in knowing what could be true than in knowing what something looked like, or how it might react when mixed with other things.

Roger Bacon and Thomas Aquinas: The Practical and Intellectual Uses of Aristotle's Natural Philosophy

The path of natural philosophy split after Grosseteste and Albertus Magnus. Those more attracted to the investigative side of the subject, such as Roger Bacon (c. 1214–94), began to copy the practical approach of many of the Arabic sources.

....

1. Albertus Magnus, *The Book of Secrets of Albertus Magnus*, ed. Michael R. Best and Frank H. Brightman (Oxford: Oxford University Press, 1973) 33–34.

This group included the growing number of alchemists and astrologers. Those more interested in philosophy and an adherence to the Greek intellectual tradition tended to see the subject in terms of its ability to train the mind and provide ways of gaining certain knowledge. This stream led to Thomas Aquinas (1225–74) and the scholastics. A third stream can be seen in the spread of primarily practical skills among the engineers, masons, smiths, navigators, and healers of the Middle Ages. This group has been overshadowed by the others because they were rarely part of the intellectual class and left few written records; yet, it is clear that everything from the construction of the cathedrals to the practice of midwifery was affected by natural philosophy as it filtered through European society.

Roger Bacon is a perfect example of the spirit of enquiry in the Middle Ages. He studied at both Oxford and Paris and later joined the Franciscan Order. He favoured the utility of natural philosophy, especially that found in Aristotle's more practical works, and argued that the comprehension of nature would aid Christianity. He wrote on optics, speculated about the design of underwater and flying vehicles, and supported the idea of experiment as a method of discovering things about

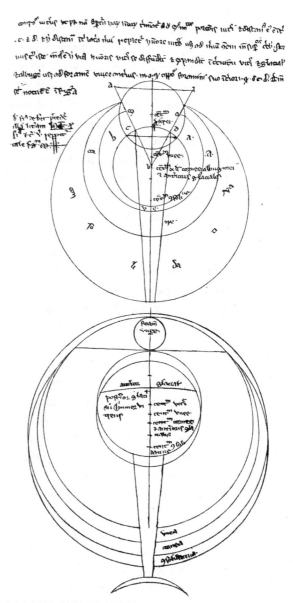

3.2 ROGER BACON'S OPTICS

Bacon's version of Alhazen's optics of the eye. In the top circle, light enters the eye from the top and passes through the vitreous humour. The bottom circle is a detail of the interior of the eye.

Roger Bacon's *Optics Diagram of the eye*, from the work of Roger Bacon/Universal History Archive/UTG/Bridgeman Images.

nature. He was the first European to mention gunpowder, but it is uncertain whether this was an independent discovery or learned from Eastern sources (having been discovered around the ninth century in China) and recreated by him. The text in which he mentioned gunpowder was his *Opus Majus*, written around 1267. The book was not published until 1733, and it is unclear how widely the manuscript circulated in his day. His speculations and defence of natural philosophy were not well received by the Franciscan leadership, but he persisted, believing that he had a duty to pursue his work. Eventually he was reprimanded, put under surveillance by his Order, and finally imprisoned for heresy in 1277.

The greatest figure of the intellectual stream was Thomas Aquinas. He had been a student of Albertus Magnus, following his teacher's lead in clarifying the interaction between theology and philosophy. For Aquinas, faith and the authority of God were primary, but in those areas not determined by revelation God had granted humankind the tools to understand nature. Thus, there could be no true conflict between religion and philosophy, since God had given us both. Any apparent contradictions disappeared when proper theology and proper philosophy were applied. Aquinas followed the philosophic path begun by Rushd (Averroes) and, in a sense, saved Aristotle by compartmentalizing his work. In one box he put Aristotle's system for gaining certain (or true) knowledge and the method of testing knowledge based on logic. If results were arrived at through the proper methods, the product of philosophy could not contradict revelation. He placed Aristotle's observations about the world in another box. These contained some erroneous material, but the big picture—such as the perfection of the heavens— was correct, and, as such, many of the observations were worth the effort of cleaning up or Christianizing. In the last box were Aristotle's ideas about theology, politics, and social structure. These, along with errors by other pagans, were disregarded as being heretical, false, or superseded.

The discussion of Aristotelian philosophy indicates in part how important Greek philosophy had become for the intellectual community of Latin Europe. Aquinas's work was situated within a serious scholarly debate about the place of philosophy (Aristotle's work in particular) in the intellectual arena, but it was also written to counter a number of specific challenges to orthodoxy. One of his chief targets was Siger of Brabant (c. 1240–84) who held a strongly Aristotelian view of the world and attempted to teach philosophy without the constraint of theology. In response, Aquinas wrote *On the Unity of the Intellect, against the Averroists*, which, while specifically attacking Siger's position, more generally argued that philosophy was dependent on theology and should not stand alone. Aquinas won, and Thomistic natural philosophy became the orthodoxy of the European scholarly world.

Aquinas's writing and reasoning were dense, even compared to contemporary medieval scholars, and this in turn made his work the focus of much study. Consider this short passage from the Introduction to his work *On Being and Essence*:

> Moreover, as we ought to take knowledge of what is simple from what is complex, and come to what is prior from what is posterior, so learning is helped by beginning with what is easier. Hence we should proceed from the signification of being to the signification of essence.[2]

While it seems reasonable to move from what is easy to understand to what is complex, Aquinas's idea of what was easy and complex has provided scholars with 700 years of debate.

Scholasticism

By the beginning of the fourteenth century the study of Aristotle in the Thomistic tradition was in complete ascendancy. While there were still evangelical members of the Church who questioned any worldly study as a distraction from faith, Aristotelianism had flowed into every aspect of intellectual life and had taken up a position alongside the Church fathers as a source of authority. The intersection of Aristotelian methodology and medieval interests including theology and certain aspects of Platonic philosophy developed into a form of philosophy known as scholasticism. The scholastics were closely associated with the universities and the more intellectually inclined religious orders such as the Dominicans.

Scholasticism represents the strongest vein of intellectualism in the Latin Church and can be traced back to Augustine in the fourth century. In the early medieval period it owed more to Plato and his idealism than to Aristotle, who was

....

2. Thomas Aquinas, "On Being and Essence," in *Philosophy in the Middle Ages*, ed. Arthur Hyman and James J. Walsh (Indianapolis: Hackett, 1984) 508.

little known except for his logic. The basic method for the scholastics was the dialectic, so that questions were posed in such a way as to establish two contradictory positions. The idea of resolving a question by presenting contradictory initial positions was well known to the Greeks and makes up the basis of the dialogue form used by Socrates, but the medieval scholars took this method to new levels of intricacy. This began with a formalized organization of the argument into thesis, objections, and solutions. Peter Abelard's (1079–1142) *Sic et non* (*Yes and No*) was one of the seminal texts for this method. Thomas Aquinas used it in his reconciliation of Aristotle with Christianity.

Herein lies the historical problem of the scholastics, since their dedication to Aristotle became so strong that, over time, their system was transformed from a method for understanding the world into an axiomatic statement about the world. The scholastics were rationalists at heart in that they argued reason was required to understand the universe, but they created a system that relied on a set of authorities that were then largely placed beyond question. This did not mean that debate ended, and in fact it remained one of the fundamental skills for scholars. Universities took up the *dissertatio* as the method to achieve higher degrees. This skill extends to the modern day with the dissertation and defence system used to obtain a PhD, a doctorate in philosophy. The system supposes that the thesis is an argument made by a student who publicly defends it against questions posed by scholars knowledgeable in the field; our continued use of the method created by the medieval scholars indicates how robust an educational system it is. There was a limit, however, to the amount of new insight that could be gained by perpetually debating the same issues. Thus, the debates were less concerned with reaching a conclusion and were seen more as a tool to train novices to understand the established answer or to bash opponents back into line.

Medieval Alchemy

While Aristotle was undergoing theological and philosophical revision on the road to orthodoxy, another conduit for investigations of the natural world was not being subjected to the same process of legitimization, because it was not a part of the school system. More than medicine, astronomy, mathematics, or philosophy, alchemy brought an interest in natural philosophy to a wide audience that included princes, physicians, teachers, monarchs, religious leaders, craftspeople, and commoners. If only by sheer numbers, alchemists were the most common proponents of the study of nature. Because the Islamic alchemists based their work on a version of Aristotelian matter theory, the works of Rhazes and Geber were the greatest conduit of Greek natural philosophy as a practical art into the Latin West. This had positive and negative effects on natural philosophy. On the positive side, it expanded the study of the material world and was instrumental in introducing skills and the concept of experiments. The negative aspects were, first, the degree of charlatanism that came to be associated with it, and, second, the alchemists' secrecy and even paranoia, which was contrary to the concept of public knowledge so characteristic of natural philosophy.

The medieval charlatans were many, and they played on the greed and gullibility of both the high and low born. The basic con was simple. The charlatan claimed to have discovered the process for creating the Philosopher's Stone, thus persuading a rich benefactor to support the actual production of gold from base metals. During the production, the alchemist was housed, fed, and clothed by the patron and might even be given a stipend to cover other living expenses. Also, costly and exotic materials were needed. Since alchemical knowledge was arcane and secret, who among the victims could say that the expensive white powder was not a rare ingredient made from the feathers of a phoenix and imported from Cathay? In addition to the money made indirectly, the alchemist often required quantities of gold as a seed for the transformation of undifferentiated prime matter into the precious metal.

Charlatan alchemists found ready victims. The medieval world was full of fantastic beasts, evil spirits, and magicians, so alchemy fit with the belief in the existence of supernatural forces. In addition, transmutation of matter was preached as doctrine by the Church. In transubstantiation the Eucharist bread and wine were transformed into the body and blood of Christ, while many biblical stories hinged on transformation of matter in some way, such as Lot's wife turning from flesh to salt, Eve being created from Adam's rib, or Christ changing water into wine. While the Church outlawed witchcraft and regarded magic as dangerous and evil, it was through the Church that alchemy came to Europe, was translated and transcribed in the *scriptoria*, was studied by popes and cardinals, and was practised by monks.

What complicates the story of the alchemists was that the "true" alchemists (those who were not simply con artists) also needed patrons and funds to carry out their work. If that meant occasionally improving results to placate patrons, that was the price of research. There were also the contradictory pressures on the alchemists to keep their processes secret (for personal and financial reasons) and the necessity of making their work public in order to attract patrons. This continues to be a problem even today, when the pressure to produce results has occasionally led scientists to fabricate or adulterate results (or at least produce conclusions far beyond their evidence) in order to secure funding for their experiments.[3]

Arnold of Villanova (c. 1235–c. 1311) is a good example of a true medieval alchemist. Famous as a physician, he was also an astrologer and alchemist. He wrote a treatise on transmutation called *The Treasure of Treasures, Rosary of the Philosophers and Greatest Secret of All Secrets*, in which he claimed to have found the secret of matter known to Plato, Aristotle, and Pythagoras. He told his readers that he would hold nothing back, but that they must read other books to understand the hidden reasoning behind his work. Transmutation could be achieved through a kind of purification of metal that would leave behind only the noble elements of silver and gold. This was to be accomplished by an *aqua vitae* (water of life) made from mercury, which in turn was used to produce an elixir that could convert a thousand times its weight in base metal into gold or silver (depending on the elixir). The process was described in terms of the life of Christ, covering conception, birth, crucifixion, and resurrection. While most of the material was theoretical, there was enough practical direction (and evidence of actual work) to encourage readers to attempt to replicate Arnold's work.

Experiment and Explanation

While alchemy was one way of investigating material that was not dominated by ancient philosophy, medieval scholars were themselves quietly examining nature and finding Aristotelian observations wanting. They were not as slavishly devoted to the Aristotelian texts as they may appear to be, given the effect of scholasticism. By using Aristotelian methodology, medieval scholars challenged what was true knowledge without risking an attack on authority, especially if they concentrated on the observational material in the compartmentalized Aristotle. In a typical approach, the natural philosopher would begin by praising Aristotle and then

---

3. A modern example of this sort of wishful thinking can be seen in the attempt by Pons and Fleischmann to create cold fusion. See Chapter 13.

proceed either to explore an area that he had not covered or to demonstrate a new idea in the guise of a moderate correction to his impeccable system.

This can be seen in the work of people such as Robert Grosseteste and Theodoric of Freiberg (c. 1250–1310) who both worked on optics and the rainbow. Aristotle argued that the rainbow was the result of sunlight reflecting off water droplets in clouds that acted like tiny mirrors. Arab work on optics with its more practical aspects showed in contrast that the rainbow was created by refraction. Grosseteste began his examination as follows:

> Investigation of the rainbow is the concern of both the student of perspective and the physicist. It is for the physicist to know the fact and for the student of perspective to know the explanation. For this reason Aristotle, in his book *Meteorology*, has not revealed the explanation, which concerns the student of perspective; but he has condensed the facts of the rainbow, which are the concern of the physicist, into a short discourse. Therefore, in the present treatise we have undertaken to provide the explanation, which concerns the student of perspective, in proportion to our limited capability and the available time.[4]

Thus, Grosseteste argued that he was not demonstrating that Aristotle was wrong about the rainbow; rather, he was merely filling in that part of the investigation that Aristotle did not cover. This was a common ploy for scholastic natural philosophers, allowing them to maintain their allegiance to the Philosopher while they presented original work without fear of being accused of hubris for placing their work above his.

Theodoric praised Aristotle and then tossed aside his theory to present his own, one based on refraction and reflection. This was likely based on material he learned from Alhazen's *Book of Optics*. He offered a method of testing the behaviour of light that falls on a raindrop by obtaining a glass globe, filling it with water, and shining a light on it. (See figure 3.3.) While Theodoric's work was not the first example of experiment in the Latin West, it is often pointed to as a precursor to experimentalism, particularly because his results are essentially those we find today. It is a good example of the kind of intellectual bridge between Aristotelian philosophy and the move to test observations that would transform the study of nature.

4. Robert Grosseteste, *On the Rainbow*, "Robert Grosseteste and the Revival of Optics in the West," in *A Source Book in Medieval Science*, ed. David Lindberg and Edward Grant (Boston: Harvard University Press, 1974) 388–89.

3.3 THEODORIC'S
RAINBOW FROM
DE IRIDE (C. 1304)
The small circles at
the right represent
raindrops that reflect
and refract the light
entering from the left
and seen at the middle.

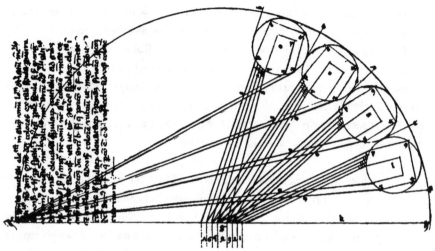

In Aristotle, the truth about nature is to be found in the intellectual construct
that results from the application of logic to observation. In other words, we know
the truth because of our ability to apply a system of classification and explanation
to sense perception. Theodoric does not deny the Aristotelian system, but he
pushes against the Aristotelian location of sense perception. The unaided eye
cannot discern the correct sense perception, so the creation of the rainbow must
be modelled in such a way that the event can be made clear to the senses. The
glass globe is not a raindrop, but Theodoric makes the implicit assumption that it
must be analogous to a raindrop and thus must represent the physical condition
of the raindrop. The truth about the rainbow no longer lies solely in the observer
(the senses and the intellect) but must also reside in the apparatus that replicates
the physical conditions.

While we have come to accept the kind of reasoning behind Theodoric's work,
it was not self-evident that certain knowledge could be gained by such a method.
One of the principal problems of reasoning from observation was the impossibility
of certainty by induction. By definition, sense perception relies on induction:
an observer noticing only white swans might reasonably go from a series of
particular observations to the general conclusion that swans could only be white.
Since observation cannot limit the possibility of a black swan, nor can the observer
know that all possible swans have been seen (since that would include both past
and future swans), the best that can be said is that all observed swans are white.

Likewise, in the Middle Ages, the argument Aristotle had made against experiments was still taken seriously. That is, forcing nature to perform unnaturally (in an experiment) does not give one insight into its natural behaviour.

Ockham's Razor

Skepticism about the possibility of certain knowledge as formulated in the Aristotelian/scholastic system was not uncommon. The primary source of attack came from mystically inclined theologians who objected to rationalism and logic altogether, but there were philosophic challenges as well. The most forceful skeptic was William of Ockham (1285–c. 1349), who attacked the Aristotelian categories of relation and substance, thereby undermining both physics and metaphysics. Ockham argued that relations were created in the mind of the observer and did not represent any underlying order in the universe. Thus, Aristotle's four spheres of elements existed only in the mind, collapsing the whole edifice of Aristotelian explanation. Ockham also challenged Aristotelian teleology, arguing that it was impossible to prove by experience or logic based on first principles that there was a final cause for any particular thing. Part of Ockham's defence of this philosophy was based on the Law of Parsimony, more commonly called "Ockham's Razor." He argued that "plurality should not be posited without necessity."[5] In more direct language this meant that an explanation of some problem would not be made better by adding arguments to it. As a philosophical device, it also suggested that, when faced with more than one explanation for a phenomenon, it was wise to choose the simplest. This idea was not originally Ockham's (versions of the idea of philosophical parsimony can be found in the work of Maimonides and even Aristotle), but it was one of his guiding principles. Much of Aristotle's elaborate system seemed to Ockham to be unnecessary or unprovable.

In addition to challenging scholasticism Ockham also challenged the hierarchy of the Church. He believed that revelation was the only source of true knowledge, and this belief set him at odds with the policy of papal authority. Although willing to accept the supremacy of the Church on spiritual matters, he objected to the extension of papal authority to secular issues such as the subordination of

---

5. William of Ockham, *Summa totius logicae* (c. 1324), in *Philosophy in the Middle Ages: The Christian, Islamic, and Jewish Traditions*, third edition, ed. Arthur Hyman, James J. Walsh, and Thomas Williams (Indianapolis: Hackett, 2010), p. 624.

monarchs to the Church's temporal authority. For his loud and public objections, he was excommunicated on June 6, 1328. At that time he was under the protection and patronage of Holy Roman Emperor Louis IV, so was protected from the wrath of the pope.

The Okhamites were few in number, in part because the position was dangerous politically, but they had a wide impact. Their philosophy was labelled "nominalism" because it denied the actual existence of abstract entities or universals. By extension, the natural world could only be described in "contingent" terms. Something that is contingent might be true or might equally be false. Consider the statement "all swans are white." This conclusion could be reached by observation and held to be universally true, but such a conclusion was proven false when the *Cygnus atratus* or Australian black swan was discovered. The matter is contingent on factors external to the proposition. If universals did not exist and nature was contingent, then the only way to discover anything about the natural world was through observation, and all general statements (such as classification) were potentially subject to revision based on further observation. This philosophy was part of a trend by some philosophers away from the study of the traditional realm of metaphysics and toward the study of experience. Moreover, the Ockhamite position suggested the independence of philosophy from theology. Although it was not the only group to do this, the nominalists opposed the majority of medieval thinkers who accepted Aquinas's position that philosophy was subordinate to theology. This was another form of the separation of the natural from the supernatural, which was a necessary step if there were to be an independent study of natural philosophy.

The Black Death and the End of the Middle Ages

Ockham's death occurred just before the greatest natural disaster of the Middle Ages: the plague or Black Death. The plague started in China in the 1330s and was carried by traders to the Black Sea, where Italian merchants, sailors, and shipboard rats were infected and passed it on to Europe in 1347. The disease was horrific, spreading through air, by touch, or by flea bite. People often died within hours of exposure. It was called the Black Death because it caused buboes (hence bubonic plague), or swellings filled with dark blood, to form on the body, especially near the lymph nodes in the groin, armpits, and throat. The Italian author Boccaccio wrote that victims "... ate lunch with their friends and dinner with their ancestors in paradise."[6] Many historians place the death toll at 25 million in five years, or one-third of Europe's population, but the figure may have been as high as 50 per cent. Many towns and villages, where the proximity of people led to a rapid spread of the disease, were totally depopulated. The effect of the plague was made worse because the people of Europe had experienced a series of bad harvests before it arrived, and malnutrition and starvation had already weakened the population.

The appearance of the plague coincided with the Hundred Years' War (1337–1453) that pitted France against England. England lost most of its continental lands, but the prolonged conflict wiped out a significant portion of France's nobility. The Great Schism (1378–1417) also followed on the heels of the Black Death and led to the central authority of the Church splitting, as competing popes in Rome and Avignon attempted to rule at the same time. All the death and destruction of the era encouraged a swing toward a more conservative theology and promoted a resurgence in mystical Christianity. The Black Death certainly seemed like a biblical curse, and no earthly action had any effect on it. Physicians often blamed disease on bad astrological events, and the medical faculty at the University of Paris concluded that the plague was the result of a conjunction of Jupiter, Saturn, and Mars that corrupted the air. The "calamitous 14th century," as historian Barbara Tuchman called the era,[7] marked the beginning of the end of medieval Europe. Although it took almost 400 years for the social structure of the Middle Ages to fade completely from Western European society, the new path was opened not by philosophers, social reformers, merchants, monarchs, or popes, but by the misery of disease.

In the plague years less original work was done in natural philosophy, since most theologians and scholars who did survive were more concerned about death and salvation than about the structure of nature. Nicolas Oresme (c. 1323–82), the Bishop of Lisieux, was one of the few who continued to work on natural philosophy. Oresme's work on mathematics was a precursor to analytical geometry as he tried to represent velocity geometrically. In *Le Livre du Ciel et Monde*, a commentary on Aristotle's *De Cœlo*, he presented the most comprehensive examination of the possible motion of the Earth to date, concluding that the evidence supported the geocentric model of Ptolemy. Oresme wrote *Ciel et Monde* and translated a number of Aristotle's works into French at the command of Charles V, and as such his work marked a shift in attitude toward the use of vernacular rather than Latin.

---

6. Boccaccio, "Introduction," *Decameron* (1351) Day I.

7. Barbara Tuchman, *The Distant Mirror: The Calamitous 14th Century* (New York: Alfred A. Knopf, 1978).

The greatest effect of the plague on natural philosophy was indirect. The death of so many people meant that when the plague years passed, the land was vastly underpopulated. For those who survived, life held many more possibilities than it had before. The survivors inherited the property of the victims, and many people grew suddenly rich as they gained the inheritance not only of their immediate family but often of distant relatives as well. Good land was plentiful, but the people to work it were scarce, so peasants got better deals from landowners and could afford to buy more luxury items. It was also easier for peasants to leave the land and enter into trades and mercantile activities. Cities, countries, and the wealthier nobles often had to compete to attract artisans and even peasants to their regions. The booming economy made rich those people who could supply the demand for luxury items such as silk and fine cloth, spices, ivory, perfume, glassware, jewellery, and a huge list of manufactured items from footwear to armour to mechanical toys. In the leading centres of commerce, particularly the Italian city-states of Genoa and Venice, this new money paid for merchant and naval fleets, public works, an explosion in patronage of the arts, and education. The people outside the trade centres saw their gold and silver flowing out of their regions and making others rich, as Italian merchants and their Arabic trade partners controlled the flow of the most expensive luxury items that came from the Far East. The Spanish, Portuguese, English, and the Italians themselves began to consider ways to get around the middlemen and trade with China directly.

To do that the Europeans needed a host of tools: better astronomy for navigation, improved cartography and geography, new and better instruments, and better mathematics to make these possible. Key to the new trade initiatives were new ships that could sail the Atlantic, so better naval engineering was required. But what they needed more than the tools were the people to devise them, build them, and take them out and use them. Natural philosophy was a key component to this drive, and, combined with Johannes Gutenberg's nifty invention of the printing press, Europe had all the elements necessary for an explosion of intellectual, economic, and cultural activity.

Conclusion

From the fall of Rome European rulers, Church leaders, and intellectuals had laboured to create a stable, hierarchical society. In intellectual terms, they began by creating a need for Greek philosophy, integrating it into their educational and theological world view. By 1300 Europe was growing slowly, and society was well ordered, carefully regulated, and somewhat inward-looking. Natural philosophy was studied by a small intellectual group primarily at the universities, while the alchemists, physicians, and artisans worked away at practical problems. For most thinkers, the demarcation had been established between philosophical and revealed knowledge. As Aquinas had shown, these two knowledge systems were not in conflict, since they dealt with exclusive areas of knowledge, with theology as the superior study and philosophy in a useful but supporting role. In other words, the Latin scholars had faced the same issue as the Greeks and had determined the separation between the supernatural world of revealed religion and the natural, rationally understood world of nature. The tension inherent in this separation was a very productive one, allowing some of the finest thinkers to create the impressive intellectual system of scholasticism. At the same time those interested in the application of this knowledge to practical ends lived more in the world than the academic scholastics. By 1450 their time had come. European society had been shaken by its encounters with the four horsemen of the Apocalypse, but in the aftermath a new sense of prosperity and freedom emerged. There was adventure in the air.

Essay Questions

1. How and why did Charlemagne support education?

2. How did Christian scholars overcome the inherent problems with Aristotelian philosophy and why did they do so?

3. How did alchemists contribute to the spread of natural philosophy?

4. In what ways did the Crusades transform the study of natural philosophy in Europe?

CHAPTER TIMELINE

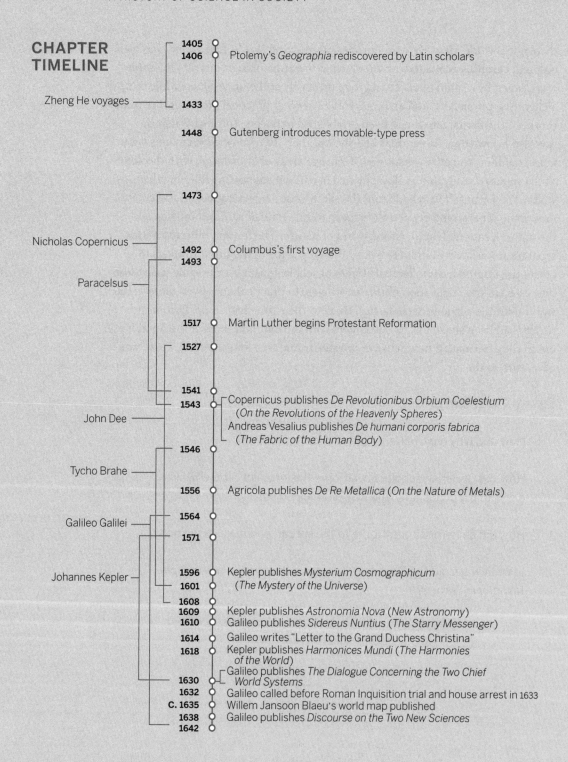

1405
1406 Ptolemy's *Geographia* rediscovered by Latin scholars

Zheng He voyages — **1433**

1448 Gutenberg introduces movable-type press

1473

Nicholas Copernicus — **1492** Columbus's first voyage
1493

Paracelsus —

1517 Martin Luther begins Protestant Reformation

1527

1541
1543 Copernicus publishes *De Revolutionibus Orbium Coelestium*
(*On the Revolutions of the Heavenly Spheres*)
John Dee — Andreas Vesalius publishes *De humani corporis fabrica*
(*The Fabric of the Human Body*)

1546

Tycho Brahe —

1556 Agricola publishes *De Re Metallica* (*On the Nature of Metals*)

1564
Galileo Galilei — **1571**

Johannes Kepler — **1596** Kepler publishes *Mysterium Cosmographicum*
1601 (*The Mystery of the Universe*)
1608
1609 Kepler publishes *Astronomia Nova* (*New Astronomy*)
1610 Galileo publishes *Sidereus Nuntius* (*The Starry Messenger*)
1614 Galileo writes "Letter to the Grand Duchess Christina"
1618 Kepler publishes *Harmonices Mundi* (*The Harmonies*
of the World)
Galileo publishes *The Dialogue Concerning the Two Chief*
1630 *World Systems*
1632 Galileo called before Roman Inquisition trial and house arrest in 1633
C. 1635 Willem Jansoon Blaeu's world map published
1638 Galileo publishes *Discourse on the Two New Sciences*
1642

SCIENCE IN THE RENAISSANCE: THE COURTLY PHILOSOPHERS

4

The intellectual life of Europe expanded in the fifteenth and sixteenth centuries. Natural philosophers found new texts, new lands, new interpretations, and new career paths. The European Renaissance, meaning "rebirth," began with a renewed interest in the discovery of classical texts. This intellectual voyaging was matched by a greater confidence and spirit of adventure that led to contacts with newly discovered peoples and places. Europeans found that they were living in a world of expanding possibilities. They encountered people who had their own knowledge, particularly of navigation. While intellectuals first looked backward, to the glorious heritage of the ancients, they soon used ancient knowledge as a stepping stone to new information and ideas. At the same time the Catholic Church lost its professed monopoly on truth with the upheaval of the Reformation, while university scholastics found themselves under attack, no longer the sole controllers of philosophic knowledge. A window of opportunity was created, especially through patronage in the princely courts and merchant halls. Because of this, different things began to be valued. Rather than syllogistic logic and theological subtleties, princes wanted spectacle, power, and wealth. Therefore, natural philosophers who were practical (or claimed to be) were valued.

The Early Renaissance: Humanists and the Printing Press

As we have seen, Europeans had never completely lost touch with Greek knowledge and natural philosophy. They had studied Aristotle intensively during the Middle

Ages, to the point where his logic and larger intellectual system had become a foundational requirement for academic and theological discourse. However, there were large sections of the Greek and Roman corpus that had disappeared from view. Plato, especially, was largely unknown to European intellectuals, as were many other works of literature and philosophy. European scholars' eyes were opened by the rediscovery of, and engagement with, these great ancient thinkers. The men and women responsible for this rebirth were called humanists.

Beginning in the fourteenth century in Italy, scholars unaffiliated with the Church or the universities began to sell their services as teachers to the children of the rich and powerful in the Italian city-states. They taught humane letters, *studia humanitatis*, and stressed the *trivium* through the study of the great Latin writing of the past. Scholars such as Petrarch (1304–74), Leonardo Bruni (d. 1444), and Guarino da Verona (Guarino Guarini) (1374–1460) looked to the wonderful prose of Cicero and Seneca in order to understand how to be the good citizen and live the good life. Because these teachers changed the venue and purpose of education, both women and men had access to the new learning, and several women became well-known humanists. For example, Isotta Nogarola (1418–66) composed the "Dialogue on Adam and Eve," a debate as to whether Adam or Eve was more responsible for their banishment from Eden (seen as an early feminist discussion). Cecilia Gallerani (1473–1536) was a friend of Leonardo da Vinci and may have held the first "salon" or meeting of artists and intellectuals. She became famous as the "Lady with an Ermine" in the portrait by da Vinci. All these humanists were convinced that good words and thoughts made wise citizens, and they worked hard to find pure versions of ancient texts in order to achieve that wisdom.

This rediscovery of ancient wisdom and a reorientation to living a good life in this world rather than only working to achieve salvation in the next is often labelled the Renaissance. Although historians today hotly debate the use of the term, the period witnessed a flowering of intellectual and artistic activity that started in Italy during the fourteenth century and was emulated in other parts of Europe over the next 200 years. While humanists stressed the language-based studies of grammar, rhetoric, and logic, this changing intellectual world affected the study of nature as well. Scholars appeared who were willing and able to ask fundamental questions about the system of natural philosophy and to develop new methods of study. What started for these scholars as an intoxicating rediscovery of Greek natural philosophy ended in an almost complete abandonment of it. Over the period the scholars themselves underwent a radical change as they increasingly moved away from the Church and from theology as the foundation and reason for the study of nature. In this there was also a rebirth of the Athenian ideal of philosophy as a study unto itself, and with the huge expansion of the universities and the patronage of the royal courts there was a way to pursue philosophy independent of theology and Church support. Natural philosophers were still called upon to justify their enterprise, something they did by calling attention to its civic and state utility.

Although most humanists were more concerned with understanding the books of the Bible and Cicero in their original languages than in predicting the paths of planets, their enterprise helped infuse new life into natural philosophy. Humanism did this in three ways: humanists rediscovered and translated classical scientific sources from the original Greek; humanist methodology treated written sources in a more skeptical manner; and humanism introduced a new purpose for, and mode of, scientific discourse.

Equally important, humanism revived Aristotelianism, both by rediscovering early Greek versions of Aristotelian texts formerly known only through Arabic translations and by forcing scholastics to make their arguments and methodology more rigorous. As a result, Aristotle's system did not give way before the humanist onslaught; rather, it incorporated much of the methodology and rigour of human-istic studies while retaining its basic framework. The Aristotelian system had proven extremely fruitful as a research program, since it provided an all-encompassing study of the physical world including physics, astronomy, and biology, and of the spiritual and social world using metaphysics, logic, and politics. Until an equally sophisticated paradigm could be established in the seventeenth century, Aristotelianism remained useful and necessary. Thus, the history of natural philosophy throughout the fifteenth and sixteenth centuries is one of the refine-ment and triumph of Aristotelianism, rather than of its defeat.

Two major factors contributed to the rediscovery of Greek natural philosophy in this period. The first was the fall of Constantinople to the Turks in 1453. Before this date, individual Greek manuscripts were traded from Byzantium or were discovered in various Italian monasteries. But with the fall of the last outpost of ancient Greek scholarship, hundreds of books, some of them literally thrown over the walls to save them from the invaders as the city fell to the Turkish army, were brought all at once to Italy. Knowledge of Greek now became absolutely necessary for scholarly work. The flooding of the intellectual market with Greek texts coincided with the second impetus to the rediscovery of Greek natural philosophy. This was the patronage of the Medicis, who were interested in a full translation of Plato. Cosimo de' Medici, head of a powerful Florentine banking family, became interested in the metaphysical philosophy of Plato in 1439 and by the 1450s encouraged humanists such as Marsilio Ficino (1433–99) to undertake translations of his work. Cosimo set up the Platonic Academy, with Ficino as its head, and in a relatively short time this group translated many of the important works of Plato into Latin. This rediscovery, combined with the discovery of mystical and magical treatises such as those of the supposed Hermes Trismagistus and the Jewish cabala helped to develop Renaissance magic as a much more esoteric study than its practical medieval counterpart, as seen in the *Book of Secrets*.

What made the rediscovery of Greek natural philosophy, and with it the growth in interest in the study of nature, a European phenomenon, rather than just an Italian one, was the invention of the printing press. In 1448 Johannes Gutenberg (c. 1397–1468) introduced movable-type printing, thereby revolutionizing communication. Movable type printing was not in itself a revolutionary idea, but it represented the perfection and combination of a number of existing technologies. Printing, using carved wooden blocks, had been around for over 1,000 years and was used by the Chinese from around 1045. The Chinese inventor Bi Sheng (990–1051) created a movable-type system using porcelain characters, but it is not clear if knowledge of Chinese printing was known in Europe. Block printing was in use in Europe by the beginning of the fifteenth century. Despite the invention of most of the components for printing in China, the development of printing for publication in that country was inhibited both by the pictographic nature of the language, which would have required thousands of characters, and by the threat it posed to the monopoly on writing of the established class of scribes. By contrast, Gutenberg worked with only 24 letters (the use of "j" and "u" had not been standardized), plus capitals, punctuation, and a few special symbols, at a time when scribes in Europe were scarce and expensive.

Gutenberg combined two Asian inventions, the screw press and paper, to develop his movable-type printing press. Paper had been invented in China around 150 BCE and was manufactured in Europe by 1189, offering a less costly alternative to vellum and parchment. Gutenberg created typographic characters by scribing each individual letter into a hard metal (steel), then using these as a punch to make a set of moulds out of a softer metal (copper). He could then cast as many letters as he needed out of a lead alloy. The letters were uniform in size and shape and could be assembled and printed, then separated and recombined repeatedly.

Gutenberg's work was meticulous, since he was attempting to replicate the typography of the written manuscript. This attention to detail and the cost of creating the actual press led him to seek financial backing from Johann Fust of

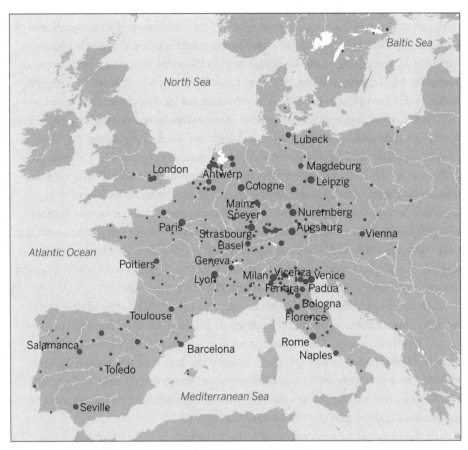

4.1 SPREAD OF PRINTING IN EUROPE

The centres of printing between 1452 and 1500.

Mainz in 1450. Gutenberg's project was the "42-line Bible" (also known as the Gutenberg Bible or the Mazarin Bible). He was not as good a businessman as he was an engineer, however, and lost much of his equipment to Fust to pay his debts, who completed the printing in 1455. About 300 copies of the Bible were printed and offered for sale at 30 florins each, which was equal to about three years' wages for a clerk.

Gutenberg's press was copied by many others, and by 1500 there were more than 1,000 printers working in Europe. (See figure 4.1.) Bibles, religious texts, and indulgences were in high demand. Indulgences were slips of paper that could be purchased from the Church for the remission of sins, and they had been used by the Church to finance everything from the Crusades to building cathedrals. With the aid of the printing press huge numbers of indulgences could be produced, sometimes as many as 200,000 in a single print run. The demand for other kinds of books also exploded as everything from Greek and Roman literature to medical texts became available. The humanist interest in first Latin and then Greek literature supplied materials for the printers and also created a demand for this classical material.

The effect of printing was enormous. Books were now available to people who had never seen a manuscript of any kind. Printing made information far more widely available, and a huge storehouse of material was opened to a growing audience. As the cost of books declined, more people could afford to own them, and reading habits changed as literacy spread. With the introduction of page numbers (which familiarized readers with Arabic numerals), tables of contents and indexes became possible. This meant that a book did not have to be read from cover to cover but could be dipped into just for information the reader thought was pertinent. Since people could now potentially own many volumes, they could compare one text with others, an impossibility in a scribal age.

The media scholar Marshall McLuhan and others have argued that the introduction of mass printing technology ultimately changed the very psychology of Western society. The change from a non-literate to a literate society changed the sense of time and space, shifted the locale of truth from human memory to written records, degraded memory, promoted dissent and a wider world view, and was partly responsible for the development of the concept of professionalization and the creation of the "expert."

In the realm of natural philosophy the introduction of print changed the discourse as well. Printing helped establish the definitive and corrected version of Greek and other natural philosophy texts, since several manuscripts could be compared for the most authoritative version. This prevented scribal drift, or the compounding of simple errors such as spelling mistakes that grew worse with repeated copying. It also allowed the insertion of illustrations, charts, and maps, items that had been so prone to scribal errors that they had usually been omitted from manuscripts or were useless. This meant that scholars could concentrate on finding new knowledge rather than constantly correcting the old. Because readers could purchase and borrow numerous books at relatively cheap prices and without personally going to monasteries housing the original manuscripts, a search of the literature was possible, as were comparisons of alternative versions, especially of star charts or botanical illustrations and descriptions. New information could also be disseminated rapidly. For example, news of Christopher Columbus's voyage in 1492 was printed immediately on his return to Spain and translated from Spanish into German, Italian, and Latin within the year, while knowledge of Marco Polo's thirteenth-century visit to China was known only to a select few, even into the fifteenth century. Finally, printing provided natural philosophers with paper calculational devices, a public forum for their ideas, and a republic of letters within which to converse with people of similar interests and aptitudes.

Copernicus, Tycho Brahe, and the Planetary System

Probably the most famous natural philosopher influenced by humanist ideas, for whom the printing press transformed his research and dissemination, was Nicholas Copernicus (1473–1543). Copernicus was born in Torun, Royal Prussia (then and now part of Poland),[1] a relatively isolated intellectual outpost. He travelled as a student to Italy, where he learned humanistic techniques and consulted original manuscripts. The most significant documents he found there were complete copies of Ptolemy's *Almagest*, the single most important source for astronomy at the time. No complete manuscript of the *Almagest* was available in all of Royal Prussia in the 1480s. By the time of his death in 1543 there were three different editions of Ptolemy's book in print, allowing Copernicus and other astronomers to compare astronomical tables, discover the discrepancies between ancient observations, and establish a new model. Copernicus also had a printed version with diagrams of Euclid's *Elements* (Venice, 1482) and a list printed by Johannes Regiomontanus (1436–76), who had printed the first edition of Ptolemy's *Almagest*, of all the important scientific works from antiquity, which became the required reading list for sixteenth-century astronomers and mathematicians. In Italy, Copernicus also encountered manuscripts that originated in Arabic sources. Historians have now shown that Copernicus's ideas owed much to these Islamic astronomers, and that his new planetary model was a part of a conversation across cultures, rather than a result of one solitary thinker. Copernicus did not care where his concepts came from but was happy to try a number of strategies to see what would work.

When Copernicus studied Ptolemy's astronomy and compared it to medieval star and planet charts, he saw serious problems. Not only did the predicted locations of the celestial bodies differ, but Copernicus believed that Ptolemy had violated his

---

1. Royal Prussia became subject to the authority of the Polish Crown in 1454 but became part of Prussia in 1772. It returned to Poland after World War II.

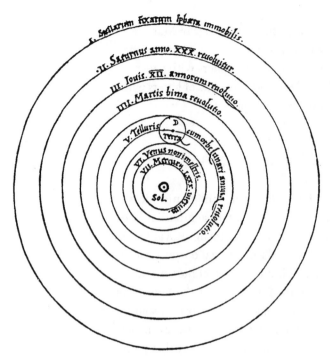

4.2 THE COPERNICAN SOLAR SYSTEM
FROM *DE REVOLUTIONIBUS* (1543)

own insistence on perfect circular motion in the heavens. Copernicus decided, as a mathematical exercise, to reverse the heavenly arrangement and place the sun at the centre with all the planets, including the Earth, revolving around it. In this schema, the sun remained stationary in the centre and the Earth now had a diurnal (daily) motion in order to account for night and day, as well as an annual orbit around the sun. (See figure 4.2.) To this, Copernicus added a third motion of the axis of the Earth's rotation that accounted for the seasons and the annual inclination of the zodiac.

Copernicus's system was just as complicated mathematically as the Ptolemaic system had been, but it did explain a number of anomalies that had been worrying astronomers for some time. For example, there was no good explanation in the Ptolemaic system for why Mercury and Venus never appear more than 45° away from the sun. Copernicus's system solved this issue by placing the inner planets between the Earth and the sun. In addition, the heliocentric model resolved the major issue of retrograde motion, which had led Ptolemy to devise the epicycle.

Moreover, Copernicus's system was aesthetically pleasing and eliminated the diurnal motion of the whole universe. It was not without its own problems, however. For example, Venus and Mercury should have phases like the moon in this new schema, but these had never been observed. More worrying, the stars did not appear to move, even though Copernicus's schema called for the Earth to move across the skies. The astronomers of the day assumed that the stars were close enough to the Earth that the angle they were viewed at would change if the Earth was orbiting the sun. This is called parallax and was not seen until 1838. Given the vast distance to the stars, it was not measurable until the development of powerful telescopes.

If the Earth was actually moving with a triple motion as Copernicus suggested, other questions of a more terrestrial nature might be asked. Why could birds fly east? Why did balls fall straight down? Why couldn't we feel the Earth moving? There existed no test that could demonstrate the motion of the Earth, and this flaw plagued astronomy for several generations.

Above all, Copernicus's system violated the whole Aristotelian ordering of the universe. Without the Earth in the centre, Aristotle's physics of "natural motion" fell apart. Catholic theology had come to depend both on Aristotelian explanation and, especially, on the centrality of the Earth as the least perfect part of the universe and therefore at the same time both the site of sin and transgression and the focal point for salvation. If the Earth was just one of many planets, could there not be other Christs and other salvations? For just such speculations, Giordano Bruno (1548–1600) was burned at the stake in 1600. Therefore, it is not surprising that Copernicus, a canon and thus an officer of the Church, delayed publishing his

Large armillary (c. 1585).

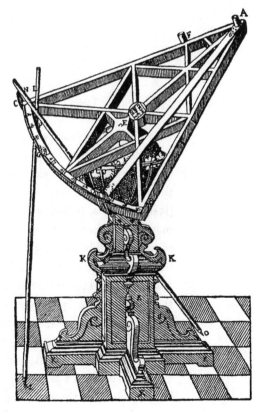

Sextant (c. 1582).

4.3 TYCHO BRAHE'S OBSERVATIONAL EQUIPMENT

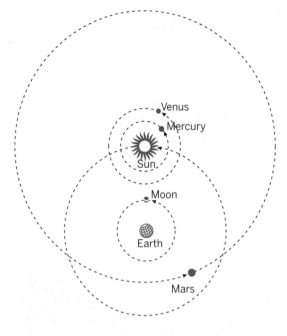

4.4 TYCHONIC SYSTEM

ideas until he was on his deathbed. He agreed, reluctantly, to the publication of *De Revolutionibus Orbium Coelestium* (*On the Revolutions of the Heavenly Spheres*) in 1543, through the persuasion of his friends, especially Georg Joachim Rheticus (1514–74). Rheticus was overseeing the publication of Copernicus's work until he was forced to leave Nuremberg. Andreas Osiander (1498–1552) took over and added an unauthorized preface claiming the whole thing was only a hypothesis. Hypothesis or no, the printing of *De Revolutionibus* allowed the whole European scientific community to learn of Copernicus's ideas, and a century of controversy began.

Scholars who read Copernicus's work fell into two categories: philosophers who were interested in the overarching cosmology and mathematicians who wanted to use the calculations without worrying about the model. Many philosopher/astronomers wished to tinker with the model in order to make one more acceptable to the Church. Tycho Brahe (1546–1601) was probably the most prominent of these. Tycho was a Danish nobleman who, rather than serving as a military commander to his king, as was typical of someone in his social position, offered his heroic astronomical work as his feudal dues instead. He was deeply indebted both to the humanist rediscovery of ancient natural philosophy and to the technology of the printing press. Even more than Copernicus, Tycho was able to compare printed tables. He was, indeed, a self-taught astronomer, learning his craft initially from printed books. He was also the best naked-eye observer in Europe. He built a huge observatory and the largest pre-telescopic astronomical instruments ever seen. (See figure 4.3.)

Tycho devised a planetary system, often called the Tychonic system, that was halfway between those devised by Ptolemy and Copernicus. In this system, the sun and moon revolved around the Earth, while everything else revolved around the sun. This saved the Earth as the centre of the universe and of God's grace while also explaining the problems of Mercury and Venus. (See figure 4.4.)

Using his impressive astronomical equipment Tycho also made some of the most important comet and new star sightings of the sixteenth century. He observed the comet of 1577, for example, and showed that its path sliced through the orbits of other planets. This was a major discovery, since it forced people to think about the physical reality of the solid transparent spheres of Aristotelian cosmology. Where did these comets come from? How could they be imperfect (transitory) and yet supralunar (above the orbit of the moon)? Tycho and others proved that their paths were supralunar and thus discredited the traditional physical explanations of the universe. But Tycho had no alternative physics to propose, which may help to explain the reluctance of astronomers and natural philosophers to abandon Aristotle.

Another of Tycho's discoveries was the sighting of several new stars—stars appearing and continuing in the skies where none had been before. Again, this made a case against the unchangeability of the heavens, and, because Tycho's observations were so good, the new stars could not be ignored. The sighting of the new star of 1572 was, in fact, a completely different event than earlier supernovas, since many people, following Tycho's lead, were able to observe the phenomenon simultaneously and report within the year to the academic community. This was one of the first instances of community agreement rather than scholarly authority as the basis for establishing a scientific "fact." The making of scientific facts increasingly became a public enterprise.

Some historians have pointed to Copernicus's work as the beginning of the scientific revolution or at least a Copernican revolution in astronomy. If by revolution we mean a rapid shift from an old to a new model, it largely did not happen. Despite Copernicus's radical reordering of the universe and Tycho's impressive observations, people were reluctant to abandon the Ptolemaic system and embrace Copernicanism. It was never fully accepted until the heliocentric schema was modified by Kepler and Newton, and only in the late seventeenth century did it become the generally accepted model. During the sixteenth century mathematical astronomers took up the technical aspects, philosophers the descriptive. Still, a number of astronomers saw the benefit of at least considering this new framework, and gradually—at different times in different places—the sun was accorded its place in the centre of the universe. Astronomers' reasons for moving from one system to the other were complex. The historian Thomas Kuhn argued, for example, that some people found the Copernican system aesthetically pleasing. A few isolated thinkers such as Englishman Thomas Digges (1546–95) and German Michael Mästlin (1550–1631) accepted Copernican cosmology, while others, such as the close-knit scholarly community at Wittenberg, adopted a hybrid system very early in the 1550s. Galileo's championing of Copernicanism had much to do with patronage, as we shall see, although Galileo was also influenced by aesthetics and concerns with bringing astronomy and physics into accord.

The Age of Exploration

The debate about the correct model of the heavens was not just a scholarly squabble. All over Europe rulers and entrepreneurs had an urgent need to understand and predict the motions of the skies, since increasingly they were interested in long sea voyages of trade and discovery. From the Crusades on, Europeans had been interested in the exotic goods available through trade with the Middle East and Asia. By the fifteenth century this trade was completely controlled by the Turkish Empire, especially after the fall of Constantinople, and so enterprising European nations decided to circumvent the bottleneck of the Bosphorus and go around the middlemen. The Portuguese began by coasting down Africa and found, despite the closed Indian Sea depicted in Ptolemy's famous maps, that they could reach the East via the Cape of Good Hope, although they still had to pass through waters controlled by Islamic people.

Islamic traders had been sailing extensively in the Indian Ocean for many years before the Portuguese started their project of exploration and trade. Arabic traders moved between the coasts of Africa and India from at least the twelfth century, trading, establishing outposts, and interacting with the Indian population. The Portuguese were interested in voyaging for a complex mixture of reasons, including curiosity and imperial expansion, but most especially commercial concerns. They were very interested in working with Arabic traders in order to develop new routes to the East. The Portuguese were happy to use any information they found and often appropriated the techniques, maps, and matters of fact from other travellers in the region. Their maps of the Indian Ocean, for example, made use of the knowledge of Arabic traders, veterans of those waters for several hundred years. When Europeans published these maps, they erased the Muslim sources, leaving behind a tale of heroic European adventure.

The Chinese had also developed the necessary navigational and mapping skills to undertake significant oceanic voyages, long before the Europeans started their "age of exploration." Most famous are the voyages of the imperial eunuch Zheng He

(1371–1433/35). Zheng He was born into a Muslim family of the Hui people in Yunnan. When Yunnan fell to the Emperor's forces, Zheng He was captured and castrated. He became a powerful member of the imperial court under the Yongle Emperor, who sponsored seven naval expeditions, with Zheng He as the admiral. These voyages, with hundreds of ships and tens of thousands of troops on board, sailed all through the Indian Ocean to the Horn of Africa and Arabia. (The theory that Zheng He actually sailed around the world is without foundation, however.) Zheng He sailed from 1405 to 1433 (although he may have died on the final voyage), bringing gifts to the leaders he contacted and returning with tributes to the emperor. Most famous was the giraffe he brought back from Africa during his third voyage of 1413–15. Zheng He's achievement was considerable, but it is important to note that he followed long-established and well-mapped routes, some dating to the Han dynasty. For example, when his fleet arrived in Malacca in 1407, there was a sizable Chinese community already established there.

Historians have debated why the Chinese did not continue with this program of exploration after Zheng He. It is clear that the death of the Yongle Emperor was a key factor, since his successor immediately stopped the voyages, which he saw as expensive and unnecessary. It also seems likely that this was a political issue, since Zheng He's achievements represented the power of the eunuchs over the scholar/ bureaucrats who were less interested in these voyages. The Chinese began to concentrate more on domestic issues and were less interested in contact with the wider world, although they continued to trade along the Silk Road and interact with other peoples in the China Sea.

Others who might have had the technical ability to sail long distances were the peoples of the Americas. Jacob Bronowski argued that the "new world" did not travel out to the old world because it lacked a sense of the heavens as a wheel—an invention little used by the Maya or other South American civilizations. While this may have contributed to a lack of exploration, two simpler reasons restricted Mayan scientific activity. The first was a series of collapses caused by endemic warfare and agricultural failure because of drought and environmental degradation. Mayan social structure did not adapt well to the challenges, making the situation worse. They did not have the time or periods of peace to develop natural philosophy, and they never separated the study of nature from their religious practices. The second problem was technological. The Maya had great mathematicians and engineers, but they did not master a number of technologies, especially high temperature smelting or glass-making, leaving them with Neolithic tools.

While this new "age of exploration" had little influence on the Chinese or Mayan world view, this was a critical period in the development of European consciousness. Although the Chinese had sailed farther and many fisher folk had been traversing the Atlantic Ocean for centuries, the achievements of Vasco da Gama (c. 1469–1524) and Christopher Columbus (1451–1506), as well as those who followed, fundamentally changed the way Europeans understood the Earth and their relationship to it. These early explorers, equipped with a Christian and imperial belief in the righteousness of their cause and the superiority of their understanding, challenged the authority of the ancients, especially Ptolemy. Ptolemy's *Geographia* had only been rediscovered by humanists in 1406, providing another view of the globe that could be used and challenged. As with other natural philosophical endeavours, then, humanist rediscovery sparked an extension and eventually refutation of ancient knowledge of the globe. Columbus and those who came after demonstrated to Europeans the existence of a continent completely unknown to the ancients (though familiar to its inhabitants). More importantly for natural philosophy these explorers disproved a number of ancient and medieval theories of the Earth, most particularly by demonstrating that the globe had a much larger proportion of dry land than had hitherto been suspected, that it was possible to sail through the equatorial regions without burning up, and that people could and did live south of that equatorial region in the lands known as the antipodes. Columbus did not prove the world was round—this had been known by learned men since antiquity—but he did prove that the globe was navigable and, ultimately, exploitable by Europeans.

The prime motivating factor for these voyages was amassing great wealth, both for the individual and for the country sponsoring the enterprises. At first, the destination was the Far East—Cathay and the Spice Islands. The Portuguese were most successful at reaching these areas, setting up key trading depots in Goa (India), Malacca (Malaysia), and the Moluccas (Spice Islands). The Spanish, having reached the Americas by mistake, soon modified their mission, and although they continued to seek gold and especially silver, the *conquistadors* began to focus on colonization, seeing the natives as a useful slave population and one that could be converted easily to Christianity. Later, as the cultivation of sugar and cotton became more important to the Spanish, the African slave trade was introduced into this economic arrangement. And yet it would be a mistake to separate these imperial and mercantile enterprises from the growing interest in and study of the Earth. The study of nature was inexorably linked with religious and mercantile concerns. The "discoveries" of this age of exploration encouraged new innovations in cartography and navigation, led to a changing understanding of the terraqueous globe, spawned an interest in the effect of climates on human beings, and launched ethnographic investigations and debates concerning the New World peoples.

There was a burgeoning interest in the mapping of the world in the sixteenth century, undoubtedly influenced by that fifteenth-century rediscovery of Ptolemy. At first, charts and plots were used as aids to descriptive and experiential knowledge, but eventually European rulers, investors, and scholars wanted to visualize their world in this new graphic way. Countries such as Spain and Portugal were quick to develop state-controlled repositories of navigational maps and charts. Later, monarchs called for the mapping of their individual countries and regions, as well as creating larger maps of imperial concerns. The result was a flood of map production, including world atlases by Gerhard Mercator (1512–94) and Abraham Ortelius (1527–98), beautifully engraved in the Netherlands, and country surveys by Christopher Saxton (c. 1542–1611) in England and Nicolas de Nicolay (1517–83) in France. Working in Amsterdam Willem Jansoon (1571–1638) and his son Johannes Blaeu (1596–1673) produced a series of detailed world maps. (See plate 2 for Blaeu's 1664 map.) Maps became objects of desire for prosperous merchants, as we can see from numerous Vermeer paintings of merchant houses with beautifully coloured maps hanging on the walls. They were used to visualize and control space, to build empires, and to swell local and regional pride and identification.

One of the most troublesome aspects of the New World discoveries was the fact that there were people there. Who were they? What were they? While European scholars and explorers could use only European categories and understanding to interpret what they encountered, this contact with a previously unknown Other had far-reaching implications for European thought. Early explorers interpreted the customs and behaviours of those they encountered from a European viewpoint and tried to eliminate customs that did not suit their preconceptions, such as the lack of private property or a nomadic way of life. Sixteenth-century Spanish theorists tried to fit Amerinds into the only classification system they knew: Aristotle's. Thus, men such as Bernardo de Mesa argued that the Amerinds were natural slaves. The discovery of the Incas and Aztecs in the 1520s made this harder to believe. Clearly, in Aristotelian terms, these people were civilized. They had government and infrastructure and lived in a complex community. And so thinkers such as Francisco de Vitoria (c. 1492–1546) claimed that these people were natural children, based on the idea that they made category errors, such as engaging in cannibalism, bestiality, or eating dirt, but had the capacity to learn from their mistakes. They had to be protected because with training they might be raised up to adult (that is, European) status. This opinion was never shared by the majority, since it implied that eventually these children would grow up and would have to have their property restored to them. Another minority opinion, that of Michel de Montaigne (1533–92), had far-reaching effects. Montaigne argued that the Timpinambas of Brazil, although cannibals, were a noble race, more moral than Frenchmen, even if they did not wear trousers. This idea of the noble savage recurred most famously in the writings of Jean-Jacques Rousseau.

Paracelsus, Medicine, and Alchemy

Internal trade in Europe had been growing steadily from the time of the Crusades and by the sixteenth century had developed into a strong mercantile culture and economy. Greatly expanded by the gold and silver bullion flooding into Europe as the New World trading networks developed, manufacturing and trade among European nations expanded considerably. Mining in Europe and the New World became a growth industry, and with these economic and industrial changes, concomitant developments occurred in natural philosophy, especially in theories of mining and metallurgy on the one hand and alchemy on the other. As well, the increasing numbers of skilled artisans began to develop links with natural philosophers, asking new questions and developing new systems of investigation.

Mining of precious metals and other minerals had taken place since antiquity, but the demand for these goods soared in the sixteenth century. Coal for heat, iron for steel, tin and copper for manufacturing were all profitable minerals. There were a number of technological problems to be overcome in mining these substances, not least the water present in mines of any depth. Pumps were devised, although none were completely satisfactory. The refining of metals was also a process that had to be worked out, and Georgius Agricola (1494–1555) in *De Re Metallica* (*On the Nature of Metals*, 1556) was the first to explain some of these processes in natural philosophical terms. (See figure 4.5.) Agricola was humanist-trained and, clearly from his use of Latin, was interested in introducing the study of metals to a scholarly audience. On the other hand, he lived in Bohemia and Saxony, the richest mining lands in Europe, and he married a mine owner's daughter, so he was not exactly a disinterested party.

Mining produced serious illnesses among the miners, so it is no surprise to find a physician who interested himself in these cases. The German physician Theophrastus Bombastus von Hohenheim, known as Paracelsus (1493–1541), was influential in bringing together medical and alchemical knowledge, and he is recognized as one of the main creators of iatrochemistry or medical chemistry. His life was deeply influenced by the religious and social crises in the German states. Paracelsus was born in Zurich; his father was a physician who wanted him to follow in the profession. In 1514 he spent a year working at the Tyrolian mines and metallurgical shops of Sigismund Fugger, who was also an alchemist. It was through Fugger that Paracelsus became intrigued by the nature of metals, and he spent much time during his life trying to identify and discern the properties of metals. After he left Tyrol, he travelled widely across Europe, studying briefly with alchemists in France, England, Belgium, and the countries of Scandinavia before finally going to Italy, where he claimed to have earned a medical degree in 1516 at the University of Ferrara.

4.5 ORE PROCESSING EQUIPMENT FROM AGRICOLA'S *DE RE METALLICA*

In 1526 Paracelsus settled in Strasbourg to practise medicine. He treated miners' diseases, especially black lung. His alchemical work led him to become an advocate of the use of metals rather than traditional plant-based drugs in treatment. Most famously, he prescribed mercury for cases of the new disease of syphilis, a cure only slightly less excruciating than the original symptoms! Paracelsus's fame grew, and when the printer and publisher Johann Froben of Basel fell ill and local physicians failed to cure him, he sent for the young doctor.

Paracelsus cured him. At the time, Desiderius Erasmus, the famous Dutch humanist and biblical scholar, was staying with Froben, so Paracelsus's success was widely noted. Paracelsus was offered the position of City Physician and Professor of Medicine in Basel. He accepted, but held the position for only two years, since his radical ideas about the treatment of disease caused great controversy. He started his career as City Physician by publicly burning copies of Galen and Avicenna in order to demonstrate his rejection of the old medicine, which treated diseases with herbs. He was also radical in other ways, insisting on lecturing in German rather than Latin. He was loved by his students but hated by his associates, whom he frequently criticized.

The city officials defended their choice of Paracelsus against a clamour of protest from apothecaries and other doctors. Then the Canon Lichtenfels fell ill and offered 100 gulden to any doctor who could cure him. Paracelsus used his metallic system and Lichtenfels recovered but then refused to pay. Paracelsus took him to court, but either because the fix was in or because of some legal mistake on Paracelsus's part, he did not win his case. He left his position and spent the remainder of his life wandering through Europe, repeatedly running into trouble with authorities for his radical ideas. Lacking a powerful patron to protect him, he was in constant danger of being arrested by secular authorities or accused of heresy or witchcraft by religious officials. Finally, in April 1541 he found employment at the court of the Archbishop Duke Ernst of Bavaria. Ernst was very interested in alchemy, so it was likely the patronage position was offered for both medical and alchemical reasons. Unfortunately, Paracelsus, weakened by years of hardship, died in September that same year.

Unlike Aristotle, who had argued that there were four basic elements, Paracelsus and many of his fellow Renaissance alchemists claimed there were only three: salt, mercury, and sulphur. The careful combination of these three, with arduous, secret, and prolonged laboratory manipulations, might lead to the illusive Philosopher's Stone, the source of eternal life, the gold of the soul, and perhaps material gold as well. While Paracelsus can be seen as an alchemist, he was not really interested in transmutation. Instead, he was interested in iatrochemistry—medical chemistry. He shared the slowly evolving view that alchemy should be concerned with employing the material world for useful purposes, not with the fruitless effort to create precious metals. Although much of his work had mystical aspects, he also promoted the concept of understanding matter based on elemental composition, one of the foundational ideas of later work in chemistry.

Patronage and the Study of Nature

There was often no clear line to be drawn between the esoteric research of the alchemist and the mundane concerns of the apothecary. Both belonged to a

4.6 MAJOR SITES OF NATURAL PHILOSOPHICAL WORK IN EUROPE, 1500–1650

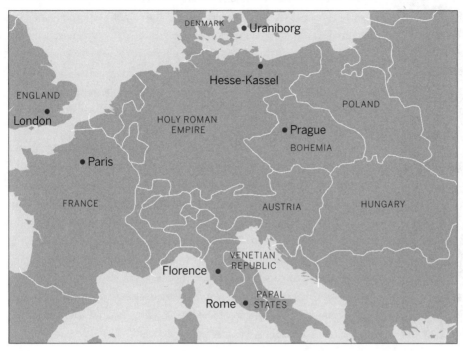

growing group of skilled artisans who plied their trade in increasingly large numbers in the urban centres of sixteenth-century Europe. Printers, instrument makers, surveyors, and shipwrights all began to ask questions about how the natural world could be used to their benefit. They often used and sometimes taught mathematics. This community of superior artisans, together with scholars trained and employed in non-traditional settings such as courts or the homes of private patrons, developed new questions about the make-up, design, and running of the world that would lead, by the next century, to a major reorientation of the scientific enterprise.

One place where natural philosophers, mathematicians, and practitioners came together was the princely court. During the Renaissance these were sites of spectacle and culture where political, cultural, and intellectual patronage encouraged some of the most glittering and opulent courts seen since antiquity. The earliest of these courts were, of course, Italian, and as we have already seen, the Medicis of Florence gathered together some of the foremost artists, humanists, and natural philosophers of their day. Other princes and courts followed suit, and

CONNECTIONS

Patronage and the Investigation of Nature: John Dee and the Court of Elizabeth I

The life of the famous necromancer, mathematician, and natural philosopher John Dee provides a fascinating glimpse into the complex and sometimes dangerous world of patronage. Dee worked hard but ultimately unsuccessfully to gain a place in Queen Elizabeth's court as her Royal Philosopher; in the process he pursued a number of practical projects that took him far from the philosophical work he valued. When he insisted on the importance of his scholarly work, his patron became less and less interested and he received less support.

John Dee received his education at Cambridge and very soon established his superior understanding of mathematics and geography. He went to Louvain to study with Gemma Frisius and Gerard Mercator, two prominent mathematicians and globe makers, and when he returned to England, he set himself up in London as an astrologer and geographical advisor. Many explorers, such as Humphrey Gilbert, asked his advice about navigational and geographical issues, including the question of the existence of a northwest or northeast passage. Dee became astrologer to the princesses Mary and Elizabeth. He was charged with treason for casting Mary's horoscope and appeared before the Star Chamber (a special law court often dealing with political trials), but was eventually able to clear his name. After Mary's death, he became Queen

soon natural philosophy became part of this patronage system, affecting the topics of investigation and how they were investigated. Hans Holbein's portrait entitled *The Ambassadors*, painted in 1533, demonstrates the importance of mathematical instruments to the self-fashioning of courtiers. (See plate 3.) These two men, French ambassadors to the court of Henry VIII, display a celestial and terrestrial globe, a quadrant, a torquetum, and a polyhedral sundial as evidence of their learning and wealth.

Patronage was a system of dependency, with personal contracts between two individuals: the patron and the client. The patron had power, money, and status, but wanted more. The client could give the patron more of these while getting some for himself. It was thus a two-way and often volatile relationship. The whole system was based on changing the balance of status. In natural philosophical relationships the client claimed special knowledge or skill, usually with some practical application, although sometimes he simply offered the patron the prestige of being able to surpass the knowledge of some other prince's natural philosopher.

Elizabeth's astrologer, advising her on the luckiest day for her to hold her coronation. Elizabeth took his advice and consulted him on many matters of astrological, geographical, and imperial importance.

Dee, however, had his sights set on higher goals. He sought to understand the underlying basis of matter through alchemy and the universal language of creation through Hermetic philosophy and magic. He hoped to develop a completely new philosophical structure for understanding the world, one that would lead to a unity of all mankind, but he could not persuade the queen to provide him with the necessary stipend that would have allowed him the freedom for such work. He sought the sort of fame and stability that Johannes Kepler had as Imperial Mathematician to Rudolph II. But Elizabeth was both practical and cheap, and although she gave Dee gifts, as was appropriate in a patron–client relationship, she never conferred on him the money or title that he sought.

Dee then moved to more esoteric and supernatural research, searching for the transcendent understanding of divine forms (the Platonic Ideals of nature) through scrying (crystal ball gazing) and angelic and demonic communications. His lack of success at Elizabeth's court caused him to try his luck at various other European courts. Unfortunately, the Polish courtiers he met were suspicious that he was an English spy, and he fared little better in Bohemia; he was forced to flee Rudolph's court, and on his return home to England, discovered his house and library had been vandalized. Although Elizabeth gave him a small position as Warden at Christ's College, Manchester, Dee was never able to achieve the success he had earlier in her reign. He died in poverty, with only his daughter to care for him. Dee's story, then, is a cautionary tale about the dangers as well as the rewards of scholarly patronage.

The philosopher sought to gain the attention of the would-be patron by dedicating a book or sending a manuscript to him or her, by circulating a letter concerning the patron's interests, or by publishing a book acknowledging the patron's greatness. Through negotiations the patron granted some court or household position to the scientist. This generally led to science at the courts that was useful, daring, and often controversial. In some cases cooperative enterprises were undertaken with the patron or other members of court.

There are numerous examples of these client–patron relationships, including that of Leonardo da Vinci (1452–1519) at the court of Charles VIII of France; German prince-practitioners such as Rudolph II and Wilhelm IV, Landgraf of Hesse; and the astronomer and mathematician John Dee (1527–1608) at the English court of Queen Elizabeth I. The best example of the patronage relationship and its effect on natural philosophy, however, is the life of Galileo Galilei (1564–1642). While modern commentators remember Galileo's final condemnation by the Roman Inquisition, he was famous in his day for his telescopic sightings. Through his astronomy and even more through his physics, Galileo constructed an abstract mathematical schema, suggesting the abstraction and mathematization of the world so integral to early modern natural philosophy. He believed that God had constructed the world using number, weight, and measure, and thus he replaced the study of causes with the study of laws. He used measurement and experiment, usually seen as part of modern scientific method. But perhaps what is most interesting about Galileo is that he did all this not within theological institutions as Copernicus had done, or in the universities as Newton would do, but at court. Galileo was every inch an early modern courtier, a kind of intellectual knight, with power to gain (and lose), and constantly looking for innovations to aid and glorify his patron.

Galileo

Galileo was born in Pisa in 1564. He moved to Florence early in his life and always thought of himself as a Florentine. His father, Vincenzio Galilei, was a famous musician who discovered a number of important mathematical musical laws. Vincenzio wanted his son to become a physician who, he said "made ten times as much money as a musician," but Galileo was more interested in mathematics. His first job, at the University of Pisa, was as a teacher of mathematics, at the bottom of the academic status ladder. His first significant post was at the University of Padua, which was under the control of Venice, and he used his patronage connections with powerful people in the Venetian elite to work his way up. Galileo was always in need of money, because he had seven brothers and sisters who relied on him for support. He needed to find big dowries for his sisters, and his errant brother Michelangelo Galilei was constantly in debt.

While at Pisa, and especially at Padua, Galileo began to study motion, although he did not publish his findings for 40 years because he could not figure out all the details to his own satisfaction. After many years he decided this was not an important question; rather than look for the cause, he developed laws of *how* motion worked. He eventually published his mechanics in *Discourse on the Two New Sciences* (1638). He rejected Aristotelian notions of motion, showing that speed does increase continuously, at least in free fall and, therefore, that impetus (the force impressed on an object, which Aristotle said would wear out with time) did not exist. Instead, Galileo argued that continuous motion once imparted, or continuing stillness, would remain forever. What separates this from Newton's later idea of inertia is that for Galileo continuous motion was circular. In Galileo's system a ball set in motion on the Earth, if unimpeded by friction or any other extraneous force, should travel continuously in an orbit around the Earth. Probably Galileo's most significant achievement in mechanics was his development of a clear picture of abstract and measurable motion.

For many years historians believed that Galileo did only thought experiments. We now know that he did practical experiments, although probably his most famous one, the Leaning Tower of Pisa experiment, was not performed by him, making this the most famous unperformed experiment in the history of science. It is possible that one of his students dropped two balls of different masses from the top of the tower, although it is not clear that Galileo witnessed this. The point of the experiment was to find out whether Aristotle, who predicted that the two balls would fall at different rates proportional to their weight, was correct. Galileo's actual experiments on motion led him to predict that the two balls would fall at the same rate. According to Galileo, if the two balls were dropped from the tower at the same time, they would hit the ground at the same instant. Allowing for a small variation due to air resistance, Galileo was correct.

The problem with the study of falling bodies was that they travelled far too fast for quantitative analysis with the equipment available at the time. There were no stopwatches in his day, so Galileo devised a method to "dilute" the rate of free fall. He rolled balls down an inclined plane, which had small notches at regular intervals. He measured the time by listening to the click as the ball hit these bumps and comparing it with someone singing Gregorian chants. He discovered that distances from rest were proportional to the squares of the elapsed time ($k = d/t^2$). This was a huge discovery, achieved by removing all real-life distractions, thereby creating an almost frictionless plane on which he could study an ideal example. Galileo was no longer asking *why* bodies fell (the cause) but rather measuring how fast they did so. This law was extremely influential for Newton's later work as he applied it to the universe and constructed a world subject to even more accurate measurement. Newton, like Galileo, avoided the question of causes.

Galileo also worked on the problem of projectile motion, important to the princes and principalities for whom he worked, since it was connected to ballistics and warfare. He determined that cannonballs move in a parabola by dividing their motion into two parts (forward motion and earth-seeking motion). He discovered that a ball shot from a cannon will hit the ground at the same moment as one dropped from the same place and determined that pointing the cannon at a 45° angle produced a maximum range. (See figure 4.7.) Since Galileo was a Copernican, he used this argument of the different vectors of motion to argue for the movement of the Earth.

As Galileo worked his way up the patronage system, he found that astronomical work was more successful in attracting patronage than mechanics or mathematics. While ballistics had been useful, it was the telescope and the discovery of new celestial bodies that brought him the greatest rewards of position, status, and authority. The telescope had been developed in the Netherlands in the first years of the seventeenth century. Galileo heard of this invention and imported a model. He worked out the optical principles and developed a more powerful version. In 1609 he demonstrated his marvel to the Venetian court, showing that his backers could see a returning merchant ship through the telescope two hours before someone searching with only the naked eye, thereby allowing a manipulation of the commodities market. Here was insider trading with a vengeance! The Venetian Senate was very impressed. They were willing to double his salary and give him a lifelong position, but in return all his future inventions would belong to the Senate and he could never ask for another pay increase.

Galileo had his eye on another prize—a position at the Medicis' court in Florence. The Medicis were arguably the most important patrons on the Italian peninsula, surpassed only by the court of the pope for power and prestige. They were possibly the richest family in Europe, with connections to the pope and business interests all over the Mediterranean and beyond. They could give Galileo the status and freedom he desired. Galileo began by teaching mathematics to the Grand Duke's son, Cosimo (descendant of the Cosimo who founded the Platonic Academy). He invented the proportional compass (which he manufactured and sold) and for a time taught practical mathematics like navigation. In 1609 he made his move.

A. Setting cannon inclination.

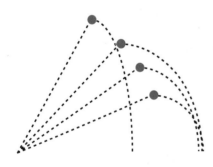

B. Various ranges by angle prior to Galileo.

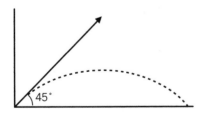

45°

C. Galileo's solution.

4.7 THE QUESTION OF CANNON RANGE

He turned his telescope on the skies and discovered that there were four moons circling Jupiter. This and other findings were all published in *Sidereus Nuntius* (*The Starry Messenger*) in 1610. He also discovered that the sun had rotating spots, that Venus had phases, and that the moon had craters and mountains. All this was highly controversial, since it showed the imperfection of the heavens, which went against Aristotelian supralunar perfection. He named the four moons of Jupiter the Medician stars, as a gift from a prospective client to a powerful patron. Cosimo, now Grand Duke, was delighted. After much negotiation, Galileo was given the position of Court Philosopher. This was a huge jump in status. But, of course, with status came risks. Galileo was now expected to take part in many intellectual wrangles as duels for the honour of his patron. Eventually, he moved from Florence to Rome and looked to the pope for patronage. These risks proved his downfall.

Johannes Kepler

Another astronomer whose career was equally influenced by these new patronage requirements was Johannes Kepler (1571–1630). Kepler was an anti-social, near-sighted man, descended from a family of misfits. Despite this, he became the Imperial Mathematician in the court of the Holy Roman Emperor Rudolph II, and thereby joined the practice of astronomy to the glory and wonder of this powerful court. Kepler is often called the first true Copernican (although several lesser-known sixteenth-century astronomers could share this title) because he whole-heartedly endorsed the heliocentric system. In the process of doing so he changed it to one that would have horrified Copernicus, since he destroyed the idea of the perfect circular motion of the heavens. He also attempted to join the physics of the heavens to a mathematical model of their motion. In other words, Kepler asked what the physical cause of the motions of the heavens was, rather than just mapping their course. His explanations were not taken up by other natural philosophers but showed astronomers that such questions were important.

Kepler was conceived on May 16, 1571, at 4:37 am and was born on December 27, 1571, at 2:30 pm, after a pregnancy lasting 224 days, 9 hours, and 53 minutes. We know this because Kepler cast his own horoscope, and these details were necessary to make accurate predictions. This demonstrates the importance of astrology to Kepler in particular and to early modern astronomers and society more generally, as well as the importance for Kepler of precision and mathematical accuracy.

Kepler had a very unhappy childhood. He grew up in a very poor Swabian Lutheran family with an abusive father and an unbalanced mother who was later tried as a witch. The high point of his young life was receiving a scholarship to the University of Tübingen, where he studied theology. When he finished his degree, he took a job as a mathematics teacher and astrologer in Graz. While teaching mathematics (to virtually empty classrooms—his pedagogic skills were low), he had a revelation that was to change his life. In a flash of insight, the structure of the universe was laid bare to him. He was circumscribing a triangle with a circle when he realized that the orbits of the planets might work this way. (See figure 4.8.)

From this Kepler developed three questions: Why were the planets spaced the way they were? Why did they move with particular regularities? Why were there just six planets? (The latter question marks him as a Copernican, since there are seven planets in the Ptolemaic scheme.) With his insight concerning the circumscribed triangle, he saw the answer to the first and last questions. He transformed his two-dimensional figure into a three-dimensional solid. Since there are only five regular solids in Euclidean geometry, the six planets fit perfectly with one solid between each orbit. This seemed to recreate the particular spacing of the planets. Kepler published this finding in *Mysterium Cosmographicum* (*The Mystery of the Universe*) in 1596. (See figure 4.9.) Later, in his *New Astronomy* (1609) and in the second edition of *Mystery*, he laid out the physical reason for the planets' motion in this particular configuration. He postulated that some sort of "magnetic" force emanated from the sun, in the centre, and was the cause of motion. That is, the sun was the prime mover, a concept that shows that Kepler was influenced by neo-Platonic ideas. Although there were many problems with this whole schema, it would provide Kepler with his life's project.

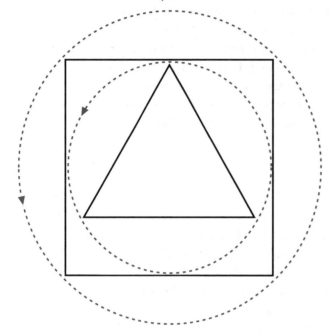

4.8 KEPLER'S ORBITS

Rotating the triangle and the square produces orbits.

Kepler recognized that in order to improve his model he needed better observations of planetary motion. He decided to go to the best observer in Europe and so became an assistant to Tycho Brahe. Kepler joined Tycho in Prague, where Tycho had recently become the Imperial Mathematician to Rudolph II. They had a very stormy relationship. Tycho insisted that Kepler work on the orbit of Mars, which he was not very happy to do. As it turned out, this was very fortunate, since Mars has the most irregular orbit of all the planets, and Kepler was forced to abandon the idea of a circular orbit in order to match observation to mathematical model. Kepler never did his own observations (he was far too near-sighted to see the stars and planets accurately), but he spent eight years calculating sheet after sheet of numbers. This was boring, repetitive, exacting work with little reward.

Tycho had hoped Kepler would prove the Tychonic system, but Kepler, as a Copernican, had other plans. After Tycho's death in 1601 Rudolph appointed Kepler Imperial Mathematician in his place. This gave Kepler status, although not much pay.

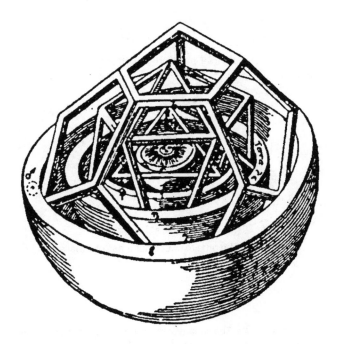

4.9 KEPLER'S NESTED GEOMETRIC SOLIDS
Based on Kepler's concept of the spacing of the planets in *Mysterium Cosmographicum* (1596).

He earned his living casting horoscopes but still had time to work on what is often seen as his greatest work, *Astronomia Nova* (1609) or, more fully, *A New Astronomy Based on Causation or a Physics of the Sky Derived from Investigations of the Motions of the Star Mars. Founded on the Observations of the Noble Tycho Brahe.* Kepler had worked on the orbit of Mars for eight years and had got his calculations to agree with the Copernican system to within eight minutes of arc. Although this is fairly accurate (Copernicus himself was only accurate to within 10 minutes), Kepler was sure that Tycho's observations were better than that. After a terrible struggle, he concluded that the orbit of the planet was an ellipse. Although this was not the first law of planetary motion he worked out, astronomers and historians came to

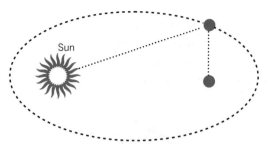

LAW 1: Planets move in elliptical orbits with the sun at one focus.

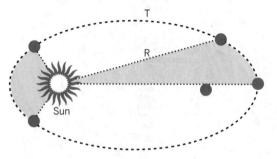

LAW 2: AREA LAW. The time to sweep out areas of equal size takes equal time.

$$(T_1/T_2)^2 : (R_1/R_2)^3$$

LAW 3: PERIOD LAW. The square of the period "T" (time to complete one orbit around the sun) of two planets is proportional to the cube of the distance "R" to the sun.

4.10 KEPLER'S THREE LAWS

call it Kepler's First Law because it underlay his other observations. He also postulated that the "magnetic" force of the sun, sweeping the planets around before it, operated in a mathematically consistent way and that a line from the sun to each planet swept out an equal area in an equal time (called the equal area law, or Kepler's Second Law). This meant that, when the planet was closer to the sun, it moved faster.

In 1618 Kepler published the third of his great books, *Harmonices Mundi* (*The Harmonies of the World*). In this work he argued that the planets, sweeping out their paths through the heavens, created harmonious music. It is perhaps telling that Kepler wrote this book that claimed to have discovered a grand scheme of harmony—in music, astronomy, and astrology— at the start of the Thirty Years' War, which necessitated his flight from Prague, and during the trial of his mother for witchcraft and the death of his daughter. As Kepler says in *Harmonices Mundi*, "In vain does the God of War growl, snarl, roar, and try to interrupt with bombards, trumpets, and his whole tarantantaran ... let us despise the barbaric neighings which echo through these noble lands, and awaken our understanding and longing for the harmonies."[2]

In the midst of discovering the major third played by Saturn and the minor third played by Jupiter, Kepler also developed what we now call his Harmonic Law, or Third Law. In this law, developed by trial and error, he demonstrated the

2. Johannes Kepler, Dedication of the *Ephemerides* (1620) to Lord Napier, in *Harmonies of the World*, as quoted in Arthur Koestler, *The Sleepwalkers* (London: Penguin, 1959) 398.

mathematical relationship between the periodic time (time for a single revolution around the sun by a planet) and the distance from the sun, so that the farther away from the sun, the greater the periodic time. He found that the ratio of the period of the orbit squared (T^2) to the mean radius of the orbit cubed (R^3) is the same value (K or a constant) for all the planets.

Despite the amount of work Kepler did, his explanations for planetary motion had little impact on other astronomers of his day. As the Imperial Mathematician to Rudolph's court, he was an important representative of natural philosophy. His books were certainly taken seriously but, except for the *Rudolphine Tables*, seem to have been seldom read. In his time his work was regarded as difficult and even dangerous. His place in the history of science depends more on his relationship to later ideas than his effect on astronomy of the day. Historians have selected the three "Laws" that accord with more modern astronomical ideas, particularly as identified by Newton, but they were mixed in with dozens of other laws created by Kepler and now forgotten. Galileo, Kepler's contemporary, saw him as a dangerous person to know, and their correspondence was polite and unenthusiastic. Kepler was suspect: as a Protestant, as a rival court astronomer, and as someone known to travel close to "occult forces," both because of the witchcraft accusation levelled at his mother and because his physical explanation for the motion of the heavens relied on action at a distance. Action at a distance required things to interact without some material connection between the objects, and Kepler's speculation about a kind of magnetic force moving the planets was seen as a magical explanation.

Newton gave credit to Kepler for a number of ideas but later asserted that he got nothing from Kepler's work. On the other hand, Kepler shows us how astronomy worked in the sixteenth and early seventeenth centuries. His years of calculating demonstrate the importance of mathematics to the study of the universe, and his place at Rudolph's court reminds us of this new site of natural philosophical knowledge.

The Protestant Reformation and the Trial of Galileo

The courts allowed men of practical knowledge, sometimes skilled artisans and mathematical practitioners, to mingle with university-trained or self-taught natural philosophers. They brought together these different ideas and interests and in doing so created new questions and goals for natural knowledge. Most natural philosophers attached to princely courts gained their reputations both for intellectual acuity and for practical applications. For example, Kepler and John Dee cast horoscopes for Rudolf and Elizabeth respectively. Dee advised Elizabeth on the most propitious day for her coronation, as well as consulting with navigators searching for a northwest passage. Likewise, Galileo's activities as a courtier were both esoteric and applied. These men walked a fine line between theory and practice, since all three were interested in large philosophical systems and desired court patronage not simply for creating improved telescopes or new armillary spheres. But monarchs wanted results, and all investigators of the natural world with court connections were compelled on occasion to dance for their supper. So claims to utility and the search for topics interesting to those princely patrons changed the orientation of natural philosophy away from philosophical specula-tion toward how things worked.

One good reason for natural philosophers to avoid philosophical speculation or the more traditional career path of Church positions was the other huge upheaval of the sixteenth century, the Protestant Reformation. While protests against various perceived inadequacies of the Catholic Church had flared up in the fifteenth century, Martin Luther's decisive stance in 1517 against indulgences split the Catholic Church in two. Just as with natural philosophy, religion was affected by the printing press—the printing of those indulgences flooded the market and made the venality of the Church more obvious, while the pamphlets printed by Luther's supporters and detractors ensured that there was not a corner of Europe that didn't know about the conflict within a few years.

The Reformation changed the intellectual, social, and institutional worlds in which natural philosophers lived. No longer did the Catholic Church have a monopoly on truth, which was either wonderfully liberating or terrifying, depend-ing on your religious position. There were new career possibilities and new places where a study of nature might be useful, such as merchants' houses, princely courts, and more secular private schools. While leaders on both sides of the religious divide called for a return to salvational concerns rather than secular ones, a window had been opened for alternative thinking and careers.

There has been much debate among historians as to the effect of the Reformation on science. Some have pointed to the flourishing of science in strongly Calvinist or at least Protestant areas as evidence of the support for science in Protestant attitudes. Others have pointed to the Catholic Church's treatment of Galileo to show the devastation caused by "superstition." The truth is that the impetus for people to investigate natural philosophy often was a way of removing themselves from sectarian strife, of finding a middle way of worshipping God through his works. The crisis of Galileo was nothing as clear as Catholics against science. A large part of Galileo's modern fame comes from his image as the "Defender of Science." However, he got into trouble not because he defied the Catholic Church but rather because he attempted, unsuccessfully, to reconcile science and religion and because his patronage choices proved too risky. All his life Galileo remained a staunch Catholic. He believed that his real enemies were not the Church authorities but "philosophers"—the Aristotelians who argued that only they had the right to make truth claims about the world.

In 1614, shortly after Galileo's astronomical discoveries, which had been hotly disputed, Galileo, and with him Copernicus, were attacked from the pulpit. Galileo, although sick, entered the fray. He wrote a letter explaining the division of knowledge between nature and scripture. When this letter fell into the wrong hands, he sent a longer version to Cardinal Bellarmine and went himself to Rome to explain the situation. Bellarmine was a humanist and moderately sympathetic to Galileo's situation. While some churchmen, especially Dominicans, believed that the motion of the Earth was unprovable, Bellarmine held it to be unproven. This was a softer position, although it is highly unlikely that Bellarmine believed that such a proof could be found.

Galileo decided to make his position clearer. In an extended version of this earlier letter, the "Letter to the Grand Duchess Christina," which was written to be circulated, Galileo claimed (following Augustine) that there must be a separation between science and religion in order to maintain the dignity of both. In an argument that stood Thomas Aquinas on his head, he argued that scripture can never be used to disprove something that has been proved by observation and right reasoning; rather, scripture must be reinterpreted to take this into account. He did not make this argument against the Church or Christianity. Instead, he was concerned that Catholic natural philosophers would lose status to Protestant ones and that the true wonders of God, as understood in his work, would not be observed and interpreted. Galileo quoted an early Church father, to very different effect: "That the intention of the Holy Ghost is to teach us how one goes to heaven, not how heaven goes."[3]

Galileo took this risky and public stand in part because of his loyalty to the Catholic Church and his desire for a strong natural philosophical community in Italy. Equally, this could be seen as a move to be noticed by the pope to whom he was looking for patronage. A client had to take risks to maintain client visibility if

---

3. Galileo Galilei, "Letter to the Grand Duchess Christina" (1615), in *Discoveries and Opinions of Galileo*, ed. Stillman Drake (New York: Doubleday Anchor Books, 1957), p. 186.

he hoped to be successful. So he went on the offensive. Hearing rumours that both his Letter and the works of Copernicus were about to be placed on the Index and therefore unavailable for good Catholics to read, Galileo went once again to Rome to seek an audience with the pope, Paul v. Instead, he had a meeting with Bellarmine, at which he was instructed to stop work on the Copernican theory. The judgement of the papal tribunal was that this theory was "foolish and absurd in philosophy," and an Interdict was produced in 1616, which told Galileo that he was no longer to hold or defend the Copernican theory.

In 1623 three new comets appeared in the heavens, and Galileo was drawn once again to astronomy. In the meantime Paul v had died, and in his place was a humanistic pope, Urban viii. Galileo, thinking he had an ally in the papacy, visited Urban viii in 1624, asking to be allowed to write about the Copernican system. He neglected to mention the earlier Interdict. Galileo left the audience believing he had received permission to write about it in a hypothetical manner. Ultimately, this turned out not to be the case.

In the 1620s Galileo began to develop his defence of Copernicanism, resulting in *The Dialogue Concerning the Two Chief World Systems* (1632). The dialogue form allowed him to present both sides of the argument (Ptolemaic and Copernican) without definitively choosing one, but since the character espousing the Ptolemaic system was named Simplicio, it was not hard to see Galileo's inclination. He used his theory of the tides as a proof of the motion of the Earth and, hence, of the Copernican doctrine. Although the theory was quite wrong-headed and convinced no one, it demonstrated to those reading the book that Galileo was indeed defending Copernicanism and so breaking the Interdict of 1616, which prohibited holding the view that the Copernican system was a proven fact. If that was not enough, the pope believed he had been personally betrayed by Galileo.

Galileo was called before the Roman Inquisition in 1632, and the trial took place in 1633, after he arrived in Rome. The trial revolved around whether Galileo had been ordered not to *teach* the Copernican system. Galileo said that wasn't part of the Interdict document he had received from the papal office but, rather, that he had been told he could not *hold* the system to be true, which was what all Catholics were enjoined to believe. The Inquisition stated that Galileo, and Galileo alone, had been told he could neither hold, teach, nor in any way defend the Copernican theory. This came from a document that was either a forgery or an unissued draft of a papal directive. Thus, this was not a trial of science *versus* religion but a matter of obedience to the Church. The Inquisition judged that Galileo had disobeyed, and they effectively silenced him. He was placed under house arrest for the rest of his life. He returned to his study of mechanics and wrote his most brilliant work, *The Two New Sciences*, also written as a dialogue. Since no Catholic was allowed to publish any of his work, the manuscript was smuggled out to the Protestant Netherlands. All his work, especially the two dialogues, became very popular and were translated into several languages.

The silencing of Galileo and the shift of scientific work to Protestant areas of Europe have suggested to some historians that Protestantism was more conducive to science. This is

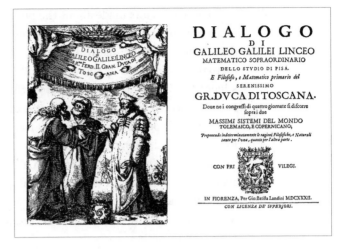

4.11 FRONTISPIECE FROM GALILEO'S *DIALOGUE CONCERNING THE TWO CHIEF WORLD SYSTEMS* (1632)

problematic in general terms, since natural philosophers continued to flourish in France, particularly within the Jesuit order, while Protestant religious leaders were often far more antagonistic to the study of nature than Catholics. In terms of the pursuit of utility, however, where knowledge of nature was seen as useful for mercantile, empire-building nations, the Protestant regions were far more willing to pursue science as a study. The whole idea of the reasonableness and simplicity of nature, although not exclusive to Protestantism, was emphasized by them, and natural philosophers also looked for the simplest answer. Protestants emphasized the idea that knowledge should be useful, either for human betterment or salvation, and natural philosophers often directed their studies to topics that had utility (or claimed it). Protestants felt that God had given them the Earth to exploit to its full extent, and exploitation became an underlying ideology of science. Puritans and Calvinists believed in personal witnessing and experience; scientific methodology increasingly employed experiment. The idea that the individual could find his own way to God through private study was borne out in science. Theories of election and vocation led not only to the idea of the investigator of nature as purer and higher but even to the cult of the scientist. Protestants rejected Church traditions; the New Science rejected traditions of Aristotelian science. Finally, Calvinism and Puritanism, especially, appealed to the urban mercantile classes, those people interested in the questions of exploration, navigation, astronomy, and mathematics, which would be the breakthroughs of the New Science.

The arguments made about the pursuit of natural philosophy in the regions controlled by Protestants and those controlled by Catholics are less about the direct relationship of natural philosophy to religion and more about the intellectual space created by the conflict. While the Catholic hierarchy may have silenced Galileo, the study of nature still remained important. Luther vehemently rejected Copernicanism, but it did not follow that Protestants abandoned astronomy. If people could question the very nature of religious faith as the Protestants did, then intellectually no question seemed out of bounds, whether for Protestants or Catholics. For a small group of people, natural philosophy seemed to offer a "third way" of worshipping God in a world where secular authority was unreliable and religious authority was wracked by dissension and uncertainty. Nature was consistent, unlike the pronouncements of monarchs, popes, and priests.

Education and the Study of Nature

Education, the other principal institution important for natural philosophers, was rapidly changing in early modern Europe. Previously, education had been largely an ecclesiastical concern. Most schools were sponsored by the Church, and many schoolmasters were clerics. From the mid-fifteenth century on, secular interest in education began to rise, first in Italy and later throughout Europe. The goal of education ceased to be only a career in the Church; government offices, secretarial positions, and eventually gentry culture and patronage possibilities all provided new incentives for achieving a certain level of education. At the same time the Protestant Reformation produced a new impetus for education and literacy, both because Protestants argued for the importance of personal and vernacular Bible reading and because the Catholic Church responded, in part, through educational strategies. Thus, education became a *desideratum* for a wider sector of the population.

A significant minority were very well educated, and increasingly during the early modern period these well-educated men were in positions of social, political, and economic power. As well, some men and women were self-taught or continued their education on an informal basis throughout their lives. Because of this increasing market for educational currency, institutions such as the universities developed less formal curricula, designed for those not interested in a credentialed profession. Even the more traditional subjects such as medicine began to look toward greater applicability of their knowledge. Other kinds of academies sprang up all over Europe to cater to specialized learning. Self-help books became more and more popular, and educational entrepreneurs, both humanists and others such as mathematical practitioners, began to sell their educational wares through individual lessons and books. Thus, this early modern period witnessed a change in the status of education among the governing classes across Europe, especially in the north and west, and thus in the demand for both educated advisors and information itself. In this climate, the utility of the subjects studied became important.

Andreas Vesalius

Andreas Vesalius (1514–64) was trained and pursued his career in this new university structure, as well as being influenced by humanism. Born in Brussels, Vesalius was the son of an apothecary of the Emperor Charles v. In 1530 he attended the University of Louvain and then moved to Paris to pursue a medical degree. In 1537 he enrolled at the University of Padua, renowned for its medical school. Almost immediately he received his Doctor of Medicine Degree and became an anatomy lecturer, a rather low-status occupation. He caused a sensation by insisting on performing dissections himself. This was almost unheard of, as anatomy lecturers traditionally had read from Galen while their assistant pointed to the pertinent parts. Vesalius soon began travelling around Italy and the rest of Europe performing public dissections. He rapidly found problems with the traditional Galenic anatomy, since it did not correspond to what he was seeing. This observation was possible only because of his personal interaction

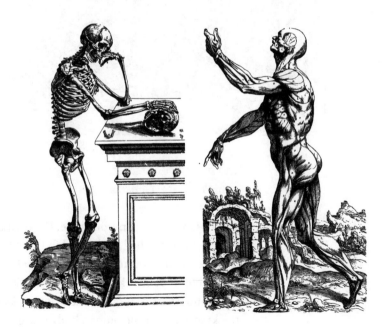

4.12 BONE AND MUSCLE MEN FROM VESALIUS, *DE HUMANI CORPORIS FABRICA* (1543)

ANDREAE VESALII
BRVXELLENSIS, INVI-
ctissimi CAROLI V. Imperatoris
medici, de Humani corporis
fabrica Libri septem.

CVM CAESAREAE
Maiest. Galliarum Regis, ac Senatus Veneti gratia &
priuilegio, ut in diplomate eorundem continetur.

4.13 FRONTISPIECE FOR VESALIUS'S *DE HUMANI CORPORIS FABRICA* (1543)

with the bodies. In 1543 he published the results of his disagreements with Galen, a new method, and a new philosophy in *De humani corporis fabrica (The Fabric of the Human Body)*. Vesalius produced a beautiful book and in the process disproved a number of Galen's ideas.

Vesalius showed that the liver was not five-lobed but one mass, that men did not have one less rib and women one more, that nerves were not hollow, and that bones were a dynamic foundation of the human body. For the first time he pictured the muscles in a rational methodical way. Perhaps his greatest achievement, however, was his method. He began with humanism, since he compared alternate texts of Galen in order to find the purest and least corrupted. Once he had questioned the text, he turned back to its source, the human body. He then used observation, seeing personal experience as fundamental for the natural investigator. Rather than relying on authority, Vesalius prescribed first-hand dissection for all would-be anatomists.

There is, of course, an irony here, since Vesalius's book soon achieved the same level of authority he had derided in Galen!

Equally important, Vesalius's dissections were not done in a closed academic forum but in public. He helped to establish public demonstration and witnessing as an important part of natural science. As the frontispiece to *De fabrica* shows, the knowledge of the human body was gained because everyone saw personally, yet together, and all agreed on what they had seen. (See figure 4.13.) This idea of public demonstration as the creation of knowledge, of matters of fact, and the need to have a group of like-minded individuals to agree on closure, became a necessary ingredient to scientific practice and discourse in the seventeenth century and beyond.

Conclusion

The establishment of natural philosophy as an enterprise to be conducted in public, in the universities, the courts, the merchant halls, and the instrument makers' shops was an innovation of the Renaissance and early modern period. The rediscovery and printing of ancient knowledge had, ironically, allowed early modern scholars to claim that they were now developing new knowledge rather than conserving what existed. The changing social, political, and religious world gave these scholars new venues to investigate nature and new claims to the secular utility of their task. All natural philosophers of this period believed that to study nature was to study God's work and that this was a sacred task, but equally they believed that the point of this enterprise was firmly rooted in the present, with the goal of human betterment or personal advancement (and maybe both). In an era of great adventure and discovery, Western Europe, especially those countries on the Atlantic, began to believe in the possibility of boundless progress. Had they not already surpassed the ancients in exploration of the Earth with Columbus's voyages and of the heavens with Galileo's telescope? The courtly natural philosophers were men of creativity and action, not austere and academic theologians. This made all the difference to their attitudes and to the face of science.

Essay Questions

1. What influenced Copernicus to develop a new heliocentric system?

2. How did voyages of exploration influence scientific understanding of the world?

3. What role did patronage play in the development of science in the sixteenth and seventeenth centuries?

4. Why did Galileo come into conflict with the Roman Catholic Church?

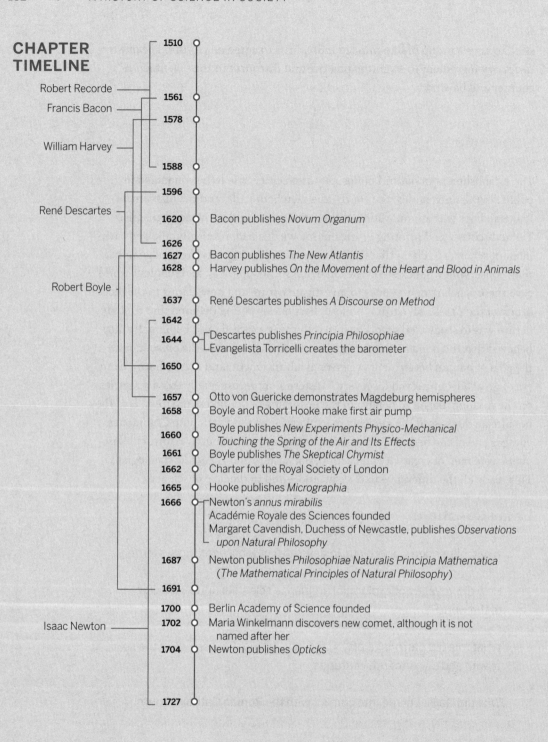

CHAPTER TIMELINE

Robert Recorde

Francis Bacon

William Harvey

René Descartes

Robert Boyle

Isaac Newton

1510

1561

1578

1588

1596

1620 — Bacon publishes *Novum Organum*

1626
1627 — Bacon publishes *The New Atlantis*
1628 — Harvey publishes *On the Movement of the Heart and Blood in Animals*

1637 — René Descartes publishes *A Discourse on Method*

1642

1644 — Descartes publishes *Principia Philosophiae*
Evangelista Torricelli creates the barometer

1650

1657 — Otto von Guericke demonstrates Magdeburg hemispheres
1658 — Boyle and Robert Hooke make first air pump

1660 — Boyle publishes *New Experiments Physico-Mechanical Touching the Spring of the Air and Its Effects*
1661 — Boyle publishes *The Skeptical Chymist*
1662 — Charter for the Royal Society of London

1665 — Hooke publishes *Micrographia*
1666 — Newton's *annus mirabilis*
Académie Royale des Sciences founded
Margaret Cavendish, Duchess of Newcastle, publishes *Observations upon Natural Philosophy*

1687 — Newton publishes *Philosophiae Naturalis Principia Mathematica* (*The Mathematical Principles of Natural Philosophy*)

1691

1700 — Berlin Academy of Science founded
1702 — Maria Winkelmann discovers new comet, although it is not named after her
1704 — Newton publishes *Opticks*

1727

THE SCIENTIFIC REVOLUTION: CONTESTED TERRITORY

5

From 1543 to 1687 some of the giants of science lived, contemplated the natural world, and produced the underpinnings of modern science. Given the accomplishments of the period, it is no wonder that historians have agonized about the idea of an era of scientific revolution. Indeed, the twentieth-century discipline of the history of science began by focusing on the problem of the origin of modern science. The work of some of the discipline's founders concentrated on what this important transformation was and how it came to take place. In recent years, historians have begun to question whether or not such a revolution happened at all. Obviously, the answer depends on how revolution and science are defined: whether there was a gradual transformation of ideas, a gestalt switch, or a sociological innovation. We argue that there was a transformation in the investigation of the natural world, in which new ideas, methods, actors, aims, and ideologies vied with one another for a newly secularized role in the developing nation-states. This, indeed, was a scientific revolution.

The scientific revolution can be understood as a series of overlapping innovations, all important in the creation of modern science. First, natural philosophers took up the epistemological challenges of the ancients and developed a new methodology for uncovering the truth about the natural world. Second, in many different areas, but particularly in physics, astronomy, and mathematics, new theoretical models of the universe were developed. Further, those interested in the investigation of nature formed new institutions and organizations that began to perform the now largely secular tasks of evaluating scientific fact and determining who could be a natural philosopher. Finally, and perhaps most significantly, men interested in the underlying truths of nature developed a new ideology of utility and exploitation, a new structure for scientific practice, and a gentlemanly coterie of scientists who applied their social standards of behaviour to the ideology of modern science.

The New Scientific Method: Francis Bacon and René Descartes

The rediscoveries of ancient natural philosophers and the challenges to that ancient knowledge in the sixteenth century caused scholars in the sixteenth and seventeenth centuries to turn to the epistemological question of how to determine truth from falsity. Perhaps spurred on by the religious turmoil of the sixteenth century, philosophers began to ask the question still fundamental to us today: How can we know what is true? This led to the development of a new form of scientific inquiry—a new "scientific method"—and a new way of articulating this search for certainty. In England this methodology was most fully articulated in the writing of Sir Francis Bacon (1561–1626). Bacon, although not himself a natural philosopher, proposed a reform of natural philosophy in the *Novum Organum* (1620) and *The New Atlantis* (1627). This program of reform was part of a grander scheme to transform all knowledge, especially legal knowledge and moral philosophy. Bacon believed that all human knowledge was flawed because of the Idols that all men carried with them. The Idols were the prejudices and preconceived ideas through which human beings observed the world. Bacon felt that the only way for natural philosophers to disabuse themselves of these Idols was to look at small, discrete bits of nature. The only way to be certain one understood these small bits was to study them in a controlled setting, isolated from the larger (uncontrolled) environment. Using this assumption, he introduced what has come to be called the inductive method. He suggested that increments of information could be gathered by armies of investigators, put together in tabular form, and explained by an elite cadre of interpreters.

Bacon described this in a section of *The New Atlantis* known as "Solomon's House." His methodology, although it appeared to be more democratic than earlier scholastic methods, proposed a means of controlling truth and knowledge by a small elite group who determined what could be studied and what answers were acceptable. In this attitude, he was probably influenced by the fact that he was trained as a lawyer and that he spent much of his political career as an advisor to Elizabeth I and then as Lord Chancellor for James I. He was thus accustomed to the idea of testing evidence in the public venue of a court. As the person most concerned with treason and heresy, he had a distrust of free-thinking and believed that ideas should be controlled by those whose position as custodians of the peace and security of the commonwealth best assured their credibility. One aspect of Bacon's job as Lord Chancellor included overseeing the use of torture in an age when evidence obtained by torture was considered reliable. For Bacon, knowledge was power, and thus an understanding of nature was important precisely because of the practical applications such knowledge would have. In many ways Bacon was a courtly philosopher, so the rhetoric of utility so well employed by Galileo was also present in his work.

This methodology was challenged on the continent by René Descartes (1596–1650) and his followers, who preferred a deductive style based on skepticism. Descartes, like Bacon, came from an influential family and was trained as a lawyer. Unlike Bacon, he worked as a mathematics teacher and practitioner, rather than as a courtier, although in the end the temptation for patronage overcame the struggle involved in living by one's wits. In 1649 Descartes, at the age of 53 the most famous philosopher of his time, accepted the post of Court Philosopher to Queen Christina of Sweden. This was a lucrative post that, like the patronage of the Medicis for Galileo, offered both financial support and status in exchange for glorifying Christina's court and providing philosophical services. Unfortunately for Descartes, whose health was poor, Christina's idea of using his philosophical services was to have him call on her three times a week at five in the morning to instruct her. He was dead of pneumonia before the winter was over. The Swedes sent his body back to France but kept his head. This led to low-grade tension between France and Sweden for close to 200 years, until in 1809 the Swedish chemist Berzelius somehow managed to get Descartes's skull and return it to the French scientist Cuvier, who reunited it with the body.

In *A Discourse on Method* (1637) Descartes offered the first early modern alternative to Aristotle's epistemological system: his method of skepticism. He began by doubting everything, peeling away all layers of knowledge until he came to the one thing he knew was true: that as a thinking, doubting being, he must exist in order to think the doubting thought. He encapsulated this idea in the famous declaration *Cogito ergo sum: I think, therefore I am.* From this starting point he developed through deduction from first principles a series of universal truths that he knew to be self-evident. This deductive method owed its origins to geometric proof, which starts with a small set of sure premises or axioms and proceeds to more complex conditions. Interestingly, although Descartes used this mathematical model and developed new mathematical methodology, most of his scientific theories were explicitly non-mathematical. Also, he was not interested in using experimentation as a means to discover knowledge about nature. Since the senses could be fooled, right reasoning was a much more reliable arbiter in natural philosophical debate than any crude experiment or demonstration might be.

Both Bacon and Descartes attempted to find ways of reasoning that would produce certain knowledge in an age where certainty was giving way to probability. Bacon answered the question of how we can know what is true in a careful, conservative way, involving a hierarchy of knowledge made to appear as a democratic republic of scholars. Descartes answered in an individualistic anti-communal way, which gave more power to individual thinkers but did not, in the final analysis, create a community of scholars.

Mathematics as the Language of Natural Philosophy

One result of this new methodology with its debate about reliable and sure knowledge was the gradual ascendancy of mathematics as the language of natural philosophy. Galileo had been convinced that God created the world in number, measure, and weight, and many other scholars interested in nature echoed this sentiment. For Aristotle, weighing or measuring a substance did not tell you anything interesting about it, but for those investigating nature in the sixteenth and seventeenth centuries knowing how heavy something was or how fast something went was surer knowledge than searching for final causes. They claimed that they only measured and observed rather than imposing underlying hypotheses they sought to prove. Sir Isaac Newton, for example, famously said *Hypotheses non fingo: I feign no hypotheses.* Following Galileo and Newton, natural philosophers increasingly looked for certainty through measurement rather than the analysis of cause.

Significant mathematical developments included the rediscovery of algebra and the development of the calculus; in addition, new and easier notation systems were devised. For example, Descartes instituted the use of *a*, *b*, and *c* for known variables and *x*, *y*, and *z* for unknowns. Mathematicians used these systems as they vied with one another for solutions to increasingly complex algebraic equations.

The creation of analytic geometry, primarily by Descartes but also by François Viète (1540–1603) and Pierre de Fermat (1601–65), placed a powerful tool in the hands of mathematicians and natural philosophers. By combining geometry and algebra, it became possible to transform geometric objects into equations and vice versa. This also opened the door to mathematizing nature, as everything from the trajectory of a cannonball to the shape of a leaf could be turned into a mathematical expression.

In turn, mathematicians began to look for new ways to measure areas under curves and to describe dynamic situations. The result was the invention of the calculus, independently developed by Isaac Newton in England and Gottfried Wilhelm Leibniz (1646–1716) on the continent. The calculus added infinitely small areas together under a curve or described the shape of the curve in formulaic terms. This allowed natural philosophers to accurately describe dynamic situations such as velocity and the motion of acceleration, something not possible with the older geometric and algebraic systems. Some mathematicians were concerned with the philosophical implications of the calculus, since it produced finite answers from the addition of infinite quantities, and the infinitely small and the infinitely large could, paradoxically, be equal. However, in an age that increasingly looked to the practical applications and utility of mathematics, the calculus proved to be an extremely fruitful device and was quickly taken up by the scientific community.

Leibniz was an influential German polymath, trained as a lawyer and employed most of his life by several German princes, especially the three dukes of Hanover. Unfortunately, his relationship with Duke Georg Ludwig (1660–1727) deteriorated, and when Ludwig became King George I of England, Leibniz was forbidden to enter the country. He was very critical of both Descartes's and Newton's work and became embroiled in an acrimonious dispute with Newton over who had invented the calculus first. This dispute, carried on by Leibniz in Germany and Samuel Clarke (1675–1729) in England, was one of the most famous philosophical disputes of the seventeenth and eighteenth centuries, as it ranged over issues of Newtonian natural philosophy and theology. In the end the participants died, and the matter was never resolved. It seems clear now that Newton and Leibniz did develop the calculus independently, with completely different notation and mathematical bases (Newton's was geometrical, while Leibniz's was analytical). In a sense Leibniz won, since the notation that has been used from the eighteenth century to the present is his rather than Newton's.

Mathematical Practitioners

One of the reasons mathematics became such a powerful tool in seventeenth-century natural philosophy was the presence of a new category of scientifically inclined men: the mathematical practitioners. Mathematics had been a quite separate area of investigation, and those interested in its issues had usually tied their studies to practical applications such as artillery, fortification, navigation, and surveying. In the early modern period, these mathematical practitioners provided the necessary impetus in the transformation of nature studies to include measurement, experiment, and utility. Their growing importance was the result of changing economic structures, developing technologies, and new politicized intellectual spaces such as courts, thus relating changes in science to the development of mercantilism and the nation-state. Mathematical practitioners claimed the utility of their knowledge, a rhetorical move that encouraged those seeking such information to regard it as useful.

Mathematical practitioners professed an expertise in a variety of areas. For example, Galileo's early work on physics and the telescope were successful attempts to gain patronage by using mathematics. Descartes advertised his abilities to teach mathematics and physics. Simon Stevin (1548–1620) claimed the status of a mathematical practitioner, including an expertise in navigation and surveying. William Gilbert (1544–1603) argued that his larger philosophical arguments about the magnetic composition of the Earth had practical applications for navigation. Leibniz used his mathematical power to act as an advisor on engineering projects for the dukes of Hanover. As well, many practitioners, including Thomas Hood (fl. 1582–98) and Edward Wright (1558–1615), demonstrated an interest in mapping and navigation.

This new interest in mathematics and in quantifying the behaviour of the world sparked interest in probability. Mathematicians did not believe the world was capricious but that our incomplete knowledge of it limited our comprehension. The introduction of a mathematical evaluation of probability was a step toward understanding complex systems in which not all the determining factors could be known with certainty. Blaise Pascal (1623–62), Pierre de Fermat, and Christiaan Huygens (1629–95) all investigated the mathematical basis of prediction of games of chance, which were popular pastimes in the seventeenth century. Pascal's interest in the geometry of chance had wider implications than gambling at cards and dice, since it led him to develop his probabilistic argument for belief in God, now known as Pascal's Wager. He concluded that, although one cannot know with complete certainty if God exists, by using four possible conditions, one's best probable outcome would result from belief in God. If God did not exist, one lost nothing by believing in Him, but if He did exist, and one believed, one would be saved. Conversely, one lost a great deal by not believing in God if He did exist, while gaining nothing if He didn't exist. (See figure 5.1.)

By the end of the century Jacob Bernoulli (1654–1705) had codified the mathematics of probability, arguing that mathematics gave us the greatest certainty possible in an uncertain world. The concept of probability was not well accepted in physics, however, where Newton's universal laws seemed to provide certain, rather than probabilistic, answers. The shift of the foundation of physics from certainty to probability was one of the most traumatic transitions in modern science, but it would not happen for almost 200 years.

5.1 PASCAL'S WAGER

	Do Not Believe in God	Do Believe in God
GOD DOES NOT EXIST	Nothing gained or lost	Nothing gained or lost
GOD DOES EXIST	Damnation	Salvation

New Models of the Universe

All these new attempts to find a path to certain knowledge were crucial, since natural philosophers were making some radical suggestions about the make-up of the cosmos. The scientific revolution is most clearly identified with the development of a heliocentric model of the universe. This began with Copernicus, who claimed that the Earth revolved around the sun and who developed a mathematical model to explain the movement of the planets. Natural philosophers had been slow to accept Copernicus's theory because it lacked a proper physical justification, such as Aristotle had provided through the concept of natural motion for his cosmological schema. Thus, the so-called Copernican Revolution (historians love to label innovations as revolutions) was incomplete until Sir Isaac Newton (1642–1727), the great English astronomer and mathematician, devised a mathematical model of motion that explained heavenly and earthly movement in a single physical system based on his concept of universal gravitation.

Isaac Newton: The Great Polymath

Isaac Newton was born on Christmas Day, 1642, coincidentally the same year that Galileo died. His father died before he was born, and his early life was not happy, spent in conflict with a stepfather he disliked and a mother who expected him to

CONNECTIONS

Science and the Marketplace: Mathematics for Sale

During the sixteenth century, many mathematical practitioners made their living by giving lectures and instruction, selling instruments, and performing a variety of mathematically based activities such as surveying or casting horoscopes. The booming metropolis of London provided an excellent marketplace for these skilled men, whose work demonstrated the interconnection of natural philosophy, mathematics, and the mercantile society of the time.

The first mathematical practitioner in London was Robert Recorde (1510–88). Trained at Oxford, he was largely responsible for introducing arithmetic and mathematics to a wider audience in England and for re-establishing a mathematical language and discipline into English scholarship. Recorde wrote a number of foundational books on mathematics in English. He was commissioned by the Muscovy Company (a merchant company that traded in northern Europe and was seeking the Northeast Passage to Cathay) to give lectures and to write a series of books on geometry, spherical geometry, astronomy, and navigation for use by their navigators. At the same time, he introduced Euclidian mathematics and algebra to an English audience, allowing natural philosophers to use mathematics in ways they had not done before.

Recorde was soon followed by other mathematical lecturers in London, sponsored by guilds, companies, or the City of London itself. Thomas Hood (1556?–1620), for example, was the first mathematics

run the family farm. Newton had no aptitude for farming, and his mother despaired of finding him a livelihood. Fortunately, the local vicar noticed his scholarly potential and helped procure him a scholarship to Trinity College, Cambridge. Newton was a relatively undistinguished scholar, except for mathematics; he taught himself geometry from Descartes's works and algebra from Viète's. In due course, he was made a Fellow of Trinity College in 1664 and was appointed Lucasian Professor of Mathematics in 1669. The latter was a prestigious position, although he often lectured to empty rooms. His teacher, Isaac Barrow (1630–77), had to pull strings in order to get him this position, since Newton did not believe in the special divinity of Christ or in the Trinity. This made him a potential heretic, since he would not take the required oath of uniformity to the Church of England, which was normally required for any high academic or government position.

lecturer paid by the City of London. In 1588, Hood petitioned William Cecil, Lord Burghley, to support a mathematics lectureship in London, to educate the "Capitanes of the trained bandes in the Citie of London."* Hood identified himself on the title pages of all his books until 1596 as Mathematical Lecturer to the City of London, sometimes advising interested readers to come to his house in Abchurch Lane for further instruction, or to buy his instruments. His books explained the use of mathematical instruments such as globes, the cross staff, and the sector, suggesting that his lectures and personal instruction would have emphasized this sort of instrumental mathematical knowledge and understanding. He was also known to have cast horoscopes, another way to make money.

Those who gave and attended these mathematical lectures had some expectation that they would be able to buy, sell, and use the instruments that were being discussed there. It is no surprise, therefore, that just as these lectures were being presented to the London community, mathematical instrument makers were beginning to ply their trade in increasing numbers in late-sixteenth-century London. Thomas Gemini, a goldsmith, was probably the first English instrument maker starting in the 1550s, followed by Humphry Cole in the 1580s. After that, many men set up shop and sold navigational instruments, surveying instruments, maps, globes, and astrolabes, as well as providing the informal instruction for those who were interested in using them.

By 1610, there was a strong practical mathematical community living in London. A number of mathematical lectures had been sponsored, attended by a variety of audiences. Books and individual lessons explaining the use of mathematics and mathematical instruments had been produced, all leading to an increasing number of men trained in and sensitive to mathematical tools and explanations. A variety of men met in the instrument shops and at the mathematical lectures—gentlemen, scholars, merchants, and navigators. Mathematics was becoming both a language of commerce and of natural philosophy.

* British Library manuscript, Landsdowne 101, f. 56.

While Newton would not compromise his religious beliefs, he kept his views very private for his entire life.

In 1665 the Great Plague returned to England. It swept through Cambridge, and Newton was forced to return to his mother's home. The enforced isolation allowed him the opportunity to put together a number of ideas he had been developing through intensive reading during the past four years. Although it is unlikely that an apple really fell on his head, during the following year, his *annus mirabilis* (miraculous year), he worked out theories about gravitation, physics, and astronomy. As if that were not enough, he also created the calculus and began his investigations into optics and theories of light.

Although Newton studied planetary motion during that year, he did not publish his results until 22 years later. He was dissatisfied with his mathematical results to the question of why the orbits of the planets were circular, or nearly circular, and so put them away for a time in order to concentrate on alchemy and theology, including his long-standing interest in the books of Revelation and Daniel. He spent many hours over the next 13 years reading commentaries on scripture, studying the Bible, and constructing his personal theology. This theology was complex, most closely resembling an extreme form of Unitarianism or Arianism, which argued that Christ was not divine but the highest of God's created beings. Over his lifetime, Newton spent more time studying theology than any other subject.

In 1679 Robert Hooke (1635–1703), Corresponding Secretary of the Royal Society of London, wrote to Newton to find out what he was doing. It was Hooke's job to act as a kind of intellectual pen pal, communicating with members of the Royal Society and putting people with similar interests in touch. Newton and Hooke had a strained relationship, however, since Hooke had criticized Newton's earlier optical work. What, Hooke asked in 1679, would be the path of a body (a rock, for example) released from a high tower down to a rotating earth? Newton replied that he was not presently engaged in natural philosophical investigations, but he suggested that it would spiral east to the centre of the earth because the angular velocity from the top of the tower was greater than on the earth's surface. Hooke argued that this was not so: the path would be the result of a horizontal linear motion at constant speed combined with an attractive force toward the centre and varied inversely with the square of the distance between the body and the earth. This caused Newton to wonder whether his original question in 1666 as to why planetary orbits were circular was misdirected.

Newton set to work on mathematical models of elliptical orbits. In 1684 his friend Edmund Halley (c. 1656–1743, and the discoverer of the comet that bears his name) rode out to Cambridge to try to get Newton to communicate some of his mathematical work to the Royal Society, a difficult task since Newton was very secretive and refused to publish anything. Halley said that he and his friends, the architect and natural philosopher Christopher Wren (1632–1723) and Hooke, had wondered what motion an orbiting body would traverse if attracted to a central body by a force that varied inversely as the square of the distance between them. Newton replied that in 1679 he had proven to his own satisfaction that it would be an ellipse, although he did not have the demonstration available. Halley was deeply impressed, seeing this as a breakthrough in mathematical astronomy, and urged Newton to publish. He even promised to finance the publication. Newton agreed and in an astonishingly short period produced the book that changed astronomy and physics. Published in 1687, *Philosophiae Naturalis Principia Mathematica* (*The Mathematical Principles of Natural Philosophy*, usually known simply as the *Principia*) laid out Newton's theory of universal gravitation and established a mathematical and mechanical model for the motion of the whole universe. In this work he used the mathematical models of Galilean physics and combined them with the planetary models of Copernicus and Kepler.

Newton had produced a universal physics, a Grand Synthesis, which finally allowed astronomers to move away from Aristotle with confidence. He had produced a real model of the universe rather than merely a mathematical set of calculations, such as Copernicus had put forward.

In the *Principia* Newton defined force for the first time, adding it to matter and motion as the third essential quality of the universe. Force, he said, was necessary to compel a change of motion. Without force, a mass (another term he originated) would continue at rest or in rectilinear motion, due to inertia. This is often called Newton's First Law. His Second Law showed the way to measure force mathematically, now expressed as $f = ma$, and his Third Law asserted that for every action there is an equal and opposite reaction. In order to understand the motion of the bodies in the universe, Newton insisted that they operate in absolute time and space. In the Newtonian system, absolute time and space were independent and unchanging aspects of reality. Absolute time was uniform, progressing at the same rate of change, everywhere in the universe and at all past and future moments. People could not perceive absolute time directly, but could infer it mathematically by observing the motion of the planets and stars. Absolute space was the unmoving and constant dimension of the universe. This meant that all observers saw the motion of the stars and planets (and everything else) within absolute space, and thus they should all see the same motion as measured against the unchanging dimension of absolute space. The implications of this were profound. It meant that Newtonian physics was truly universal and someone measuring the universe from a distant star would get the same results as we would.

Newton took the concept of centrifugal force (the tendency of an object to fly away from a circular path) and stood it on its head. In his early work in 1666, he had followed the older view of circular motion, which sought to quantify and explain this tendency to fly away. But, in a major insight, the *Principia* argued for a centripetal force instead, a force that pulls objects into the centre. This is gravity. Thus, the moon's motion is composed of two parts: first, inertial motion carries it in a straight line; second, gravity constantly forces it to fall toward the Earth. The balance of these two forces holds the moon in orbit. (See figure 5.2.)

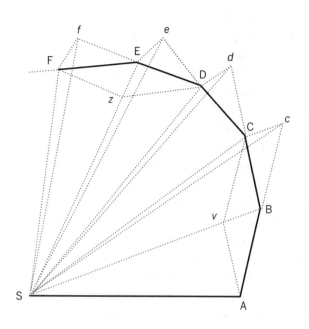

5.2 NEWTON EXPLAINS THE MOON'S MOTION

From Newton, *Principia Mathematica*, Proposition I. At each point, A, B, C, etc., the moon's motion seeks to move in a rectilinear line away from S, but is drawn back by centripetal force (gravity) toward S.

Newton set up his argument in order to disprove Descartes's theory of vortices, which Descartes had articulated in *Principia Philosophiae* (1644). Descartes had argued that the universe was like a machine, a concept called "the Mechanical Philosophy." Except for those parts filled with coarse matter (such as the Earth), the universe was a plenum, or a space filled with an element called ether. The planets moved in a sort of whirlpool of ether that carried them along in their orbits. (See figure 5.3.)

The *Principia* was in many ways an attack on Descartes's book (the title reflected Descartes's), which is why Book Two concentrated on the study of how fluids work and how things move through them. This exploration of fluid mechanics may confuse modern readers, since it seems to have little to do with forces, gravity, or the motion of the planets. Newton's objective was not just to present his own system but to discredit Descartes's by showing that the vortex model was critically flawed.

The most far-reaching and long-lasting achievement of the *Principia* was how Newton found a way to tie together through the concept of universal gravitation the orbits of the planets and the satellites (including the moon and comets), Galileo's law of falling bodies, the fixation of objects to the Earth, and the tides. He proved that an apple falling from a tree obeyed the same laws as the moon orbiting the Earth. He had a law that applied to *everything*—the moon, the satellites of Jupiter and Saturn, the Earth, rocks, and the distant stars.

The sheer power of Newton's universal laws suggested to scholars in many different fields that there should be similar laws governing human interaction as well. Philosophers from many other areas, including economics and political philosophy, searched throughout the eighteenth century for such laws and argued that any society that failed to follow the universal laws was doomed to failure.

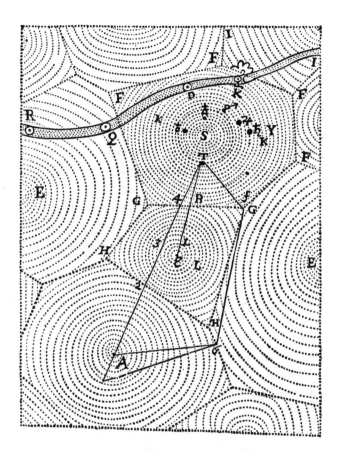

5.3 DESCARTES'S VORTEX COSMOLOGY

Points S, E, A, and ε̇ are centres of vortices. By the churning of the corpuscles within each vortex, the centre is self-illuminating, and thus is a star. Around S (the sun) move the planets, swept along by the vortex (T is the earth or terra). The stripe at the top indicates a comet's path. From Descartes's *The World or a Treatise on Light* (1664).

Ironically, the *Principia* was not a popular or financial success. The Royal Society, the pre-eminent institutional home of English science, refused to sponsor it, since their previous publishing venture, Francis Willoughby's *History of Fishes*, had been a financial disaster. They were unable to sell more than a handful of copies of this expensive illustrated book and were forced to give it to their employees, especially Robert Hooke, in lieu of salary! The members of the Royal Society, perhaps accurately, saw the *Principia* as a book with a limited readership and so forced Halley to fund it himself, which he did through the sale of subscriptions. This story reminds us that, as revolutionary as we might now consider Newton's work, at the time it held the interest of a limited audience because so few people could understand his mathematical arguments.

Newton and Alchemy

Because Newton considered his studies esoteric, accessible only to a select few, he was not overly concerned about reaching a wide audience. One aspect of his work that has been overshadowed by his physics and mathematics and so has faded from modern consciousness was his great interest in alchemy. Both Newton and Robert Boyle were involved in alchemical investigations, looking both for the Philosopher's Stone and the basic structure and functioning of nature. Although "multipliers," or alchemists interested only in gold, were looked upon as crass, many respectable gentlemen studied alchemy. Both Newton's and Boyle's investigations came from the belief in God's ordering of the universe and were a search for the active principles that animated nature, another attack on mechanical philosophy. Since Newton was interested in how the universe worked on a microscopic as well as a macroscopic level, he investigated what the prime matter of the universe was. He read both ancient and modern authors copiously and performed experiments of long duration. Because of his expertise in this area, in later years he was appointed Master of the Mint, in which position he assessed those who claimed to have produced gold. At his death, a number of people interested in alchemy were anxious to gain access to his extensive library of alchemical works.

In spite of his lack of bestseller status and his more arcane studies, Newton was the most famous scientist of his generation. He received many important marks of honour indicative of his high status, including his position as Warden and then Master of the Mint and election as President of the Royal Society. His opulent state funeral in 1727 demonstrated his exalted position, and reports of the event were transmitted around Europe, with Voltaire (who was deeply influenced by Newton's natural philosophy) providing one of the eye-witness accounts. Voltaire was extremely impressed that a natural philosopher, and one with heterodox religious views at that, was buried with such pomp and circumstance. Here was a country that understood the importance of its intellectuals!

Mechanical Philosophy

At the same time that Newton was introducing a mathematical basis for motion throughout the universe, other natural philosophers were searching for a more concrete model. Some, looking to the new instruments and machines being developed around them, argued that the world itself was a sort of machine. As men constructed more and more precision mathematical instruments, especially for navigation, astronomy, surveying, and timekeeping, and more complex machines, this suggested to them that God might have created the most complex machine of all. This conception of the world operating in a mechanical fashion became known as the Mechanical Philosophy or, alternatively, the atomical or corpuscular philosophy. It was first developed by René Descartes and Pierre Gassendi (1592–1655). Descartes especially extended the mechanical philosophy to include living things, arguing in his *Treatise on Man* (1662) that human physiology operated just like a machine. Corpuscular philosophy was later taken up by the Englishmen Robert Boyle and Thomas Hobbes, the latter of whom used it in his political theories. Basically, it was an interpretation of the world as a machine, either as a clock (implying order) or an engine (showing the power of nature). If the universe was a machine, then God was the Great Engineer, or Clockmaker. At first, Gassendi devised this interpretation of nature to give God a role as a transcendent, rather than immanent, being. That is, Gassendi argued that God could exist outside the material universe because He had set a mechanical structure in place. There was no need for God to exist within and tinker with an imperfect world. Corpuscular philosophy was eventually accused of being atheistic because, if the universe were a perfect clock, it would never stop, and there would be no need for God. The claim that natural philosophers were removing God from the universe was unfair but widespread, affecting Newton as well as Gassendi. It recurred in many forms and as a charge against many philosophers and scientists over the next centuries.

Mechanical philosophy was rooted in ancient theories of atomism, held by Epicurus, Democritus, and Lucretius, which had been rediscovered by the humanists. These ancient thinkers had posited that the world was composed of infinitely small simple particles. For the ancient Greeks, this had shown the eternity and total materiality of the world, an aspect of the theory that Descartes and especially Gassendi (a Catholic priest) set out to change. They argued that although it appeared that the world had existed forever, this could not be true, which meant that its creation was of a different kind than its operation and was, therefore, known by God, but unknowable through natural philosophy. A material universe could not be used as a proof for or against the existence of God, since God's existence was a metaphysical, not physical, question.

Mechanical philosophers reduced matter to its simplest parts, atoms, just as Descartes had stripped away ideas through skepticism. These atoms had only two qualities: extension and motion. Since extension was a definition of matter—that is, that matter must take up space and all space must be matter—a vacuum was not possible in this philosophy. Therefore, the universe was filled with a plenum of particles. All force-at-a-distance was actually motion through the plenum, which explained magnetism and the motion of the planets. It was this aspect of Descartes's theory that Newton had attacked so forcefully in the *Principia*. And the question of whether or not a vacuum could exist in nature was soon taken up in Robert Boyle's experimental program. It is a great historical irony that the close examination of the nature of matter, meant to prove mechanical philosophy, refuted the very theory that the universe was full of matter.

The Use of Experiments as Proof: William Harvey and Robert Boyle

Experimentation, as a source of sure knowledge about nature, was new to the early modern period. Aristotle, of course, had argued that forcing nature into unnatural situations would tell us nothing about how it really behaved. This attitude began to change in the sixteenth century, partly because of new attitudes toward certainty and man's power over nature, and partly because skilled instrument makers were able to create precise philosophical instruments. Francis Bacon, who as Lord Chancellor had overseen the torturing of traitors, believed that human beings, when subjected to extreme pain, would be forced to tell the truth. Likewise, putting nature on trial, including torturing nature through experiments, would force her to reveal her secrets. While the reliability of truth claims as a result of experimentation was under constant scrutiny in this period, it is fair to say that one of the significant changes to the study of nature in this period of scientific revolution was the increased use of, and reliance on, experimentation.

Probably the first extended discussion of experimentation came from the study of human anatomy in the work of William Harvey (1578–1657). Following the success of Vesalius in the sixteenth century, scholars developed a keen interest in the structure and function of living things. For example, Girolamo Fabrici (c. 1533–1619) examined the structure of veins and in 1603 found the existence of valves at particular intervals. Harvey, using keen observation, some well-thought-out experiments, and a belief in the similar structure of all animals including humans (what was later called comparative anatomy), developed a theory of the circulation of the blood that proved very influential in the years that followed.

Harvey received his medical training at Padua, where Vesalius had taught, and returned to London in 1602 to work first as a physician at St. Bartholomew's

Hospital and eventually as Royal Physician to Charles I. While in these positions he conducted a series of experiments on the blood in animals. This resulted in the publication of *On the Movement of the Heart and Blood in Animals* (1628), in which he demonstrated that the blood in animals and humans was pumped out by the heart, circulated through the entire body, and returned to the heart. He proved this through a series of elegant experimental demonstrations, some involving vivisection of animals and some less invasive demonstrations with humans. In his preface Harvey drew the political parallel between this circulation and the movement of citizens around their king, indicating his close affiliation with the Royalist side of the civil war soon to erupt in England.

One of the clearest experiments that Harvey performed was his proof that there were no passages through the septum of the heart. From Greek times through Galen and Vesalius, anatomists had argued that blood must pass through the wall separating the two ventricles or large chambers of the heart. While this explanation satisfied some aspects of the assumed purpose and path of blood, problems persisted. The first was the question of volume, since the old system required the liver to produce a constant supply of blood that was completely consumed by the rest of the body. This seemed to Harvey to be out of all proportion to the amount of matter that went into the body. The second problem was purely anatomical. While Galen had said there were pores in the septum, Vesalius, who worked with human hearts rather than the cow and pig hearts that Galen had used, could find no such passages. Rather than contradict Galen completely, Vesalius argued for a permeable septum with either sponge-like tissue or pores too small to see.

Harvey reasoned that if blood were circulated, a far smaller volume would be needed; it was simpler to conceive of a re-used supply of blood than a constantly created and consumed supply. At the centre of the system was the heart, working as a pump, but to demonstrate the two-part circulation (heart-to-lungs and back, heart-to-body and back) he had to show that blood did not move from the venous to the arterial system by way of the chambers of the heart. In a later experiment he demonstrated this, using a cow heart and a bladder of water. He tied off the passages to the heart so that he could squeeze water into one ventricle and see if it passed into the other. When it did not, he had the first proof that the blood did not pass through the septum. He then tied or untied the constraints and used the bladder of water to show that the blood must pass in sequence from right atrium to right ventricle and out to the lungs, and then through the pulmonary vein to the left atrium and left ventricle and out to the body through the aorta. While it was still unclear how the blood got from the arteries through the tissues of the body and back into the veins (and would remain unclear until microscopy developed far enough to see the microvessels at cell level), Harvey's work better explained the evidence than the older system. (See figure 5.4.)

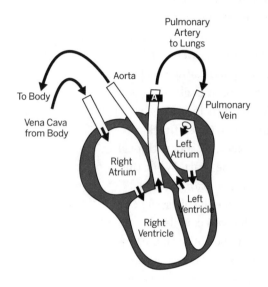

5.4 HARVEY'S MODEL OF THE HEART
Harvey closed off the artery at "A" and pumped water through the vena cava to demonstrate that no fluid passed from the right ventricle to the left ventricle.

Harvey also used careful observation and some experimentation in his embryological studies. Fabrici, in his book *On the Formation of the Egg and the Chick* (1621), had observed that in viviparous generation the embryo was created by a union of semen and blood from the parents. Harvey followed this work with a close examination of the development of ova. He examined fertilized eggs from their unformed state to birth, tracing even more closely the stages of growth. He published this work in *Exercises Concerning the Generation of Animals* (1657). This spurred further work by Marcello Malpighi (1628–94), who introduced the use of microscopic observations of ova development in 1672.

The work of Harvey and Malpighi, showing the power of observation and experiments, accorded with the inclination among many natural philosophers to use instruments and demonstrations to isolate phenomena and break down investigations into smaller components. Perhaps the man most responsible for the elevation of instrumental investigation during the scientific revolution was Robert Boyle (1627–91). Boyle, the son of an Irish noble family, came to Cambridge during the English Civil War and became a key player in the creation of the Royal Society after it. He joined the elite of London society from the 1660s onwards, living with his sister, Lady Ranelagh, in her house in Pall Mall, and entertaining royal and scholarly guests alike at his laboratory there. While at Cambridge, he set up his own alchemical laboratory to investigate the underlying make-up of matter. He denounced old-fashioned alchemical investigations and, in *The Skeptical Chymist* (1661), laid a foundation for the new study of chemistry.

Boyle, with Robert Hooke as his assistant, investigated airs, using a newly devised air pump. He employed instrument makers to attach a large, carefully blown glass globe to a pump in order to evacuate the air from the globe. In this, he

5.5 ILLUSTRATION OF VON GUERICKE'S MAGDEBURG EXPERIMENT

followed the lead of Otto von Guericke (1602–86), who had performed his own air experiments in the 1640s. Most famously, in 1657, von Guericke demonstrated that two teams of horses hitched to joined hemispheres of copper (the so-called Magdeburg hemispheres) from which the air had been evacuated could not pull them apart, since the weight of the air outside the spheres was so much greater than that on the inside. (See figure 5.5.)

Boyle and Hooke built their first air pump in 1658 and performed a number of experiments. (See figure 5.6.)They demonstrated that air had weight, that a vacuum could exist, and that some component of air was necessary for respiration and combustion. Their results were published in 1660 as *New Experiments Physico-Mechanical Touching the Spring of the Air and Its Effects.* Boyle used his air pump for less spectacular demonstrations than von Guericke, for example, placing small animals in the glass sphere, removing the air, and watching them perish, or alternately placing candles therein and watching the flame extinguish. (See plate 4 for a later depiction of these events.) From these experiments he concluded that there was something in the air that supported both life and combustion. His work was thus connected to Harvey's because it touched on what part of air was needed for life and that it seemed to be brought into the body by way of the lungs. This was part of a growing interest in "vitalism," the search for the spark of life that transformed inanimate matter into living plants and animals.

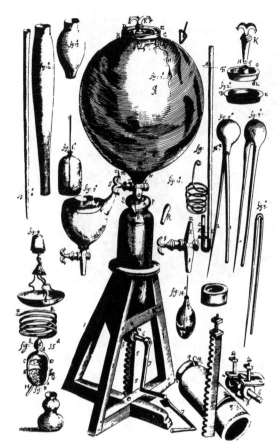

5.6 BOYLE'S AIR PUMP AND TOOLS FROM *NEW EXPERIMENTS PHYSICO-MECHANICALL* (1660)

Boyle's work also demonstrated the relationship between pressure (the "spring" of the title) and volume of air. He and Hooke used a j-shaped tube filled with mercury to show that increasing or decreasing the pressure on the short stem raised or lowered the level of mercury in the long stem. (See figure 5.6.) Boyle argued that "according to the *Hypothesis*, that supposes the pressures and expansions to be in reciprocal proportion";[1] in other words, as pressure goes up, the volume of air goes down in equal proportion and vice versa. While Boyle was making a specific argument about atmospheric air, which he considered an elastic fluid, not an element in itself, the basic relationship he pointed out was later transformed into $PV = K$, what we now call "Boyle's Law" or occasionally "Mariotte's Law" after Edmé Mariotte (1620–84) who independently found the same relationship in 1676.

Unfortunately, Boyle's air pump was plagued with problems. It leaked quite badly, so that it was not really possible to evacuate all the air from the interior. Also, although he published detailed accounts of his instrument and its operation, complete with schematic diagrams, other natural philosophers across Europe could not replicate his results. Thomas Hobbes (1588–1679), a natural philosopher as well as a moralist, severely criticized the work of Boyle and Hooke. Hobbes claimed that the air pump did not work and that it in no way represented a vacuum, as they had claimed. Although Hobbes presented many sound arguments, Boyle's rising status, both socially and scientifically, ensured that his was the winning side of this disagreement.

....

1. Robert Boyle, *A Defence of the Doctrine Touching the Spring and Weight of the Air* (London: M. Flescher, 1682) 58.

Despite equipment problems, the claim of replicability became fundamental to the experimental program from Boyle on. That is, natural philosophers began to claim that experiments were useful and produced true and certain results precisely because they were not dependent on the experimenter but could be repeated by anyone. The air pump was to be seen as transparent; nothing stood between the observer and the aspect of nature being observed. In other words, an ideology of the objectivity of instrumental experiment arose from Boyle's work with the air pump. Our modern reliance on experimentation—and on replicability—is in part based on the triumph of Boyle and the Royal Society over Hobbes and other skeptics. That "proof" (that is, evidence of the true state of nature) results from experimentation was, and is, a powerful idea, even if it has philosophical flaws. Many things make an unmediated view of nature impossible, but because of Boyle's status and the status of his witnesses, the air pump became "black-boxed." That is, the air pump and other experimental instruments were considered neutral and objective, unproblematically revealing nature's state. For example, when we look at a thermometer to help us decide what kind of coat to wear outside, we have accepted a particular concept of temperature. The thermometer is not an unbiased object but embodies a philosophical idea about the quantification of nature. To someone unfamiliar with the concept of temperature, a thermometer would be a meaningless device. This does not mean that what the thermometer reveals is false but that all scientific instruments represent a system of beliefs about the world.

The Development of Philosophical Instruments

During the seventeenth century, other instruments (often called philosophical instruments) and experimental programs were devised, all sharing this ideological stand with Boyle's work. Boyle himself used the barometer, developed in 1644 by Evangelista Torricelli (1608–47), and others followed this lead in investigating the weight of air. Torricelli filled different glass tubes with quicksilver (mercury), inverted them into a basin, and discovered that all the tubes maintained a constant level. Torricelli claimed that the space above the mercury, the "Torricellian space" as Boyle called it, was a vacuum and argued that "we live submerged at the bottom of an ocean of the element air, which by unquestioned experiments is known to have weight."[2]

....

2. Evangelista Torricelli, "Letter to Michelangelo Ricci" (1644), in *Encyclopedia of the Scientific Revolution*, ed. Wilbur Applebaum (New York: Garland, 2000) 647.

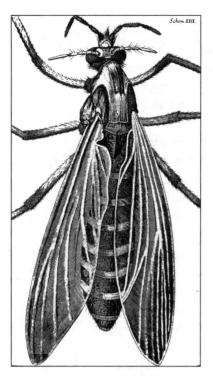

Schem XXIX.

5.7 ILLUSTRATION FROM ROBERT HOOKE'S *MICROGRAPHIA* (1665)

He also predicted that if one ascended to higher altitudes, the weight of air would be less and the column of mercury would descend further. This prediction was taken up by the mathematician Blaise Pascal. Pascal first worked to replicate Torricelli's instrumental experiment, which proved a difficult task. He eventually succeeded, both with a column of mercury and with a much larger one of water. This resulted in a heated debate about the possibility of a vacuum, strongly denied by a number of leading theologians. To avoid this discussion and concentrate instead on the question of the weight of air, in 1648 Pascal took the barometer up a mountain near his brother-in-law's home in Clermont, France. Sure enough, the higher the ascent, the lower the column of mercury and the larger the "Torricellian space" at the top of the tube. Because Pascal soon thereafter had a crisis of faith and turned from natural philosophy to spirituality, the results of this investigation were not known until after his death, with the posthumously published *Traités de l'équilibre des liqueurs et de la pesanteur de la masse de l'air* (1663). This work, with Boyle's, set up an experimental research program for the coming century, as well as demonstrating the power and "objectivity" of philosophical instruments.

Perhaps the most innovative philosophical instrument of the seventeenth century was the microscope, first developed in the first decade of the century. Following the success of the telescope to bring distant sights closer, unknown instrument makers, probably in the Netherlands, produced instruments designed to greatly magnify the very small. The five microscopists best known for their startling observations and discoveries were van Leeuwenhoek, Hooke, Malpighi, Swammerdam, and Grew. Antoni van Leeuwenhoek (1632–1723), a merchant in Delft, first turned these magnifying devices on a variety of substances, most famously male semen, where he claimed to observe small animalcules in motion. This resulted in a series of interesting letters to the Royal Society. Robert Hooke turned to the favourite subjects of microscopists—insects, seeds, and plants— and captured some stunning images of various enlarged phenomena through engravings in his bestselling, lavishly illustrated book, *Micrographia* (1665). (See figure 5.7.)

The Italian anatomist Malpighi turned the microscope on the human body and, in addition to his embryological work, discovered capillaries and their role in the circulation of the blood. Jan Swammerdam of Amsterdam (1637–80) disproved contemporary theories about the metamorphosis of insects, while the Englishman Nehemiah Grew (1641–1712) found the cellular structure of plants. All five successfully overcame early suspicions that the instrument was creating and disguising as much as it was revealing to produce theories and observations much sought after by the natural philosophical community. By the eighteenth century the vanishingly small was seen to be observable, without scientists worrying about any interference from the apparatus itself.

Newton and the Experimental Method

Isaac Newton also played a part in the development of experiment (and experimental instruments) as a legitimate methodology. Beginning during his *annus mirabilis* he developed a theory of light based on a series of simple and elegant experiments. Because of a decades-long dispute with Hooke about optics, he refused to publish, but in 1703 Hooke died. Newton's *Opticks* came out in 1704. Unlike the *Principia*, the *Opticks* was written in English, in simple language, and laid out so that the experiments could be recreated by anyone who could read the book and afford a few pieces of optical equipment such as prisms, mirrors, and lenses. Even more than Harvey or Boyle, who had likewise explained their procedures in print, Newton became the model for the new experimental method. It was a smash hit, snapped up by an eager public and translated into French, German, and Italian within the year. It has been in print almost continuously to the present day.

Newton was contributing to a long tradition of optics research, stretching back to the Middle Ages and earlier Arabic natural philosophers. He was also building on the work of Kepler, who had argued that light travelled in rectilinear rays, enabling a mathematical description of its path. This allowed several natural philosophers, including Thomas Harriot (c. 1560–1621), René Descartes, and Willebrord Snell (1580–1626), to develop the sine law of refraction, which stated that when a ray of light passes from one transparent medium to another (such as air to water), the sine of the angle of the incident (original) ray divided by the sine of the angle of the refracted ray equals a constant. While this was a useful and demonstrable relationship, it led to a disagreement about the nature of light. Was it a motion through matter (that is, a wave)? Or was light made of particles? How did light travel? Could it travel in a vacuum? During the 1670s Huygens developed a wave theory of light in which he argued that the idea of a wave front represented the path of the light.

Newton criticized this wave theory, largely because it seemed to contradict the rectilinear nature of rays put forward by Kepler. Newton argued that light was corpuscular and proved it to his own satisfaction with a series of elegant experiments. By passing a beam of sunlight through a series of prisms, he demonstrated that white light was not pure light, as had been supposed, but rather was a composite of many colours (the spectrum). His demonstration has been called the *experimentum crusis* or crucial experiment, the demonstration that confirms the hypothesis. What Newton noticed was that light passing through a prism smeared into an oblong with colours at top and bottom. It had been assumed since antiquity that such an effect was the result of some corruption of the white light. If this was the case, it seemed reasonable to assume that by passing some coloured light through a second prism, the degree of corruption would be increased. So Newton passed light through a prism, then allowed a small amount of coloured light to project through a slit, and then through a second prism. There was no change in the colour of the light. In other words, nothing was added to or taken away (see figure 5.8.). To confirm this, he also placed two prisms together so that the first separated the light and the second, an inverted prism, gathered all the bands back together, producing a spot of white light.

Newton believed that light was composed of particles and that their differing speeds resulted in a differing angle of refraction when passed through a prism. If, for example, all red light was composed of small particles of similar nature, there seemed to be no mechanism to change the nature of the red particles as they passed through successive prisms.

This was hotly debated, the French following Descartes's and Huygen's lead in preferring a wave theory of light, but Newton's *Opticks* provided a foundation for a new English school of optical research throughout the eighteenth century.

Newton's *Opticks* also laid out a research program for natural philosophers who followed him. The book concluded with a series of "Queries," topics that interested Newton but which he had not had time to fully investigate. In addition, since there was a strong bias among Newton's peers against "theoretical" science, meaning the presentation of philosophical ideas without experimental demonstration, Newton presented his ideas as a series of questions. This list extended beyond optics, covering a range of scientific areas related to light such as the relationship between light and heat, the effect of the media of transmission on the behaviour of light, and the condition of the universe. For close to 100 years many natural philosophers and scientists interested in finding important areas of research started their inquiries

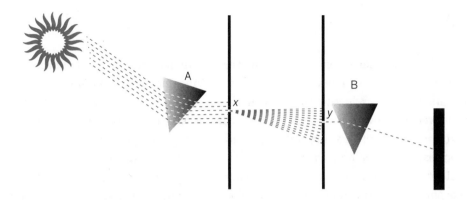

5.8 NEWTON'S DOUBLE PRISM EXPERIMENT

Sunlight passes through the prism "A" and falls on the screen. A portion of the spectrum produced (light of a single colour) passes through the screen at slit "x" and continues on to the second screen, where it goes through slit "y" and on to the second prism "B." The light is finally projected on to the wall. Newton's experiment demonstrated that light of a single colour could not be further broken up into a spectrum.

by taking on one of these questions. For example, in Query 18 (of the fourth edition of the *Opticks*), Newton noted that two thermometers, one in a vacuum and the other in a closed container of air, both seemed to heat up and cool down at about the same rate. This suggested to Newton that there must be a medium of propagation "more rare and subtle than the Air" that transmitted heat like a vibration. This observation prompted a number of scientists to look for the imponderable fluid or ether, an idea that was eventually resolved by the work of Einstein. It also got other scientists to investigate the nature of heat as separate from temperature, leading ultimately to the kinetic theory of heat and the laws of thermodynamics.

One of the most famous questions posed by Newton was Query 31 from the 1718 edition. Newton asked: "Have not the small Particles of Bodies certain Powers, Virtues, or Forces by which they act at a distance, not only upon the Rays of Light for reflecting, refracting, and inflecting them, but also upon one another for producing a great Part of the Phenomena of Nature?" Newton goes on to suggest that known forces such as gravity and electricity may play a role in the attraction of particles, but unknown forces may also be at work. By observing attraction, according to Newton, it should be possible to figure out the law of attraction that controls the way matter combines and functions. This idea contributed to the concept of chemical affinity, an idea used by most chemists to explain how matter combined. Affinity theory was one of the foundational concepts of modern chemistry until the nineteenth century when it was replaced by valence theory.

New Scientific Organizations

The new instruments, experiments, and the underlying assumptions about the relationship between humans and nature were fundamentally important to the creation of a new scientific enterprise in the seventeenth century. Equally important were the new institutional structures that developed in this period for the express purpose of fostering and supporting natural philosophy. The founding of such assemblies as the Royal Society of London and the Académie Royale des Sciences[3] in Paris contributed to a dramatic new organization of science, one that encouraged natural philosophers to develop social codes of behaviour, rules about who could do science and what counted as science, and statements about the secular usefulness of their enterprise. Just as Galileo had tried to separate religious and scientific claims to the monopoly of truth, so too did these new scientific societies. This was the beginning of the institutionalization of science, different in kind from the university-based science and distinct from the court-based science of the sixteenth century, although owing much to that earlier model. The breakdown of individual princely courts cast some natural philosophers adrift. Growing absolutism, especially in France, led to a more particular focus for patronage, as well as to a growing urban elite and intellectual culture in the capital cities. One hundred and fifty years of bloody religious wars caused people to look elsewhere than the church for salvational knowledge and for a secure group with status. The rising leisured class was looking for secular legitimation and something to do.

In about 1603 the first secular scientific society was formed, the Accademia dei Lincei (Academy of the Lynx) in Rome. This organization was founded by Federico Cesi (1585–1630), later Prince Cesi, and existed in great secrecy for the first years of its life, persecuted by the authorities including Cesi's father. Cesi was a man of huge intellectual energy and curiosity, who established a research program in natural history of great scope and imagination. When Galileo joined in 1611, the Accademia was given a major boost, especially since Galileo donated his microscope to it. After the condemnation of Copernicus, however, followed by Cesi's death in 1630 and Galileo's condemnation, the Accademia ceased to function. In 1657 a scientific society was founded in Florence, the Accademia del Cimento (of experiments). This society had neither a formal membership nor statutes and existed for ten years as a

---

3. The name of this organization changes according to informal or official use. Historians say "Académie des Sciences," although officially it is called "Académie Royale des Sciences." To make things more complicated, the name changed several times even though it remained the same group. Within the text, we will use the term commonly used by historians, that is, the Académie des Sciences.

loose collection of men interested in experimental research. It, like the Lincei, was in reality a hybrid, neither court-based nor autonomous, since it was focused on the court of Grand Duke Ferdinand II de Medici and his brother, Prince Leopold.

The most famous of the seventeenth-century scientific societies, the one that first marked this new form of scientific organization and the only one that has survived as a continuously operating body to the present day, was the Royal Society of London, established by royal charter in 1662. There is much debate about its origins. A number of people interested in natural philosophy, experimentalism, and the utility of natural knowledge met informally in England during the English Civil War and in the Interregnum that followed. The Civil War was fought between Royalists, who wished to maintain the power of the Crown and the liberal Anglicanism of the Church of England, and Parliamentarians, who argued for the primacy of political power from the people in Parliament and for the predestinarian theology of the Puritans. Historians of the Royal Society have sought founders of modern science in both camps, but most particularly among the Puritans. Samuel Hartlib (c. 1600–62), an educational reformer who came to England from Prussia to escape the Thirty Years' War and who was closely associated with the Parliamentary camp, tried to develop an educational program to bring natural philosophy and the new experimental method to English intelligentsia. The Hartlib circle was one of the groups responsible for the founding of the Royal Society after the Restoration, although on much more conservative and elite lines than Hartlib and his circle had envisaged. In fact, while earlier historians sought the origins of this new institutional structure in radical religion and politics, the truth seems to be that most natural philosophers were eager to find an alternative to the crippling political and religious controversies of their day and that natural philosophy provided just such a third way.

With the Restoration of Charles II in 1660, various groups from London, Oxford, and Cambridge came together in London to form the Royal Society. Although it received a charter from the king, it was autonomous and, therefore, different from earlier university, church, or court-based spaces of scientific investigation and discussion. It was founded with a strong inclination toward a Baconian philosophy of research, setting out to employ an inductive, cooperative method in order to discover useful information for the benefit of the nation. Here the rhetoric of utility that emerged with the courtly philosophers was carried into an urban, gentlemanly locale. Thomas Sprat, the first official historian of the Royal Society, said that the Society was founded as a way to avoid the "enthusiasm" of the Puritans and the sectarian disputes that had ripped the country apart during the Civil War. Although

Sprat was hardly a disinterested observer, it does seem that the Royal Society attempted to find a third way through the religious and civil disagreements of the period. While Royal Society members included those of many different religious

affiliations (from Catholic to Puritan), what they had in common was the desire to keep clear of religious controversy and to do natural philosophy instead of theology.

The Royal Society developed a strict method of choosing members, who had to be known to existing members and have an active interest in natural philosophy. The exception to this were some aristocrats, necessary to maintain the Society's elite nature. It also refused to allow women to join, although Margaret Cavendish, Duchess of Newcastle (1623–73), did attend some meetings and had published more books on natural philosophy than many of the members put together. They were also very hesitant to accept tradespeople, preferring instead the trustworthiness of gentry. The Royal Society developed a gate-keeping function, determining who counted in natural philosophical

5.9 FRONTISPIECE OF SPRAT'S *HISTORY OF THE ROYAL SOCIETY* (1667)

inquiry. Through the work of their first corresponding secretary, Henry Oldenburg (c. 1619–77) and through the publication of their journal, the *Philosophical Transactions of the Royal Society*, founded in 1665 and still published today, they also were able to determine what counted as proper natural philosophical work. Thus, in one fell swoop, the Royal Society became the arbiter of just who could be a natural philosopher and what qualified as acceptable natural philosophy.

The other successful seventeenth-century scientific society was established in a very different way. The Académie Royale des Sciences was founded in Paris in 1666 by Louis XIV's chief minister, Jean-Baptiste Colbert. Although there had been an informal network of correspondence centred on Father Marin Mersenne (1588–1648) since the 1630s, the Académie was a top-down organization, another element of the absolutist French state. Unlike the Royal Society, where members were elected and unpaid, the state appointed 16 academicians, paid as civil servants, to investigate the natural world as the king and his advisors required. So the Académie can be seen as the root of the professionalization of science, since this was the first instance where scholars were paid exclusively as scientists. Because their research agenda was set by the state, they could take on projects beyond the scope of individual scientists. For example, the Académie sponsored the measurement of one minute of arc of the Earth's surface, resulting in the first accurate measurement of the size of the Earth and the distance to the stars. In the long run, however, it was less successful than the Royal Society. An appointment as an Academician was often the reward for a life's work rather than an incentive to new work. Most of the massive projects came to nothing. Its journal, *Journal des Sçavans* (founded in 1665), was largely a reprint service. The Académie des Sciences did well as a promoter of science as an elite and respected activity, but it was not a place that sponsored innovation.

These new scientific societies created four enduring legacies for science as a profession. First, science was now seen as a public endeavour, although with carefully defined limits, members, and methods. Second, its cooperative nature was stressed, through projects such as the History of the Trades sponsored by the Royal Society and the History of Plants and Animals investigated by the Académie. Such undertakings led to the Enlightenment view that all was knowable if properly organized and to the sense of the utility of the knowledge gained. When Leibniz founded the Berlin Academy of Science in 1700, he chose as its motto *theoria cum praxi*, theory with practice.

Third, scientific communications were established as an essential element of the scientific enterprise. While it had been true that letters between natural philosophers (for example, between Galileo and Kepler) or within the letter-writing circle of Father Mersenne had been integral to maintaining a community of scholars, the establishment of scientific journals in the seventeenth century both broadened and controlled this community. These journals acted as a guarantor of veracity and reliability, even while issues were socially determined and highly contested. They also broadcast scientific ideas and experiments to a much larger audience, allowing ordinary people to take part in science by "virtual witnessing." This led to a wider interest in the investigation of nature and a greater acceptance of new scientific ideas and of scientists as respectable, if awe-inspiring people.

Finally, scientific societies established scientists as experts, qualified by membership to pose and judge questions about nature. This was especially true in

France, where election to the Académie was the culmination of one's life's work. But equally within the Royal Society, some natural philosophers, such as Newton or Boyle, were respected both within and outside the society as experts and significant scholars. Below them were the collectors, those who found interesting natural phenomena to report but who left theorizing to their betters, much as Bacon had laid out in Solomon's House. Thus, the seventeenth-century scientific societies established ideologies about science and its practice still with us today.

During the sixteenth century, connections between natural philosophers and skilled artisans aided greatly in the development of new ideas about nature and new problems to be investigated. This was also true during the seventeenth century, although the focus became the urban mercantile centre, rather than the princely court. Just as the scientists themselves were forming associations with their fellow scholars in these new, secular, non-court settings, often in major trading centres, they were also closer to shipyards, print shops, instrument makers, and chart makers. However, projects like the Royal Society's History of the Trades, an unmitigated disaster, shows us that the communication between these artisans and scholars was more complicated than one might think. The History of the Trades attempted to find out how all the different manufacturing trades in England were performed, so that natural philosophers might find a more rational scientific way to manufacture goods. Not surprisingly, the tradesmen were remarkably unforthcoming about their trade secrets, and the suggestions made by the bewigged gentlemen were at best unhelpful and at worst positively dangerous. It would take some time for a new collaborative approach to bring together the skills of craftspeople and the precision of scholars.

The Place of Women in the Study of Science

Most of this discussion of seventeenth-century science has focused on the contributions of men of science; this was, however, a period when women were attempting to make their mark on the study of the natural world. Both Anne Finch, Viscountess of Conway (1631–79), and Margaret Cavendish, Duchess of Newcastle, were interested in breaking into this previously clerical and male preserve. The same social and intellectual upheaval that made science a gentlemanly pursuit gave women a brief window of opportunity to become involved in natural philosophy. Bethsua Makin (c. 1612–c. 1674) wrote *An Essay to Revive the Ancient Education of Gentlewomen* (1673), arguing for the right and ability of women to study natural philosophy. Anne Conway corresponded with Leibniz and shared with him her theory of "monads" that became the basis of his particulate philosophy of the universe. Noblewomen such as Christina of Sweden engaged in natural philosophical conversations. Margaret Newcastle wrote many books of natural philosophy and attended a meeting of the Royal Society. She also composed what has been called the first English work of science fiction, *The Description of a New World, Called the Blazing World* (1666). These women, however, were exceptions. The seventeenth and eighteenth centuries saw restrictions on women's sphere in science as in much else.

Such restriction came from a general change in attitude to a gendered nature and from changing theories about women's role in society and in reproduction. In a pre-industrial society the vast majority of people existed in a close and symbiotic relationship with nature, which was perceived as female, a nurturing mother. The ideal was coexistence, not control. Mining, for example, was either the rape of the Earth or the delivery of a child, since minerals developed in the Earth's womb. Therefore, sacrifices, prayers, and apologies were necessary before mining could begin. The vitalism of Paracelsus, the natural magic of the neo-Platonists, and the naturalism of Aristotle all gave the female contribution of nature and to nature its due.

In early modern Europe, however, this began to change. As scholars began to view the Earth as exploitable, its image changed to a wild female who must be tamed. At the same time women were losing socio-economic status, becoming less autonomous and less able to earn wages or operate independently in craft guilds. Increasingly, accusations of witchcraft were levelled, especially against women. Scientific theories of sexuality and procreation changed. Where during the Middle Ages women had been recognized as providing something to initiate procreation (matter for Aristotle, female semen for medieval authors), during the sixteenth and seventeenth centuries theorists such as Harvey claimed that women were merely a receptacle, the incubator of offspring, and, as such, were totally passive. Likewise, women were no longer seen as equal partners in intercourse but rather seducers, enticing men into sexual activity that was deleterious to their health and intellectual well-being. Women were also losing their professional role in reproduction, as female midwives were replaced by licensed male surgeons or male midwives, who used their new technology of forceps to manage nature. Finally, with the introduction of the Mechanical Philosophy, introduced partly to deal with perceived disorder in the world, the soul of nature was stripped away, leaving only inanimate atoms; nature was dead, and women's claim to vitality through reproduction was rendered void. In other words, the natural philosophy of the seventeenth century articulated an ideology of exploitation, an image of the world that could be constructed according to man's specifications.

We can trace this change in attitude toward women as natural philosophers through the careers of two scientists: Maria Sybilla Merian (1647–1717) and Maria Winkelmann (1670–1720). Their careers demonstrate, on the one hand, the possibility of women's involvement in natural inquiry and, on the other, the restrictions to their participation through the creation of the new institutions of scientific societies.

Maria Merian's career illustrates the success a woman could have in a scientific field, particularly one based on an entrepreneurial model. Merian was born in Germany into a family of artists and engravers. From an early age she was interested in drawing and painting insects and plants. After marrying her stepfather's apprentice she became a renowned insect and plant illustrator, publishing well-received and beautifully engraved books. In 1699 the city of Amsterdam sponsored her travels to Suriname, where she observed and recorded many new plants and animals. On her return she created a bestselling book of these findings, which was published posthumously. Merian's career thus followed a very successful path in the older apprenticeship and mercantile model. However, after her death her work came to the attention of the new community of natural historians and philosophers, and her reputation suffered. Her Suriname book was strongly critiqued, condemned for her classification system and more particularly for her credence of slave knowledge about the use of these plants and insects. Her reputation, therefore, was greatly diminished in the increasingly misogynistic and racist attitudes of the eighteenth-century scientific community.

Maria Winkelmann was the daughter of and later wife of astronomers (she married Gottfried Kirch in 1692) and worked with both her father and husband on telescopic observations in Berlin. In 1702 Winkelmann independently discovered a comet and published her findings. She was a full participant in her chosen scientific field, but after Kirch's death in 1710 her status fell sharply. The Royal Academy of Sciences in Berlin refused to allow her to continue in her husband's position as official astronomer to the Academy, eventually appointing her less capable son instead. Even Leibniz's support was insufficient to help her maintain her position; the Academy was unwilling to set a precedent by allowing a woman to hold such an important job. In this they followed the lead of the Royal Society, which had likewise debated allowing Margaret Cavendish to join and had resisted the undesirable precedent. Eventually the Academy, embarrassed by her presence at the observatory, forced her to leave the premises; without access to large telescopes, she was unable to continue her observational work. The new organization of science, the Academy, proved itself to be more restrictive for women than the earlier apprenticeship model had been.

Seventeenth-Century Scientific Ideology

The new ideology of exploitation of and superiority toward nature reflected a changing attitude toward knowledge and nature. As crucial as the new knowledge and methodologies proved to be, of even greater significance in the creation of modern science were the new locales of scientific discussion and the new ideology and code of conduct the seventeenth-century societies established. Science had previously been the property of clerics and academics, but the upheaval of the seventeenth century—its religious and political wars, its economic strife—created an opportunity for a new group of gentlemen practitioners to develop a new standard for scientific conduct and a new place to practise science. This was particularly true in England, where the move away from absolutism at the end of the century allowed a certain freedom of association among the leisured classes and where the controversies associated with the Civil War and its aftermath gave gentlemen a desire for civility and an alternate way to establish matters of fact.

Robert Boyle was particularly instrumental in this transformation of scientific ideology. First, he set up laboratories, spaces for scientific experiment and investigation, in private locations, particularly in his sister's townhouse in fashionable Pall Mall in London. The privacy of this space was paramount, since Boyle was able to control access and behaviour within the site. That is, he could allow in people with the proper credentials, worthy to witness his experiments and guaranteed to know the proper way to behave. Similarly, private museums of natural history developed in this period all over Europe as private spaces, also belonging to aristocrats and gentry, who controlled visitors and bestowed on those visitors status through their permission to observe.

Because this space was private, Boyle could decide who had the proper credentials to observe, to take part in experiments, and to participate in the making of natural knowledge. He developed a number of criteria, later used by the Royal Society and other scientific bodies. The person seeking entry had to be known to Boyle or his circle and, thus, was a gentleman. This person should also be a knowledgeable observer, one who was able to validate the experimental knowledge and to witness matters of fact, rather than just gawk. Still, it was more important that the observer be of the leisured classes than that he or she be philosophically knowledgeable. This was clear from von Guericke's famous hemisphere experiment, which gained its status as one that created natural knowledge about the weight of the air by the presence of a large group of gentry onlookers (as can be seen in figure 5.5). Similarly, one of the important roles of the

SIR ISAAC NEWTON.

5.10 NEWTON

Engraving of Newton. From Sarah K. Bolton, *Famous Men of Science*. NY: Thomas Y. Crowell & Co., 1889.

Royal Society was to provide a credible audience for various experimental demonstrations, thereby establishing their veracity.

Because these experiments and demonstrations of knowledge were performed in spaces created and used by gentlemen, gentlemanly codes of behaviour were adopted as the codes of behaviour for scientists. For example, gentlemen argued for the openness and accessibility of private space at the same time that they were carefully controlling access to such places. Likewise, scientific laboratories as they developed claimed to be open public space while limiting access to those with the knowledge and credentials to be there. Gentlemen were very concerned with issues of honour and argued that their word was their bond. A gentleman never lied, which was why there was much outrage (and duels fought) at any suggestion of cheating. This was why matters of fact could be established through the witnessing of a few gentlemen, who would, of course, see the truth of the investigation and report it accurately to others. A scientist's word was also his bond. Scientists would not cheat or lie, and thus they claimed the role of a completely trustworthy enclave in society. The Royal Society in concert with Boyle created a community of scientists who could decide what constituted truth and reality and who were allowed to make pronouncements on that reality.

Conclusion

By the time Newton died in 1727, the place of the natural philosopher had changed considerably. The image of the natural philosopher had shifted even further away from the "pure" intellectuals of ancient Greece, the Islamic wise men, or even the courtiers of Galileo and Kepler's era. A bust commemorating Newton sits in the entrance hall of Trinity College, Cambridge. (See figure 5.10.) It does not present him as an erudite scholar or as an important member of society, as had earlier portraits from his student days, or as Master of the Mint. Rather, it presents him as a modern Caesar who, with firm gaze and noble brow, has conquered all he surveyed. While Caesar captured Rome and gained an empire, Newton conquered

Nature and made it man's dominion. The poet Alexander Pope said of Newton's life: "Nature and Nature's Laws lay hid in night / God said, 'Let Newton be,' and all was light."[4]

By the end of the seventeenth century many aspects of modern science had been established. Philosophers of this period of scientific revolution had wrestled with questions of epistemology at the beginning of the century and decided on a behavioural model of truth-telling by the end. Some of these thinkers developed new ideas, theories, and experimental discoveries, setting in place a series of research programs for the coming century. They had also introduced a new methodology of science, which included the mathematization of nature and a new confidence in, and reliance on, experimentation. New secular scientific institutions sprang up; with them came an articulation of the utility of their knowledge to the state and the economy. Finally, the domination of natural philosophical enquiry by secular gentlemen from the leisured classes ensured a code of behaviour that was gendered and class-based. These ingredients led to a new scientific culture that rapidly assumed a recognizably modern face. All these changes together constituted a scientific revolution.

Essay Questions

1. Was there a scientific revolution? If so, of what did it consist?

2. In what way can Newton be seen to have completed the Copernican revolution in astronomy?

3. In what ways were the Royal Society and the Académie des Sciences similar and different?

4. What role did women play in the development of natural philosophy and did this change over time? Why or why not?

---

4. Alexander Pope, "Epitaph Intended for Sir Isaac Newton." In Westminster Abbey, 1735.

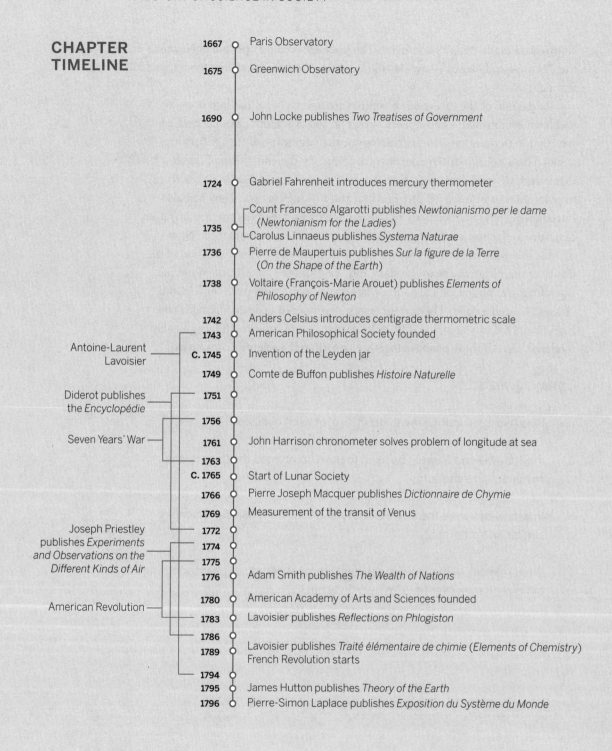

CHAPTER TIMELINE

1667 — Paris Observatory

1675 — Greenwich Observatory

1690 — John Locke publishes *Two Treatises of Government*

1724 — Gabriel Fahrenheit introduces mercury thermometer

1735 — Count Francesco Algarotti publishes *Newtonianismo per le dame* (*Newtonianism for the Ladies*)
Carolus Linnaeus publishes *Systema Naturae*

1736 — Pierre de Maupertuis publishes *Sur la figure de la Terre* (*On the Shape of the Earth*)

1738 — Voltaire (François-Marie Arouet) publishes *Elements of Philosophy of Newton*

1742 — Anders Celsius introduces centigrade thermometric scale

1743 — American Philosophical Society founded

Antoine-Laurent Lavoisier

C. 1745 — Invention of the Leyden jar

1749 — Comte de Buffon publishes *Histoire Naturelle*

Diderot publishes the *Encyclopédie*

1751 —

1756 —

Seven Years' War

1761 — John Harrison chronometer solves problem of longitude at sea

1763 —

C. 1765 — Start of Lunar Society

1766 — Pierre Joseph Macquer publishes *Dictionnaire de Chymie*

1769 — Measurement of the transit of Venus

Joseph Priestley publishes *Experiments and Observations on the Different Kinds of Air*

1772 —

1774 —

1775 —

1776 — Adam Smith publishes *The Wealth of Nations*

American Revolution

1780 — American Academy of Arts and Sciences founded

1783 — Lavoisier publishes *Reflections on Phlogiston*

1786 —

1789 — Lavoisier publishes *Traité élémentaire de chimie* (*Elements of Chemistry*) French Revolution starts

1794 —

1795 — James Hutton publishes *Theory of the Earth*

1796 — Pierre-Simon Laplace publishes *Exposition du Système du Monde*

THE ENLIGHTENMENT
AND ENTERPRISE

6

I n 1727 François-Marie Arouet (1694–1778) witnessed the opulent state funeral of Sir Isaac Newton and reported to the intellectual world the status and importance of science in English society. Arouet, better known as Voltaire, was one of the leading lights of the French philosophical and social reform movement and was well placed to encourage the dissemination of Newtonian science to eighteenth-century continental circles. Voltaire and many other reformers were deeply influenced by the work of Newton and by the freedom of the English political system and sought to bring the best of rational thought from natural philosophy to the question of human relations. Voltaire wrote about the freedom of the English in his *Philosophical Letters on the English* (1734) and about Newton in his *Elements of Philosophy of Newton* (1738), written with Emilie de Breteuil, the Marquise du Châtelet-Lomont (1706–49). For many Enlightenment thinkers natural philosophy was both a model for, and a tool of, reform.

The spirit of reform worked in both directions. Just as the *philosophes* hoped to transform society, the new natural philosophers hoped to reform science, revolutionizing the practice of investigation, changing the language of scientific discourse, and placing their discoveries and expertise at the service of the state. Many natural philosophers took their research agenda and methodology from Newton, and so the mathematical and physical sciences of the eighteenth century, especially those concerned with force and matter, owed much to the venerable English scientist.

The eighteenth century was a time of burgeoning mercantile growth, as many European nations expanded their trade empires and colonies. With this expansion, Europeans encountered cultures and peoples with different knowledge and understanding of the world, and so this period marks an important moment of the circulation of knowledge from the wider world back to Europe, as well as bringing European ideas of science and philosophy to other countries. Europeans were interested in products made from newly discovered flora and fauna, since such products could be very profitable in this new commercial revolution. It is, therefore, no surprise that many of those interested in studying nature increasingly did so in the service of commercial ventures or the state. Thus, two contradictory elements of scientific study developed in this period. On the one hand, the philosophical issues of freedom, democracy, and toleration led many natural philosophers to espouse radical political positions. On the other, scientists increasingly took their directions from the state or commerce and added their expertise to the exploitation of worldwide resources, with the result that they exerted a greater and greater influence on the wealth of nations and the power of the state. Many of the more radical efforts were overshadowed by the violence at the end of the century as reason fell to tyranny during the French Revolution, but by the nineteenth century science was a powerful and necessary ingredient for the operation of the modern state.

Universal Laws and Progress

The definition of the Enlightenment has been as hotly contested by historians as that of the scientific revolution. Different countries experienced different degrees of reform and resistance to reform. Still, at the heart of the Enlightenment were two monumental concepts. The first was a reappraisal of the human condition that led to the conception of universal human rights. The second was a belief in the inevitability of progress. Both led to cries for social, economic, and governmental reforms, and both owed much to conceptual changes in natural philosophy. One of the most obvious Enlightenment debts to the New Science, particularly following the Newtonian approach, was the belief in an underlying universality of natural laws. Newton had demonstrated that the law of gravitation and the laws of motion united the universe. There were no privileged realms and no exceptions; the laws of nature were the same for kings and peasants. It followed that similar universal laws existed and could be found to govern human relations. A number of philosophers argued for universal laws of human interaction in law, government, economics, social life, and religion. They were more than social critics; they called for action to amend the errors of the past. Like Francis Bacon, they challenged the very concept of traditional authority and hoped to replace it with rule based on reason and universal laws of human rights.

The philosophical tradition that led to the Enlightenment took shape when Newton was still alive. Thomas Hobbes (1588–1679) debated the nature of scientific knowledge with Robert Boyle and argued that a sovereign was necessary to protect people from each other and that to gain this security people had to give up personal rights. Boyle's view on scientific method won, while Hobbes's political views made him someone dangerous to know. He may have been rejected for membership in the Royal Society because of his political stance. John Locke (1632–1704) also refuted Hobbes's position on government by arguing for a natural and unalienable right to life, liberty, and the enjoyment of property. In his famous work *Two Treatises of Government* (1690), Locke defended the rights of the people to reject unwanted government, arguing that it was the duty of government to protect the rights of people and that the people had a right or even duty to depose any government that failed to protect their inherent and universal rights.

Historians have now cast doubt on whether Locke had much of an immediate impact on the course of English politics in the seventeenth century, but his ideas gained currency in the eighteenth century as more philosophers sought fundamental structures in society that paralleled Newton's structure of the universe. When Adam Smith (1723–90) wrote on economics and social interaction, particularly in *The Wealth of Nations* (1776), his analysis was based on laws of the marketplace that, like the Newtonian universe, were self-regulating. The "invisible hand" of the marketplace was not a description of a ghostly spirit of commerce but a mechanistic model of economics that he hoped would be as reliable and certain as the Newtonian model of the solar system.

A more subtle debt that the philosophers of the Enlightenment owed to natural philosophy was a belief in progress. At its simplest level, this was a belief that tomorrow would be better than today. Leibniz's formulation of this optimistic view was ridiculed by Voltaire, in his novel *Candide*, as that of a pedant believing in "the best of all possible worlds." However, the true natural philosophical articulation of this concept was much more complex. Rather than a vague belief in a better world, natural philosophers believed that they would understand the rules that governed nature better in the future than they did in the present. In fact, they believed that they would know everything about nature, identifying all the components of the machinery of the universe just as a clockmaker could identify and reassemble all the parts of a clock. This image of a clock fit well with Newtonian mechanics while at the same time was indicative of the growing power of industry, which could manufacture the precise and delicate mechanism of the mantel clock and the naval chronometer.

Diderot's *Encyclopédie*

The most ambitious and far-reaching intellectual project of the Enlightenment was Denis Diderot's (1713–84) *Encyclopédie*. When the first volume was published in 1751, it caused a sensation. French censors banned the book and revoked the publisher's licence. France's attorney general called it a conspiracy against public morals, and the pope issued a declaration that anyone buying or reading the work would be excommunicated. The outcry from authority was so strong that it led some of Diderot's contributors to drop out of the project. In fact, his co-editor Jean d'Alembert (1717–83), a physicist who had worked on celestial mechanics, resigned over the controversy. There is nothing like a storm of controversy and whiff of scandal, however, to increase the desirability of such a work. Because of the furore, the *Encyclopédie* was soon the talk of Europe and reached a wide audience. Diderot worked on, often under difficult circumstances and the monumental work eventually reached 28 volumes, 17 volumes of text appearing by 1765 and 11 volumes of illustrations finishing the series in 1772. It sold over 20,000 full sets, with thousands of abridged, partial, and pirated volumes in circulation as well.

Diderot's aim was nothing less than providing to all educated Europeans the most current information on every subject of modern thought. The spirit of the *Encyclopédie* was one of universal knowledge and a belief in progress; many of the articles contained the latest understanding of science, technology, and philosophy, complete with wonderful illustrations. Diderot believed that by making this knowledge available, the whole of civil society could be transformed. Just as the humanists had believed that education, based on learning the classics, made a person a good citizen, Diderot held that understanding science and the new knowledge it revealed would create a just society. The *Encyclopédie* was infused with social criticism, the promotion of modern industry, and a faith in the power of scientific discovery to bring enlightenment to society. It was both a compendium of the state of arts, crafts, and letters and a bold call for reform and progress. It also suggested order in nature and order in the study of nature.

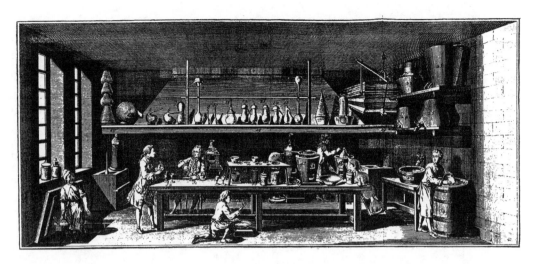

6.1 ILLUSTRATION SHOWING A CHEMICAL LABORATORY FROM DIDEROT'S *ENCYCLOPÉDIE* (1772)

Scientific Societies and the Popularization of Scientific Study

The age of Enlightenment was also the period when natural philosophy began to flow into the general culture beyond a small elite of the intellectual class. The courtly philosophers had long provided a connection between natural philosophers and the seats of power, but with a few exceptions these connections existed only in aristocratic and elite social circles. The founding of the Royal Society in London, however, and the Paris Académie des Sciences had created a space for natural philosophy outside the university, the church, or the courts. With a declining number of princely courts in the eighteenth century, the creation of this new site for natural philosophical discussion and enquiry offered both a market and a reservoir of talent for an increasingly popularized scientific enterprise.

The two major scientific societies were copied across Europe, so by the end of the eighteenth century there was a scientific society in almost every major city and official national bodies in most countries. (See figure 6.2.) On the advice of Leibniz, the Prussian Academy of Sciences was created in 1700, making it one of the oldest in Europe. The Austrian Academy of Sciences began in 1713, while the Russian Academy of Science was founded by Peter the Great in St. Petersburg in 1724. The Royal Danish Academy of Sciences and Letters traces its roots back to 1742, and the Imperial and Royal Academy of Sciences and Letters of Brussels was founded

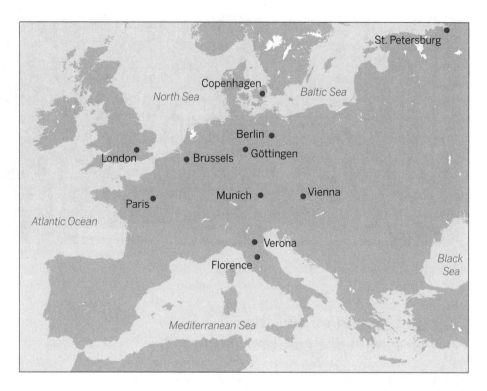

6.2 SCIENTIFIC SOCIETIES IN EUROPE 1660–1800

during Austrian rule in 1772. A number of German societies including the Bavarian Academy of Sciences and Humanities (1759) and the Academy of Sciences of Göttingen (1751) were created in mid-century. Italian societies included the Tuscan Academy of Sciences and Letters established in 1735 and the National Academy of Sciences (known as The Forty), founded in Verona in 1782.

The intersection of Enlightenment ideology, revolution, and science was particularly evident in the founding in North America of the American Philosophical Society in 1743 by Benjamin Franklin (1706–90) and others, including George Washington and Thomas Jefferson. The American Academy of Arts and Sciences, founded in 1780, counted among its founders John Adams and John Hancock and later listed Benjamin Franklin, John J. Audubon, and Louis Agassiz among its members. The connection between the reform of natural philosophy through these new scientific organizations and the reform of government by the men soon to be the founding fathers of the new American republic was integral to their overall philosophy and attitudes.

Due to rising levels of literacy and the spread of Enlightenment ideology, natural philosophy became a much more fashionable subject. In Britain, the locales of scientific discourse were the coffee houses and lecture halls, gathering places for the growing urban middle classes, at first in London and later in the provincial cities spawned by industrialization. This new audience was fascinated by the information natural philosophers provided, especially by the message of progress and the utility of such knowledge. Demonstrators like Jean Desaguliers (1683–1784) told people about Newton and his mechanics and exhibited new scientific principles using machines and models, which were closely tied to similar industrial innovations. These demonstrations and lectures provided not only a polite and rational language with which to think about the transformations taking place in society, both political and, especially, technological, but were also a morally acceptable diversion for both men and women. For religious dissenters, like Unitarians and Quakers, these introductions to God's rational power showed the utility of such knowledge and spurred the growth of science teaching at dissenter colleges, an important source of science education by the end of the century.

In France, this same interest in natural philosophy as a subject of polite and rational discourse was encouraged by the development of salon culture. The salons mimicked certain aspects of the princely courts but included people of a much wider range of backgrounds. They were often created by influential members of society and brought together artists, writers, musicians, and philosophers with the elites of commerce and government. The intellectuals were invited both to entertain and to inform. As such, the salons were also useful as conduits for patronage, especially for the bright and ambitious young men who lacked the resources or titles necessary to attend the king's court. It was for this audience that Voltaire's friend Count Francesco Algarotti (1712–64) wrote *Newtonianismo per le dame* (*Newtonianism for the Ladies*), published in 1735.

One of the leading salons of the period was presided over by Madame Geoffrin. Born Marie Thérèse Rodet (1699–1777), she had married a wealthy husband and, particularly after his death, used her wealth and home to bring together interesting and important people. Madame Geoffrin was a close friend of Montesquieu, corresponded with Catherine the Great of Russia, and entertained both Benjamin Franklin and Thomas Jefferson. She was a friend to Diderot and d'Alembert and supported the *Encyclopédie* financially and socially. Most of the intellectuals of the day who lived in or visited Paris attended her regular Monday or Wednesday meetings.

CONNECTIONS

Science and the Revolutionary Spirit

The Philosophical Society was founded at Philadelphia in 1743, making it the oldest learned society in North America. The founders included Benjamin Franklin, William Alexander, and Francis Hopkinson, all of whom would be major figures in the American Revolution. Franklin said of the reason for creating the Society that, "The first drudgery of settling new colonies is now pretty well over and there are many in every province in circumstances that set them at ease, and afford leisure to cultivate the finer arts, and improve the common stock of knowledge."* In 1769 the Philosophical Society joined with the American Society for Promoting Useful Knowledge and was renamed the American Philosophical Society Held at Philadelphia for Promoting Useful Knowledge, which was quickly shortened to the American Philosophical Society (APS).

What was established as a forum for people to discuss ideas and discoveries about nature turned out to be home to some of the most radical and revolutionary thinkers of the era. The connection between the study of nature and a desire for rationality and order in the human world proved to be a strong motivation to action.

The list of early members reads like a roll call of the Revolution, with George Washington, John Adams, Thomas Jefferson, Alexander Hamilton, Thomas Paine, and James Madison all joining. What these

Electricity

A good example of the interest in natural philosophy evident in salon culture in France and natural philosophical demonstrations in Britain was the fascination with electricity. The existence of this mysterious substance had been known for generations, since William Gilbert identified it in the sixteenth century, but it was not easily subjected to investigation. Otto von Guericke in the seventeenth century had created a machine that generated an electrical charge by spinning a ball of sulphur and rubbing it with the hands or a cloth. This allowed for a controlled production of a static charge but was still of limited use as an experimental device. In the 1750s a number of people designed generators that used spinning glass disks to generate the charge, the ultimate example being a massive device built by John Cuthbertson (1743–1821) that could generate a charge of about 500,000 volts.

people had in common was a belief in reform and an abiding interest in natural philosophy and reason. To the members of the APS, reason and natural philosophy, particularly following the Newtonian model as presented in the *Principia* and in the more practical *Opticks* (both of which Franklin and Jefferson are known to have owned), were the basis for reliable knowledge. Just as nature functions under a single, universal physics, so too should human law be based on a single, universal law. Paine, for example, argued that society existed in a state of "natural liberty" (meaning that human rights originated in nature, not human convention) in which all people were equal, not divided into higher and lower classes. This was very much in the Enlightenment tradition and it was made manifest in the documents of the American Revolution. This is especially true of the Declaration of Independence (the writers of which included members of the APS), which includes direct reference to the laws of nature in the Introduction:

When in the Course of human events, it becomes necessary for one people to dissolve the political bands which have connected them with another, and to assume among the powers of the earth, the separate and equal station to which the Laws of Nature and of Nature's God entitle them, a decent respect to the opinions of mankind requires that they should declare the causes which impel them to the separation.

The Preamble presents human rights as axiomatic just as physical laws are axiomatic and states that there should be one universal, natural, and equal law for all people, just as Newton's law of gravity was universally applicable. Although natural philosophy was not the cause of the American Revolution, it was part of its intellectual heritage. The APS would continue to foster revolutionary ideas, including among its membership Charles Darwin, Louis Pasteur, Elizabeth Agassiz, Linus Pauling, and Margaret Mead.

....
* www.amphilsoc.org/aboutQ4

Smaller static generators were frequently used for demonstrations such as the one featuring an electrified boy. In it, a child was suspended from the ceiling by silk cords, a charge was generated and communicated to him, and he then moved small objects using the static electricity or delivered mild shocks to visitors. The electrified boy was clearly an entertainment, as were games based on passing a shock from person to person or electrifying chairs to zap unsuspecting visitors. (See figure 6.3.) But there was a point to such demonstrations: they were often used to illustrate concepts such as electrical flow, circuits, and insulation.

The initial research on electricity was qualitative, concentrating on finding electricity, observing its behaviour, and learning how to manufacture it. Franklin's 1752 kite experiment is the most famous episode in the hunt for electrical phenomena and has risen to the status of popular mythology. It is likely that Franklin did fly a kite as an experiment to determine whether lightning was electricity, but he

6.3 THE ELECTRIFIED BOY

Charged with static electricity (rod at "n") and suspended by silk cords, the boy attracted bits of paper to demonstrate the communication of electricity. From Johann Gabriel Dopplemayr, *Neu-entdeckte Phaenomena von bewunderswürdigen Würkungen der Natur* (1774).

left no direct account of the event, nor any indication of the date it might have taken place. The clearest record comes from Joseph Priestley's reporting of Franklin's account, but this was published 15 years after the fact. What is certain is that Franklin did not fly a kite in the middle of a thunderstorm as is popularly depicted. He flew a kite at the approach of storm clouds, reasoning that the accumulation of electricity must precede the discharge of lightning. He broke the path of the electricity by attaching the kite rope to one side of the loop on top of a key and a silk ribbon to the other side, which he was careful to keep dry (and thus insulated). He ran an iron wire from the key to the ground. When he noticed the fibres of the hemp twine of the kite rope stand up and repel each other, he put his knuckle near the key and got a shock. While this was convincing evidence, he also used the wire to charge a Leyden jar to demonstrate that lightning charge was identical to that collected by conventional generation.

While qualitative work was a necessary first step in understanding electricity, it was of limited value in understanding how electricity worked or what could be done with it. In particular, the problem of complete discharge made electricity difficult to study. The creation of the Leyden jar around 1745 was a major step toward controlling the mysterious energy. It was not discovered by one person but by three at almost the same moment: Ewald Jürgen Kleist (1700–48) in Germany and Pieter van Musschenbroek (1692–1761) and Andreas Cunaeus (1712–88) in Holland. While Kleist was probably the first to use a jar filled with water to try to collect an electrical charge, it was Musschenbroek who described the instrument. Since the work was conducted in Leyden, over time that was the name adopted by most people (although it is still sometimes called the *von Kleistiche Flasche* in Germany). A glass jar was wrapped in metal foil (this was thought to prevent the electricity leaking through the glass) and filled with water. A spike or metal rod pierced a stopper that sealed the jar. Electricity from a static generator was conducted to the interior through the rod and stored. When the rod touched a conductor, the electricity was discharged. (See figure 6.4.)

Franklin used the Leyden jar to explore the behaviour of electricity in a very practical manner. He used the discharge to mimic lightning and designed the lightning rod to protect buildings. He made a model called the Thunder House to demonstrate the effectiveness of the lightning rod. Inside the model house was a small amount of flammable gas in a closed container. When electricity was applied to the top of the house, it produced a spark that ignited the gas and blew the lid off the container and the roof off the house. With the lightning rod attached to the house, the charge was conducted away from the model and the house was safe from damage. Franklin was later called to London to discuss the effectiveness of lightning rods with King George III. The king remained convinced that such rods were more effective with a round globe at the top, and this rather inefficient design was adopted by the British.

Theory of the period described electricity as a fluid, which in effect "flowed" into the Leyden jar from the electrostatic machine. The jars also helped to clarify the concept of an electric circuit, and it was from this that another salon demonstration arose. After filling a Leyden jar, a circle of people joined hands. One person broke the circle and with the free hand held the bottom of the jar, while the person on the other side of the break touched the rod with their free hand. If there was sufficient charge, everyone in the circle felt a shock. Because the Leyden jar could hold only a fixed amount of electricity (its capacity; hence, the name capacitors for devices based on the concept), such demonstrations

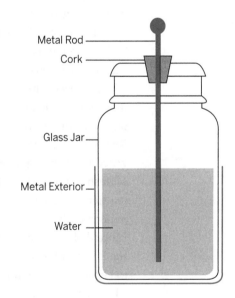

Metal Rod ——

Cork ——

Glass Jar ——

Metal Exterior ——

Water ——

6.4 LEYDEN JAR

helped to shift the study of electricity from a qualitative examination to one based on quantitative ideas, such as the relationship between the area of coated surface, the amount of charge, and its force.

New Systems of Measurement

The difficulty of doing experiments on electricity highlights one of the central problems for scientific research in the period. Electric phenomena could largely be understood qualitatively but not quantitatively. The precision of experimental results lagged far behind the potential revealed by physics and mathematics. The location of distant planets could be more accurately determined than the place of a ship at sea. The increased interest in quantification, developed from the seventeenth-century allegiance to measurement as the path to true knowledge, led in the eighteenth century to the creation of new instruments and, in keeping with the age of reform, the introduction of a new system of measurement.

Chief among these new instruments was the thermometer. Although there had been a number of attempts to measure temperature, including work by Galileo, the first instrument accurate enough for scientific work was not created until 1714 when Gabriel Fahrenheit (1686–1736) used the expansion of alcohol to measure temperature. He also introduced the use of mercury as the expanding liquid and published his method of construction in the *Philosophical Transactions of the Royal Society* in 1724. He went on to produce a scale based on the freezing points of water, establishing 0° as the freezing point of salt water, 32° as the freezing point of fresh water, and 212° as the boiling point.

Setting zero as the freezing point of salt water made sense in an era when understanding the sea was crucial to trade and power. In the same year that Fahrenheit created his first mercury thermometer, the English Parliament through the Board of Longitude offered a reward of £20,000 (equal to about $500,000 today) for anyone who created a system for determining longitude at sea. This was a long-standing problem that had plagued navigators for generations. Both the Paris Observatory (created in 1667) and the Royal Observatory at Greenwich (established in 1675) had the improvement of navigation as one of their main projects. Longitude could be calculated from celestial observations on land but not on shipboard. Over the years a number of solutions had been proposed, but all had failed as either impossible to carry out on a moving ship, such as the observation of the moons of Jupiter, or simply irrelevant, such as measuring magnetic variation. What prompted the 1714 prize was the ongoing danger of navigation, now more a concern than ever due to the expansion of trade and volume of shipping. The necessity of better navigation was made very clear when in 1707 Admiral Sir Cloudesley Shovell lost four ships and died along with between 800 and 2,000 men in a shipwreck off the coast of England. While finding latitude was relatively easy using celestial observations, what was needed to find longitude was an accurate method of timekeeping. This had been understood since at least 1530 when Gemma Frisius (1508–55), mathematical geographer and teacher of Gerardus Mercator (1512–94), argued that he could determine longitude if he had an accurate clock.

Therein lay the problem. Measurements of time accurate enough to determine longitude were possible using astronomical events such as the eclipses of the moons of Jupiter, as suggested by Galileo, but such observations were impossible on a ship at sea. Even after Christiaan Huygens (1629–95) developed a pendulum clock in 1656 that was accurate enough to keep good time, the instrument was useless on board moving ships. The problem was finally solved by the clockmaker John Harrison (1693–1776), who in 1761 had his timepiece tested on two voyages to Jamaica. Despite the success of Harrison's chronometer, he was awarded only £5,000 by the Board and did not receive the whole prize until King George III interceded on his behalf in 1773. As a craftsman he did not fit the gentlemanly profile of natural philosophers developed by the Royal Society. The leadership of the Royal Society had expected one of their Fellows would win the prize.

Navigation was vital to European nations competing for colonial holdings and operating global trade networks. This race to empire dictated the subject of some research programs, like that for longitude and the cure for scurvy. At the same time, the expanding reach of these nations made possible a number of experiments and observations that could have been conceptualized but not undertaken in earlier times. Take, for instance, the expedition to Lapland of Pierre de Maupertuis (1698–1759). Maupertuis was a member of the Académie des Sciences and in 1736 was chosen to lead the expedition to measure the precise length of a meridian degree. He was a fervent Newtonian and had been one of the first to introduce Newton's theory of gravity to French scientists. The resulting report, *Sur la figure de la Terre (On the Shape of the Earth, 1736)*, confirmed Newton's theoretical prediction that the Earth was not spherical but flattened at the poles.

Mathematical Physics

Maupertuis was one of a growing group of continental thinkers who turned their work to mathematical physics. They bridged the disciplines of mathematics and experimentalism, often following the direction set by Newton both mathematically and in practical terms by the research suggested in the "Queries" section of Newton's *Opticks*. Among this group were Joseph Lagrange (1736–1813), Pierre-Simon Laplace (1749–1827), Augustin Jean Fresnel (1788–1827), and Leonhard Euler (1707–83), one of the most prolific mathematicians of all time, who published over 1,000 articles and books on everything from number theory to mathematical cartography. Although the interests and studies of these men were diverse, ranging from the most esoteric aspects of the calculus to chemistry, what ties their work together was their effort to put into mathematical terms everything that was around them. Fresnel worked out a wave theory of light, challenging the corpuscular theory of the Newtonians, but he also developed the Fresnel lens, which was built in a series of concentric rings. This was used in lighthouses (and later in theatre lighting) because it allowed a much greater concentration of light than traditional metal reflectors had permitted, but was considerably more compact than a regular lens.

Lagrange began his serious mathematical work at the age of 17 by looking at optics. He sent a proof of his results to Euler, who encouraged him to continue in this endeavour. Lagrange did continue and essentially founded the field of astrophysics. While Newton's mechanics had set a framework, the precise details about why the solar system moved as it did or the nature of the gravitational relations of the Earth and its Moon were complex problems that needed to be worked out. Lagrange did so in his book *Mécanique analytique* (1788), which presented mechanics in strict mathematical terms. The Preface stated:

> One will not find figures in this work. The methods that I expound require neither constructions, nor geometrical or mechanical arguments, but only algebraic operations, subject to a regular and uniform course.[1]

Lagrange's work on gravity led him to predict that there would be five points in space—the Lagrange points—where the gravitational forces of two bodies (such as the Earth and the Moon) balanced each other. Of the five points, two (L_4 and L_5) were more stable than the others, and the forces of attraction worked in such a way that a

---

1. Joseph Lagrange, "Preface," *Mécanique analytique* (Paris: Gauthier-Villars et fils, 1889), p. 1.

third body placed there could remain in stable orbit indefinitely. The L_1 spot in the Earth–Sun system is currently occupied by the Solar and Heliospheric Observatory Satellite, which is, as the name implies, observing the sun. (See figure 6.5.)

Like Lagrange, Laplace also worked on celestial mechanics. He wrote many technical papers, as well as *Exposition du Système du Monde* (1796), a popular account of how the universe functioned. In practical matters, he was an assistant to the great chemist Antoine Lavoisier, aiding him in his work on heat and energy. These scientists added new vigour to physics, turning their attention to a wider and wider range of natural phenomena. One of the grandest schemes was offered by Roger Joseph Boscovich (1711–87), who claimed that all matter in the universe could fit inside a nutshell and who tried to encapsulate the structure of the universe in a single physico-mathematic idea. His *Theoria Philosophiae Naturalis* (1758) eliminated matter altogether and dealt only with points of force, presenting one of the first attempts at a grand unified theory (such theories are sometimes called guts) based solely on physical principles.

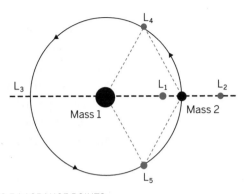

6.5 LAGRANGE POINTS

The Lagrange points are points in space where satellites would have their most stable orbits.

Scientific Expeditions: The Transit of Venus

Grand mathematics stirred the hearts and minds of a number of scientists, but it was often the adventures of investigation that attracted public notice. Among the most famous scientific expeditions of the period were the investigations of the transit of Venus. Kepler had earlier recognized that both Mercury and Venus must pass in front of the Sun and had calculated when these events would occur. He determined that the transit of Venus happened in pairs every 105.5 years (with the second eight years later), followed by an interval of 121.5 years (and another eight years). Just a year after Kepler's death, Pierre Gassendi had observed the transit of Mercury in 1631. It was now time to observe the next transit of Venus, due in 1761 and 1769. The importance of this astronomical event was recognized by Jeremiah Horrocks (1619–41), who realized that if observations were made simultaneously at different points on the Earth, the information gained could be used to calculate the distance to Venus and the distance from the Earth to the Sun.

In 1761 Joseph-Nicolas Delisle (1688–1768) organized a major effort to measure the transit of Venus by sending observers to Pondicherry in India, St. Helena, Isle de Rodrigues in the Indian Ocean, and Tobolsk, Siberia. Observations were undertaken by many other observers as well, but unfortunately poor weather and the Seven Years' War (1756–63) prevented some of the observers from recording the event, and the results were not as good as astronomers had hoped.

The expeditions of 1769 were more successful. Captain Cook's voyage to Tahiti and the adventures of Guillaume Le Gentil (1725–92) captured the public's romantic imagination. Tahiti had only recently been discovered by British explorers and was portrayed as an earthly paradise. Because it had a superb tropical climate and the islanders were extremely friendly, reports about Tahiti were widely spread by the European press. Cook's mission was in large part scientific—mapping, observing the transit, testing the chronometer for longitude, and looking for a way to deal with scurvy—but he was also under orders to claim any previously undiscovered land for the British.

Le Gentil went to Pondicherry, undertaking a difficult trip overland and then across the Indian Ocean, but missed the first transit of Venus. He stayed in the region until the second transit, only to be thwarted again because of bad weather. He wrote of his adventures in *A Voyage in the Indian Ocean* (Volume I published in 1779, Volume II in 1781). So began the age of heroic science, when facing extreme danger in the name of the expansion of knowledge gained a significance almost equal to the religious crusades of long ago. The Académie des Sciences team, sent to Mission San Jose del Cabo in Baja, California, all died from an epidemic except for a single survivor. This man crossed the territory that would become Texas on foot in gruelling conditions and returned to French territory and home only after great travail. It took several years to compile the data (Cook, for example, did not get back to Britain until 1771, since he was searching for the Great Southern continent), but the astronomer Jerome de Lalande (1732–1807), the director of the Paris Observatory, used the combined transit data from all these teams to calculate a distance of 153 to 154 million kilometres from the Earth to the Sun. While this was not as accurate as astronomers had hoped, it was a major accomplishment. It also demonstrated how scientific knowledge, as well as guns and colonies, confirmed the imperial power of European nations such as France and Britain.

Other Scientific Encounters

Europeans were interested in discovering more about the vast resources of other parts of the world, for commercial, political, and scientific reasons. This was the period that saw the first concerted efforts to map countries and regions largely unknown to Europeans. For example, the first British maps of the subcontinent of India were made in the eighteenth century. James Rennell (1742–1830) produced maps of the Mughal Empire, relying on assembling in London the diverse field reporting of members of the East India Company. Europeans were also interested in the uses of plants and animals in other regions and were reliant on native knowledge to give them this information. Maria Merian's investigation of the properties of insects in Suriname was only possible because the Surinamese shared their knowledge with Merian. Europeans and non-Europeans were often on an equal footing when it came to knowledge of the natural world, although information not seen as useful to Europeans was frequently not transmitted home. For example, knowledge of the abortificant (abortion-causing) properties of the peacock flower of South America were not communicated to Europe, although the flower itself became known and prized for its beauty.

Perhaps the event that best demonstrates this negotiation between scientific world views was the diplomatic mission of George McCartney (1735–1806) to China in 1793. McCartney was sent by the British to establish better trading relations with the Qing dynasty. He brought diplomats and soldiers, but he also brought scientists and scholars, along with scientific devices he felt would astonish the Chinese (for example, a world map and an ornate clock). Joseph Banks advised him to find as many tea plants as possible, with an eye to diversifying the new tea plantations Banks had already founded in India in 1778. The mission was a complete failure. McCartney got into a difficult situation because he refused to kowtow to the Qianlong Emperor and the Chinese were not impressed by these marks of European scientific knowhow. They had their own maps of the world and their own time keeping devices. The British had expected to encounter a back-ward people, eager for instruction. Instead, they met a sophisticated and educated culture, with its own scientific understanding.

The Industrial Revolution and the Study of the Earth

Geographers, mathematicians, cartographers, astronomers, and clockmakers all worked to solve the complex problems of navigation and astronomy, thereby contributing to the power of their nascent empires. Just as importantly, their success aided the exploitation of the natural resources and captive markets of the colonies. The wealth so generated in turn helped finance the Industrial Revolution, particularly in Britain, where there was a burst of creative industrial activity starting around 1750. By 1780 the introduction of the factory system of production and James Watt's (1736–1819) condensing steam engine had begun to transform the lives of many people as they moved from the farms to the factories, creating on one hand the terrible slums of early industrial Glasgow, Birmingham, Manchester, and London, and on the other the incredible wealth of the industrial barons. Historians of science have long debated the relationship between the growth of science and the technological and economic changes in this period. Rather than scientific breakthroughs contributing directly to new technologies, however, science was instrumental in creating a culture of progress and a discursive claim for the utility of the scientific enterprise.

As merchants were sailing to and trading with new parts of the globe, natural philosophers came to view the Earth itself as a legitimate topic of scientific investigation. Studies of the Earth—cosmography and geography in the scientific revolution, adding geology in this Enlightenment era—were motivated by economic considerations to do with mining and land formations, political issues dealing with newly discovered peoples and the navigability of the globe, and religious questions of the evidence for God's handiwork. All contributed to geology and earth studies in the eighteenth century, which were focused on several questions: the process of creation and subsequent history of the Earth, the age of the Earth, the configuration of strata, and the place of fossils.

Natural philosophers were first concerned with the material reality of God's creation. The Englishman Thomas Burnet (c. 1635–1715), for example, in *Sacred Theory of the Earth* (1691), wanted to explain creation in mechanical (Cartesian) rather than miraculous terms, following the trend of mechanical philosophy to avoid supernatural explanations. He started with the Genesis story, but used natural, mechanical explanations to account for the events of biblical creation. He claimed that the Earth in the beginning was very hot; as it cooled, a thin shell of smooth land formed above the waters. Burnet did not eschew all biblical explanations, however, since he argued that the coming of sin into the world caused the shell to break, sending floods over the land to create the imperfectly spherical world, with mountains and oceans.

Burnet was criticized for his irreligious stance, despite his use of biblical reasoning. This did not prevent another English natural philosopher, William Whiston (1667–1752), from making a similar point in *New Theory of the Earth* (1696). Whiston was a Newtonian—he had been at Cambridge with Newton—and therefore sought to place Earth history in the context of Newtonian thought. He argued that the Earth was formed by a comet that condensed into a solid body because of gravity and that the deluge was caused by another comet passing close by, depositing water, and knocking the Earth out of circular orbit. Hence, the imperfection after the Fall of Adam and Eve.

These late-seventeenth-century creation stories were coupled to the heated religious conflicts of the age. In the eighteenth century it was more possible to move away from biblical stories and look instead for rational explanations. This is not to say that eighteenth-century geologists were atheists, however. Rather, they saw God's work as essentially rational and therefore explicable. As well, the rise of the amateur gentlemanly philosopher brought a new purpose to geological investigation—for status and political power rather than for ideological or religious reasons. Geology, for instance, concentrated on the observation of actual phenomena, including fossils, rather than on large speculative systems, such as those of Burnet and Whiston. In the early seventeenth century, fossils were considered to be figured stones rather than biological in origin, but now scientists such as Nicolaus Steno (1638–86) and John Woodward (1665–1728) interpreted them as the remains of living creatures that had been petrified in rocks.

Recognizing fossils as biological remains, however, caused more problems than it solved. Most of the fossils found in Europe seemed to be of sea creatures and did not resemble contemporary living forms. Moreover, many of the best fossil deposits were high up on mountains, where people could not believe there had been an ocean or lake at an earlier time. While the biblical flood might provide some explanatory options, extinction and submerged mountains required a major leap of scientific faith. This crisis of explanation led to the development of two competing geological theories by the end of the eighteenth century: Neptunism and Vulcanism, named after the Roman gods of the seas and fire.

Neptunism was so named because its proponents believed that water was the fundamental agent for the formation of the Earth. George Louis Leclerc, Comte de Buffon (1707–88), argued in his *Histoire Naturelle* (1749) that a cooling Earth had gone through six epochs of earth formation, corresponding roughly to the six days of creation. In these six epochs, the Earth cooled, water condensed, and oceans retreated; the Earth evolved in a clear direction toward the present day. The true founder of Neptunism, however, was Abraham Gottlob Werner (1749–1817), which is why this theory is sometimes called the Wernerian theory. Werner was a mineralogist teaching at a mining school, one of the new educational innovations of the Enlightenment. He decided that strata must indicate the order of rock formation and that the layers had been precipitated out of primordial oceans. His explanation was based largely on his findings of sedimentary rock formations; he assumed that volcanic activity was local and relatively unimportant. He was probably influenced in his formulation by the lack of such volcanic activity around Freiberg, where he worked. Werner also insisted that forces now in action were too feeble to have made the world as it now is, and so forces in the past must have been far stronger and more powerful. In other words, he believed that geological formation operated in only one direction, toward the present. Werner was extremely influential, because he taught many mineralogists, miners, and geologists, and because his scheme fit with a classification scheme for minerals.

Soon, however, Werner was challenged by an alternate theory, which came in response to the big holes in his own. For example, where had all the Flood water gone? How were igneous rocks formed? And, the most vexing question, why were sea fossils found on the top of mountains? Gradually, a number of mineralogists began to argue for some contribution of the inner heat of the earth, sometimes called Vulcanism. There was only one dedicated vulcanist, Jean Étienne Guettard (1715–86), since most other scientists acknowledged a role for water as well. Guettard lived in France near Auvergne where there were a number of mountains with cone shapes, indicating they had once been volcanoes. He argued that volcanic forces in the past had been much stronger than now and widespread enough to account for rock formation, fossil placement, and igneous rocks. He was the first person to create geological maps of France, his *Atlas et description minéralogiques de la France* (*Mineralogical Atlas and Description of France*) appearing in 1780. Along with a young Lavoisier, he contributed to a project to do a complete geological survey of France, of which only a tiny portion was completed.

The more people began to investigate rocks and their layers, however, the less satisfying both these accounts seemed to be. James Hutton (1726–97), a Scottish geologist, developed a new theory that combined the inner heat of the Earth with the actions of water. In *The Theory of the Earth* (1795), he suggested that sedimentary rock had been formed from earlier igneous rock and that the Earth was in a constant state of change from one form into the other and back again. Volcanic activity forced igneous rocks up to the surface, where water caused erosion and the formation of sedimentary rocks, some of which were fused by heat to form igneous. And so the cycle went. As Hutton said, "We find no vestige of a beginning, no prospect of an end."[2]

With this statement, Hutton attacked all the mineralogists and geologists before him, since he argued that forces now in action did have sufficient strength

....

2. James Hutton, "Theory of the Earth," *Transactions of the Royal Society of Edinburgh* IV (1788): 304.

to cause the evolving nature of rock stratification and that the inner heat of the Earth was the same now as before. In other words, this was a model *without direction*, either biblical or material. This is called "uniformitarianism." Hutton's was a wildly radical theory, and although his earlier papers were favourably reviewed when they first appeared in 1785, with the publication of his book in 1795 he was widely condemned. This was largely because the French Revolution in France had whipped up conservative concerns, and Hutton's "atheist" and "Jacobin" theories had no place in a patriotic country at war. The liberal legacy of the Enlightenment was over, at least for a time.

Museum Collections and Scientific Expeditions

The importance of geology in the eighteenth century was mirrored by the passion for collection and classification. Collectors in the sixteenth and seventeenth centuries had assembled their collections in museums, first as private places of study or contemplation (a *studio*). These early collections by natural philosophers such as Konrad Gesner (1516–65) and Ulisse Aldrovandi (1522–1605) included books, engravings, and artifacts. The selecting and organizing principles in these museums were usually based on the singularities, freaks, and discontinuities of nature rather than some underlying order or continuity. Collections might include stones in bizarre shapes or two-headed calves, rather than a complete set of the flora of a region. In the late seventeenth and early eighteenth centuries princes developed collections as part of the spectacle and wonder of courtly science. These *studiolo* or cabinets of curiosities (which were *de rigueur* for gentlemen by the nineteenth century) were not working collections, as Gesner's and Aldrovandi's had been, but rather were meant for public display and ostentation.

The first truly public museum, meaning that the public could pay a fee and tour the collection, was the Ashmolean Museum, established in Oxford in 1683. Its model in terms of public access was probably the Bodleian Library, also in Oxford, the first European public library, which people could enter for a fee if they had a reason to use it. Not all visitors were pleased with the open-door policy, since it violated the privileged status of the gentlemanly witness. Count von Uffenbach of Germany, for example, visited the Ashmolean in 1710 and was not at all impressed, since he was forced to rub shoulders with the hoi poloi! However, public museums, or at least museums that could be accessed by an educated public, soon became a mark of enlightened Europe. Museums in Russia, Italy, Spain, and France all opened their doors to the public in the eighteenth century (although opening the Jardin du Roi in France required the storming of the Bastille). The British Museum opened in 1753, following a bequest by the late President of the Royal Society, Hans Sloane (1660–1753), from his own collection.

Sloane, a physician to the wealthy in London, amassed an enormous collection. At first he collected Old and New World plants, and then branched out to include a huge array of other artifacts, including shells, insects, fossils, minerals, antiquities, coins, books, and manuscripts, most collected for their rarity rather than to create a coherent classification system. He began his collection, as well as his path to riches and power, on a trip to Jamaica in 1687 under the patronage of Christopher Monck, Duke of Albemarle. When the duke unexpectedly died on the voyage, Sloane had a somewhat grotesque opportunity to practise his preservation technique, embalming Albemarle's body to preserve it on the trip home. On his return Sloane made use of one of the new species he had identified in South America by marketing it to the English public. He had observed South American natives consuming chocolate but found it too bitter for his own taste. However, he discovered that mixed with sugar and milk, the result was very pleasant, and so he made his fortune importing and manufacturing milk chocolate.

Sloane's collection was visited by most of the important people of the day— Carolus Linnaeus, Benjamin Franklin, and Georg Handel, to name a few. Sloane himself was a somewhat touchy and anti-social man who got into prolonged fights with nearly every important scientist in early-eighteenth-century Britain. For example, he had a huge row with the geologist Woodward, whom he accused of grimacing at him at a Royal Society meeting. In 1742 he retired to Chelsea, outside London, and set up his Physic Garden. At his death he established a trust that soon formed the British Museum, which became the largest public museum of the eighteenth century.

Sloane's successor as President of the Royal Society, and an extraordinary collector in his own right, was Sir Joseph Banks (1743–1820). Early in his career Banks signed on as ship's botanist for James Cook's (1728–79) first voyage to the Pacific in 1768. Indeed, on this first voyage, Banks, who represented the Royal Society on the trip, was more famous than Cook, who was a relatively unknown naval officer. Banks's entourage threatened to overwhelm the ship, and, in fact, his demands for space and resources were so great that he was not asked to participate in Cook's next voyage. Banks was overwhelmed by the wealth of new flora and fauna he observed on this journey. He named their first landfall in Australia Botany Bay, because the vegetation was so luxurious and exotic. He returned determined to collect all of the world's botanical species.

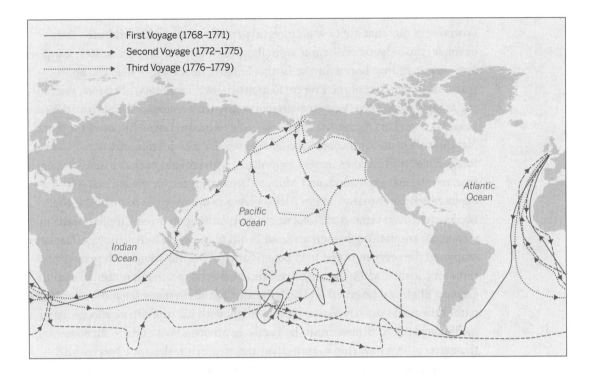

First Voyage (1768–1771)
Second Voyage (1772–1775)
Third Voyage (1776–1779)

Atlantic Ocean

Pacific Ocean

Indian Ocean

6.6 CAPTAIN JAMES COOK'S THREE VOYAGES (1768–1779)

Cook's voyages themselves demonstrated an intersection of empire-building, heroic science, and the collecting spirit. Cook set out to prove the existence or non-existence of *Terra Australis Incognita*, the theoretical continent that should have balanced the Eurasian continent on the other side of the globe. He proved that it did not exist, while at the same time discovering a number of new lands and islands (and claiming them for Britain), testing the chronometer, and establishing that eating limes could prevent scurvy. These voyages were symbolic of the British Empire, collecting and laying claim to parts of the world and establishing the power of British science and technology. Although Banks did not go on Cook's next two voyages and therefore did not witness Cook's murder on Hawaii nor travel to Vancouver Island, he did issue instructions, with the approval of the Admiralty, for the collection of flora and fauna and the observation of natural phenomena.

Banks became the greatest botanical collector in Europe, as well as the most powerful natural philosopher in Britain, controlling appointments and influencing patronage. He established Kew Gardens, not only collecting specimens but growing them and distributing them to other gardens. Most of the oriental flowering plants that are now an integral part of European horticulture—for example, rhododendrons—came from Banks's collection. Banks saw Kew Gardens as a great exchange house for the British Empire; he used it to help transport plants from one part of the Empire to another—with exempla, of course, remaining at Kew. Here, then, was museum-collecting with a new imperial thrust. As a personal friend of George III, Banks was able to instruct expeditions of the Royal Navy to collect and transport specific samples. He also hired professional collectors, such as Archibald Menzies (1754–1842), who brought back the monkey puzzle tree and the giant redwood, and Mungo Parks (1771–1806), the famous African explorer, who eventually died in Africa trying to discover the route of the Niger. Banks's two most famous exploits were transplanting tea from China to India in 1788 and transplanting breadfruit from Tahiti to the West Indies in 1787. The latter employed the services of the HMS *Bounty*, and sailors impatient with poor conditions for the men while the breadfruit received good treatment mutinied against Captain Bligh and resettled on bleak Pitcairn Island. Perhaps the most astonishing part of the mutiny on the *Bounty* is that Bligh, with his supporters, succeeded in navigating back to Britain from the Pacific in an open boat and set out soon thereafter with a new ship to transplant the breadfruit, this time successfully.

Following the British example, other European nations strove to create an empire through scientific observation as well as military coercion. The French, particularly, made an imperial bid for the Pacific just as the British were doing. For example, Louis-Antoine de Bougainville, Comte de Bougainville (1729–1811), landed at Tahiti, brought back a new plant named in his honour, *Bougainvillaea*, and set in motion the philosophical paradigm of the noble savage, as articulated by Jean-Jacques Rousseau. Jean-François de Galaup La Perouse commanded another voyage of discovery, exploration, and empire building, but his ship was lost in the Pacific in 1787, leading to many romantic speculations as to his fate. Much time, imagination, and money were spent trying to find this lost expedition, but since the French Revolution broke out almost immediately, the first ship sent to search mutinied and returned home, and La Perouse was never heard from again.

Amateur Scientific Societies

For those who remained at home, the curiosity for collecting and understanding natural phenomena and for attending natural philosophical lectures, as well as the legitimating power of scientific knowledge, led to the formation of a number of amateur-based scientific societies. Perhaps the most famous was the Lunar Society of Birmingham.

The intersection of scientific utility with industrial and economic interest was illustrated by the collaboration of the members of the Lunar Society. Based in Birmingham, it started in 1765 when a small group of men met informally to discuss natural philosophy and issues of the day. They called themselves the Lunar Circle. The group expanded in 1775 at the instigation of William Small (1734–75) and Benjamin Franklin and was renamed the Lunar Society, as a reference to the fact that the members met monthly on the Sunday or Monday evening closest to the full moon so there was more light by which to travel the unlit streets of Birmingham. Meetings of the "Lunatics," as they called themselves, often took place at Soho House, Matthew Boulton's home. The Society faded around 1790, largely because the French Revolution made them seem a potentially subversive group, but sporadic meetings may have continued until 1809.

While the Lunar Society was never as formal as the various scientific societies of the day—in fact, many of its members were also Fellows of the Royal Society (FRS)—it took the concept of putting the utility of scientific knowledge into practice far more aggressively. Knowledge of the natural world was not simply for a better understanding of God's creation or to improve the mind. Knowledge was to improve the human world. For the most part, these men saw no distinction between personal profit and social benefit. Many of the members were businessmen first and natural philosophers second, and most worked in a wide and diverse range of occupations during their careers. While some aspects of reform were of interest to them, they might be better characterized as "improvers," restless and constantly tinkering and looking for ways to make things work better, whether it was an engine, a business practice, education, or the production of sulphuric acid. In particular, the collaboration of James Watt with the industrialist John Roebuck and his even more fruitful partnership with Matthew Boulton were instrumental in the development of the steam engine that literally powered the Industrial Revolution.

Other philosophical societies sprang up across the country. For example, the Manchester Literary and Philosophical Society was founded in 1781 in order to promote polite knowledge, rational entertainment, technological instruction, and professional occupation. The pursuit of scientific knowledge was seen as both transcendent (as a path to God beyond the material world) and as an intellectual ratifier of a new world order based on industrialism and the exploitation of the material world. Manchester was a new manufacturing city, and its elite citizens sought legitimation and an ideology of progress and utility that could represent their lives. The Manchester Lit and Phil (as it was called) was founded by medical men and included Quakers, Unitarians, and manufacturers. This was a society where provincial and mercantile concerns were more important than London directives. London scientists were not telling manufacturers how to run things; equally, manufacturers were not telling scientists what to do. Rather, these societies provided a controlled outlet for radical agitation in the first generation and a means of maintaining the power won in the second. Soon, Lit and Phils were founded in Bristol, the Potteries, Newcastle, and Edinburgh.

Classification Systems

One of the recurring concerns of those collecting artifacts for museums or fossils for geological theorizing was the question of how to categorize or classify their findings. Classification was a goal of the encyclopedists, matter theorists, and imperialists. Control of the world and its resources could come only from knowing the names of everything and where each type or individual fit into the larger scheme of things. Through the Middle Ages and Renaissance, classification had been based on the Great Chain of Being, which was a strict hierarchy descending from God, angels, people, animals, plants, and ending with the inanimate world. Scholars were no longer satisfied by this system and therefore sought new, more rational classification schema. These thinkers devised rational systems to classify first plants and animals and later chemical elements, participating in the great Enlightenment project to know everything.

Carolus Linnaeus (Karl von Linné) (1707–78), a Swedish botanist, was one of the most successful systematizers in the eighteenth century. He amassed a massive botanical collection (dried, rather than living, as opposed to the collection at Kew), receiving plants from other collectors all over the world. Linnaeus developed a classification system, first articulated in *Systema Naturae* (1735), based on increasing specificity: Kingdom, Class, Order, Genus, Species, Variety. In order to classify specific plants, he used "artificial classification"; that is, he based his system on attributes that were easily counted and measured but probably did not link the species in nature. All systematizers in the eighteenth century used artificial systems but hoped one day to find the natural basis for connections among species. As it turned out, it was only through adding time, as Darwin did, that such natural classification became possible. Linnaeus based his system on the sexual characteristics of plants, specifically by counting the number of stamens and pistils. This was not random choice, since sexual organs seemed fundamental in passing on certain basic characteristics. He then developed a binomial nomenclature of genus and species, which is still used today.

Linnaeus looked at the relationship of clearly related genera and species to express the concept of branching classification. That is, he argued that there was no clear linear progression or hierarchy from simplest to most complex organism, as the Great Chain theory posited, but that each genus, each species, was equally complex. He pictured the relationships of species laid out like countries on a map, with each species touching many others. At first, in *Philosophia Botanica* (1751), he argued for the fixity of species, seeing no transformation from one to another. By 1760, he had decided that a number of species might have common ancestors but believed that this was caused by hybridization only. Linnaeus saw classification as a closed system in which basically no new species could appear; therefore, all were theoretically knowable. He envisaged the map of nomenclature as a finite table, where systematizers could work to fill in the gaps, eventually producing full knowledge of all living beings. (This was very similar to the development in the nineteenth century of the chemical periodic table.) Thus, Linnaeus showed the natural affinity of species, based on similarity of characteristics, and mapped out an ambitious research program of naming all living things.

Although Linnaeus was first and foremost a botanist, he did transfer his nomenclature and his system to the naming of animals. Until then,

REGNUM VEGETABILE. 837

CLAVIS SYSTEMATIS SEXUALIS.
NUPTIÆ PLANTARUM.
Aĉus generationis incolarum Regni vegetabilis.
Florefceatin.
PUBLICÆ.
Nuptiæ, omnibus manifeftæ, aperte celebrantur.
Flores unicuique vifibiles.
MONOCLINIA
Mariti & uxores uno eodemque thalamo gaudent.
Flores omnes hermaphroditi funt, & ftamina cum piftillis in eodem flore.
DIFFINITAS.
Mariti inter fe non cognati.
Stamina nulla fua parte connata inter fe funt.
INDIFFERENTISMUS.
Mariti nullam fubordinationem inter fe invicem fervant.
Stamina nullam determinatam proportionem longitudinis inter fe invicem habent.

1. MONANDRIA.	7. HEPTANDRIA.
2. DIANDRIA.	8. OCTANDRIA.
3. TRIANDRIA.	9. ENNEANDRIA.
4. TETRANDRIA.	10. DECANDRIA.
5. PENTANDRIA.	11. DODECANDRIA
6. HEXANDRIA.	12. ICOSANDRIA.
	13. POLYANDRIA.

SUBORDINATIO.
Mariti certi reliquis præferuntur.
Stamina duo femper reliquis breviora funt.
14. DIDYNAMIA. | 15. TETRADYNAMIA.
AFFINITAS.
Mariti propinqui & cognati funt.
Stamina coharent inter fe invicem aliqua fua parte vel .cum piftillo.

16. MONADELPHIA.	19. SYNGENESIA.
17. DIADELPHIA.	20. GYNANDRIA.
18. POLYADELPHIA.	

DICLINIA (a δις bis & κλίνη thalamus f. duplex thalamus.)
Mariti & Feminæ diftinĉtis thalamis gaudent.
Flores mafculi & feminei in eadem fpecie.

21. MONOECIA.	23. POLYGAMIA.
22. DIOECIA.	

CLANDESTINÆ.
Nuptiæ clam inftituuntur.
Flores oculis noftris nudis vix confpiciuntur.
24. CRYPTOGAMIA.   CLAS-

6.7 PAGE FROM LINNAEUS'S *SYSTEMA NATURAE* (1735)

animals had been categorized by the Aristotelian groupings of four-footed versus two-footed, plus birds and fishes. Linnaeus worked to find similar characteristics in order to link different species. Perhaps his most controversial move was the identification of one group of animals by the mammaries of the female half of the class (that is, mammals). This may have been influenced by the fact that Linnaeus was himself an influential member of the anti-wet-nursing campaign in Sweden, which encouraged upper- and middle-class women (like his own wife) to breast-feed their own children. By grouping together horses, dogs, apes, and humans into a single class, he set in motion a huge debate about the place of human beings in nature, both in God's plan and in the biological world.

The Study of Matter: Chemistry and the End of Alchemy

An interest in classification merged with issues of power over nature through understanding its inner forces in the study of matter. Contrary to the trend of mathematical physics, which had been organized and simplified by Newton into a series of axiomatic rules, chemistry lacked a central organizing conception or even a commonly accepted language and nomenclature; as a result, matter theory was in a state of chaos. More substances and processes were being discovered, but without a set of central organizing principles these simply meant there was more to be confused about. Industrial chemistry was still largely a craft or guild system of production, but the demand for materials as Europe began to industrialize pushed producers to expand production and look for new methods. Some materials—gunpowder, dyestuffs, acids, and shipbuilding materials such as pitch—were in such high demand that their production became concerns of national importance.

Alchemy, while still common, was under increasing attack by natural philosophers who sought to put the study of matter on a rational and experimental basis. Since Boyle's publication of *The Skeptical Chymist* (1661), these researchers had been working to establish methods for examining the material world that did not depend on hidden or occult forces. This was reinforced by the 1732 publication of the *Elementa chemiae* (*Elements of Chemistry*) by Herman Boerhaave (1668–1738). Translated into most major European languages, it was considered to be one of the foundational texts for chemistry until it was supplanted by the work of Lavoisier 50 years later. A further attempt to bring some order to chemistry was introduced by Pierre Joseph Macquer (1718–84). Trained as a physician and appointed to the

Académie des Sciences in 1745, his work moved away from Boyle's approach and adopted a more Newtonian and corpuscular system. In 1751 he published two influential textbooks, *Elémens de chymie pratique* and *Elémens de chymie théorique*, and in 1766 produced the *Dictionnaire de chymie*. Although certain conceptions from Aristotle still lingered in the work of Macquer and his contemporaries, the emphasis was increasingly on experimental procedure and quantitative analysis.

Many of the leading chemists of the eighteenth century worked on pneumatic chemistry, studying "airs," or, as we call them, gases. There were a number of reasons for the primacy of the study of various airs. First, they were intimately linked to life, and the investigation of life was understandably popular. They were also linked to other processes—for example, combustion, rusting, and calcination—and were the by-product of many important operations such as brewing, smelting, and dye making. On an intellectual level, airs were the finest type of matter, so it was believed that understanding their structure and behaviour would open the door to a more general understanding of matter. This followed the tradition of the corpuscularians, particularly Descartes, Boyle, and Newton, who had all argued that matter and even light (in the case of Newton) were made of extremely small particles.

Only one aspect of this growing list of substances was in any way unified, and that was the principle of combustion. The study of combustion (which included calx or rusting and respiration) had been investigated by a number of philosophers in the seventeenth and early eighteenth centuries; most had agreed that fire was a substance liberated from compound bodies when they were heated. In the older alchemical tradition, fire was considered an essence or spirit. In the later tradition of iatrochemistry, people such as Joachim Becher (1635–82) argued that it was a *terra pinguis*, or oil earth (somewhat akin to the Aristotelian fire element) that was combined with other matter to form the materials of the world. The more *terra pinguis* a material contained, the greater its potential combustibility.

Becher's earlier system was revamped by Georg Ernst Stahl (1660–1734), who had corresponded with him. In 1718 Stahl replaced *terra pinguis* with phlogiston, using the Greek root *phlogos*, meaning flame. He argued that metals were composed of calx and phlogiston, so that when the metal was heated, the phlogiston was released into the atmosphere and left behind the calx. All inflammable substances contained phlogiston, but phlogiston itself had properties that set it apart from other forms of matter. This system allowed chemical reactions to be charted.

Metal (*calx* + *phlogiston*) $\xrightarrow{\text{heat}}$ calx + phlogiston (*released to atmosphere*)

The phlogiston theory neatly explained many of the aspects of combustion. A substance would stop burning if it saturated the surrounding air with phlogiston (such air was then "phlogisticated"), as in the case of a candle in an enclosed jar. (See figure 6.8.) Combustion would also stop if all the phlogiston was expelled from the substance in the complete formation of a calx. Phlogisticated air would not support respiration, but the atmosphere did not become saturated because plants absorbed or fixed phlogiston from the atmosphere, returning it to the flammable wood. Supporters of the phlogiston theory found that substances were often heavier after combustion or calcination than before, suggesting that phlogiston had negative weight or positive lightness and was not attracted to the Earth like matter in the Newtonian system.

A number of researchers continued the work on airs throughout the eighteenth century. A technical development by Stephen Hales (1677–1761) made the collection of airs much easier. Hales introduced the pneumatic trough in which gases were collected by the displacement of water. While the method may have predated Hales, his use of it was quickly taken up by other experimentalists such as Joseph Black (1728–99) and Henry Cavendish (1731–1810). Black, working in Scotland, began his chemical investigations while researching his doctoral dissertation in medicine. He became interested in the relationship between acids and alkalis and their relation to "fixed air," or what is today called carbon dioxide. Black demonstrated that fixed air would not sustain combustion by pouring the invisible gas onto a lit candle in a container, thereby putting out the flame. Using carefully measured experiments, he proved that fixed proportions of chemicals combined in chemical reactions and that fixed air was a component of atmospheric air, as well as being one of the gases produced when exhaling. By these demonstrations, he showed that atmospheric air was a mixture of gases and not elemental as it had been considered in the Aristotelian and many later systems.

Cavendish, after whom the great Cavendish Laboratory in Cambridge is named, identified the properties of "inflammable air" (now called hydrogen) in 1766 and distinguished it from a number of other gases that were known to burn. The properties of inflammable air suggested to a number of people that it was phlogiston. Around 1784 Cavendish was the first to demonstrate clearly that water was a compound, undoing another of the Aristotelian elements.

While the work of Black and Cavendish was wide-ranging, the greatest researcher on the nature of "airs" was Joseph Priestley (1733–1804), who isolated and studied more new gases than any other investigator. He had little formal training in natural philosophy, but he came into contact with a number of people

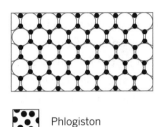

 Phlogiston

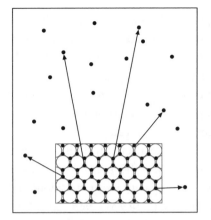

Dephlogisticated air supports combustion
by allowing phlogiston to move from the
area of high concentration in the element
to low concentration in the air.

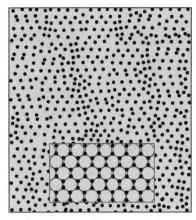

Phlogisticated air is saturated,
so no combustion is possible.

Element

6.8 PHLOGISTON

who were interested in the subject such as Matthew Turner (d. 1788), a physician
who had lectured on chemistry, and Benjamin Franklin. When Priestley took a
position as librarian to Sir William Petty, second Earl of Shelburne, he had both a
patron and a position that allowed him to work on his research. Between 1774 and
1786 he produced six volumes entitled *Experiments and Observations on the Different
Kinds of Air.* He investigated what we would now call nitric oxide, hydrogen
chloride, ammonia, sulphur dioxide, and oxygen, among others. He investigated
the properties of phlogisticated air in water (seltzer or soda water) as a cure for
scurvy. His most important work, as it would turn out, was on what he called
"dephlogisticated air." When he heated a calx of mercury, he obtained what he
thought was a new gas that was very combustible. He reasoned that this air
contained so little phlogiston itself that the phlogiston in the combusted substance
rushed to fill the void. The reverse of this, phlogisticated air, was so full of phlogiston that no more could enter, and thus combustion could not take place.

Priestley's chemical interests were closely tied to his radical political and
religious views. He was a Unitarian, connected with the dissenter academy of
Warrington Academy, later Manchester College, which offered science training as
part of its goal of progress and rationality. Priestley believed in the idea of progress,
the perfectibility of man, and the ability of humans to find out the truth about
everything, starting with the natural world. He was, therefore, in favour of the

abolition of repressive laws and was sympathetic to both the American and French Revolutions. Just as in the case of Hutton, Priestley became a more and more suspect character, especially once the British declared war on the French. On July 14, 1791, his house in Birmingham was vandalized by a mob (sometimes called the Priestley Riots). He left Britain in 1794 and settled in Pennsylvania, where he continued to support the phlogiston theory until his death.

While important work on airs was done in Britain, the centre of this scientific research was in France, particularly Paris. By 1750, the Académie des Sciences in Paris had attracted some of the most influential scientists on the continent. Because there was a government salary for full positions and the number of places in the Académie were limited, there was much competition and political intrigue associated with membership. Into this world of salon society, scientific development, and reform movements came Antoine-Laurent Lavoisier (1743–94). He was a reformer, aware of the political implications of his work within science and for French society. He began his higher education in law, following in his father's footsteps, but was attracted to science, particularly chemistry. He gained an associate position in the Académie in 1768, and, although this position was the lowest rank and unpaid, he took it, deciding to devote his life to science. To finance his work, he took a share in the Ferme Générale, a private tax farm, an organization that collected taxes for the government. Although his eventual elevation to Academician status provided him with a government salary, his personal fortune and his share in the Ferme made him independently wealthy. He eventually became a Farmer-General, one of the high officials in the organization. While this supported his research, his association with tax collection led to his death on the guillotine in 1794 when he was tried and convicted by a revolutionary court of being anti-revolutionary.

In 1771, Lavoisier married Marie Anne Pierrette Paulze (1758–1836), the daughter of one of the Ferme's partners. Marie was integral to Lavoisier's scientific and political life. She managed his affairs, learned English so that she could translate materials for him, and attended to laboratory work. She studied art with Louis David and was responsible for the engravings that accompanied Lavoisier's work. In keeping with salon culture, she hosted the twice-weekly meeting of intellectuals that Lavoisier entertained in their home. She continued this practice even after his execution. In this, she fits a pattern of women in science who work behind the scenes and who were often unnamed partners in the scientific enterprise.

Lavoisier, as a member of the Académie, was also a public servant, expected to place his intellectual abilities at the service of the state. Given his energy and interest in reform, he was more than willing to undertake reports on a wide variety of subjects, including a review of the water supply of Paris, the condition of prisons, adulteration of food, ballooning, and a range of industrial concerns such as the ceramic industry, glass making, and ink manufacturing. He also worked on the reform of agriculture, becoming a member of the Committee on Agriculture in 1785. In connection with his work for the state and his prowess as a chemist, he was made director of the Royal Gunpowder and Saltpeter Administration and was set the task of improving French gunpowder, which was often of very poor quality and whose raw materials were difficult to obtain. In this position he was given a fine house at the Paris Arsenal and ample space for a laboratory.

When Priestley visited Paris in 1774, he discussed his work on dephlogisticated air with Lavoisier, who had been looking for the component part of the atmosphere that supported respiration and combustion. Ironically, Lavoisier used Priestley's discovery to destroy the phlogiston theory. By 1777 he had concluded that "eminently respirable air" was converted to "fixed air" by combustion and respiration. Because of its relation to acids, he named this air *oxygène* ("acid former" in Greek). By 1778 he demonstrated

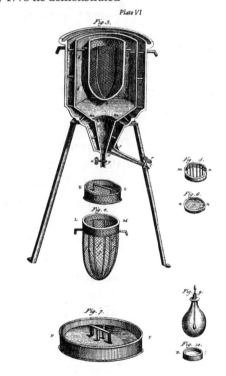

that atmospheric air was a combination of this respirable air and an inert air. This opened the door for a number of further experiments, including a demonstration that water was composed of oxygen and Cavendish's inflammable air. These experiments convinced Lavoisier that the phlogiston system could not work, and in 1783 he submitted a paper to the Académie entitled *Reflections on Phlogiston* that set out the problems of the old theory and how his oxygen system solved them. This controversial position gained Lavoisier not only a number of supporters, including Joseph Black, but also stiff resistance from prominent chemists, including Macquer and Priestley himself.

Discarding phlogiston left a problem concerning heat. If calx and respiration were simply a chemical combination, what was heat? With the mathematician/physicist Pierre-Simon Laplace, Lavoisier reformulated the place of heat in the system, introducing the concept of caloric to replace the principle of phlogiston. To quantify the production of heat, they created the ice calorimeter, which used the latent heat of fusion to demonstrate the expiration of heat from chemical reactions and respiration. (See figure 6.9.) Latent heat had been identified by Joseph Black

6.9 LAVOISIER AND LAPLACE'S ICE CALORIMETER FROM *ELEMENTS OF CHEMISTRY* (1789)

in 1760 when he noticed that at the melting point additional heat melted ice without raising its temperature (until all the ice was melted when the water began to heat up). Thus, a specific quantity of heat was needed to melt a certain volume of ice into water. By measuring the amount of water, the amount of heat could be calculated. Lavoisier considered caloric to be an imponderable fluid that caused other substances to expand when it was added to them. For example, during the combustion of carbon and oxygen, the resulting carbon dioxide had the combined weight of the original substances and gave up the caloric as heat and light. Although the concept later proved to be wrong, Laplace and Lavoisier's measurement of energy is the basis for the calorie of food energy.

Lavoisier and a number of his supporters decided that to put chemistry on a rational and useful foundation—in this case meaning his system and the rejection of the phlogiston theory—they would have to reform all of chemistry. In keeping with humanist and Enlightenment philosophy, they started with language, and in 1787 in collaboration with Claude Louis Berthollet (1748–1822), Antoine François de Fourcroy (1755–1809), and L.B. Guyton de Morveau (1737–1816), Lavoisier published *Méthode de nomenclature chimique*. This work attempted to unify and systematize the naming of chemicals and elements, replacing old common names with Latin and Greek roots. Thus "vitriolated tartar," "sal de duobus," and "arcanum duplicatam" all became potash. In addition to the root names, compounds were distinguished with various suffixes to indicate their classes. Salts formed by sulphuric acid were called sulphates, while those formed from sulphorous acid were sulphites. Using the system of nomenclature in effect meant accepting Lavoisier's underlying oxygen theory, thereby ensuring that his system would be the new path for chemical research.

Initially, opposition to Lavoisier's new system was very strong, especially among older chemists. His campaign of persuasion was carried out not only in the formal world of publishing and the Académie des Sciences but also in the salons. In the twice-weekly meetings hosted by Marie he entertained virtually every important natural philosopher who visited Paris. As Director of the Académie from 1785, he manipulated the organization of the chemistry section so that it was made up only of anti-phlogistonists. When the editorial control of the *Journal de physique* was taken over by phlogistonists in 1789, Lavoisier and his disciple Pierre Adet (1763–1834) founded the *Annales de chimie* in an effort to improve both the reporting on and quality of chemistry. This journal continues to be a leading scientific publication today.

In 1789 Lavoisier published his most influential book, *Traité élémentaire de chimie* (*Elements of Chemistry*). This brought together all aspects of his work, introducing his nomenclature, his experimental system and apparatus, and his methods and standards of measurement, and including an extensive compilation of all elements and compounds recognized under his system. Widely read and quickly translated, it gave the death blow to the phlogiston theory. Essentially, no young chemist or person interested in matter theory could claim to be current in the field without being acquainted with Lavoisier's system. His nomenclature was so functional that it quickly came to be the most widely used system, carrying his theory of chemistry with it. Equally, measurement and experimentation became integral to chemical research, paving the way for very different projects than were possible with the older qualitative practice.

The first stage of the chemical revolution ended with the French Revolution. Lavoisier, who was a reformer and political moderate, had hoped to use science to support a new and more progressive France. He had worked hard to make France powerful, both intellectually and in practical terms with his system for producing gunpowder, agricultural reform, and geological work. When the Revolution degenerated into the Terror, Lavoisier's link to the old regime through the tax farm made him suspect in the eyes of the radicals. He was denounced and arrested along with 27 other members of the Ferme. Their trial lasted just a few hours, and, although the charges were without substance, the convictions were really a foregone conclusion, and nothing Lavoisier or his friends said could sway the outcome. When Lavoisier went to the guillotine on May 8, 1794, France and science lost one of its most powerful minds. Lagrange said of Lavoisier's death that "It took only a moment to cause this head to fall and a hundred years will not suffice to produce its like."[3]

With the rise of chemistry as a public and systematic study, inquiry into the transmutation of materials faded from the realm of serious consideration by natural philosophers and the increasing number of academicians studying science. One final episode effectively closed the door on alchemy and alchemists. This was the affair of James Price, born James Higginbotham in London in 1752. Although he graduated from Magdalen Hall, Oxford, as a medical doctor in 1782, he had been elected a Fellow of the Royal Society a year earlier, in part because of his work on chemistry. After his graduation, he invited some important people to his home, claiming to have transmuted mercury to gold. He added a mysterious white powder to 50 times its weight in mercury, mixed in some borax and nitre, and

3. J.J. O'Connor and E.F. Robertson, "Joseph-Louis Lagrange," *MacTutor History of Mathematicians*, available at http://www.history-mcs.st-and.ac.uk/~history/Biographies/Lagrange.html.

heated the mixture in a crucible. It produced an ingot of silver. When Price followed the same procedure with 60 times the weight of mercury, the product was gold. The ingots were found to be genuine metal and shown to the king.

This created a sensation, and Price introduced a new twist to the old con of alchemy. Rather than gaining patronage directly for his work, he wrote a pamphlet on his experiments that became a bestseller. By 1783 Price said that his supply of white powder was exhausted, and the cost of making more would be too great in expense, labour, and drain on his health, since he hinted that some spiritual effort was required to make it.

Joseph Black reviewed Price's work and said that it was a mass of errors. He was astonished, he said, that Price had ever received a medical degree. Supporters and detractors rushed into the fray, until a committee was struck to investigate the situation. Price refused for some time to cooperate, but pressure from the Society and his friends forced him to accept the investigation. He had six weeks to prepare, but it was all for nothing. When the team of three observers from the Royal Society arrived, he showed them into his laboratory, excused himself, and left. He drank a vial of hydrocyanic acid, returned to the laboratory, and dropped dead at the feet of the investigators. In effect, alchemy died with him, since after this incident no scientific society in Europe would notice alchemical claims except to debunk them.

Conclusion

The French Revolution and the Europe-wide warfare that followed put an end to the more radical or "free-thinking" positions of a number of natural philosophers. Rational secular explanations for natural phenomena, planetary development, and the evolution of species became dangerous ideas in the 1790s. Nonetheless, natural philosophy had transformed in this Enlightenment period. Quantitative analysis, logical and rational classification systems, measuring and explaining phenomena such as electricity and heat that had once been seen as esoteric—all this was the legacy of a century of development.

One of the lasting reforms that eventually transcended the chaos of revolution and warfare was the metric system of measurement. Even within a single country, there was often no uniform set of measurements. France, for example, had around 300 different units of weight. As early as 1670 Gabriel Mouton, a French vicar, proposed to reform the systems of weights and measures using scientific principles. In 1742, Anders Celsius (1701–44), a Swedish astronomer, introduced the centigrade thermometric scale, the first important decimal (base ten) measuring system to be widely adopted in science. In the reform-minded era of the 1790s, Thomas Jefferson proposed a decimal-based system for measurement in the United States, and while this idea was not enacted, the American mint introduced decimal currency in 1792. In 1790, Louis XVI of France authorized an investigation into the reform of French weights and measures by the Académie des Sciences. Five years later, the French Republican government officially adopted the metric system. Although the measurement system of science we know today as S.I. (Système Internationale d'Unités) was the product of international conferences in the nineteenth century, the concept of a measuring system that was both uniform in units and integrated (a fixed volume of water equalling a specific mass, for example) was founded on the revolutionary reforms of the Enlightenment.

Just as the Enlightenment and the Industrial Revolution created whole new categories of jobs, so too in this age we see the creation of scientists: people who by profession, education, and association made the new science their life's work. Scientists concentrated on the utility of their knowledge about nature, investigating ways in which their study could improve the lives of citizens, the power of the state, and the bottom line of joint-stock companies. Science was now inexorably intertwined with the great imperial project of the nineteenth century.

Essay Questions

1. How did the ideas of the scientific revolution influence the Enlightenment?

2. How did geologists argue the Earth had been created? What was their evidence?

3. How did scientific collecting change the development of science?

4. What did Lavoisier revolutionize in chemistry?

CHAPTER TIMELINE

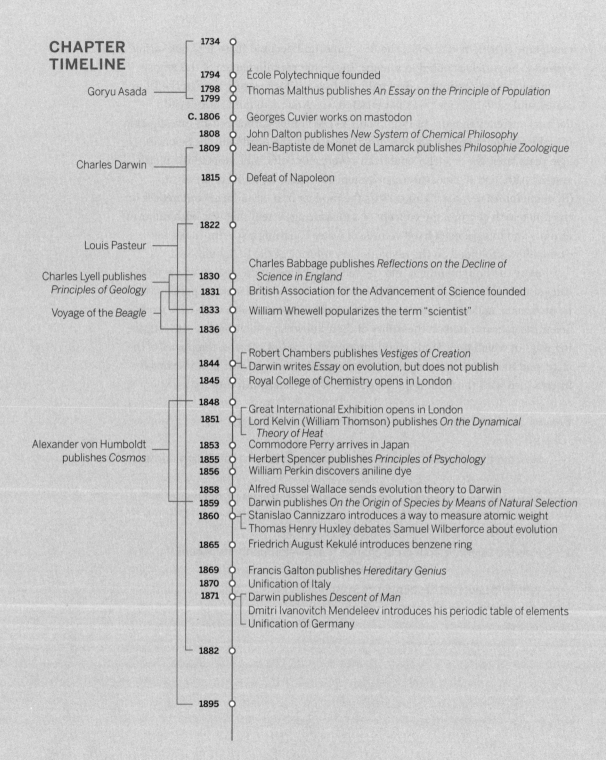

	1734	
	1794	École Polytechnique founded
Goryu Asada	1798	Thomas Malthus publishes *An Essay on the Principle of Population*
	1799	
	C. 1806	Georges Cuvier works on mastodon
	1808	John Dalton publishes *New System of Chemical Philosophy*
	1809	Jean-Baptiste de Monet de Lamarck publishes *Philosophie Zoologique*
Charles Darwin		
	1815	Defeat of Napoleon
	1822	
Louis Pasteur		
	1830	Charles Babbage publishes *Reflections on the Decline of Science in England*
Charles Lyell publishes *Principles of Geology*	1831	British Association for the Advancement of Science founded
Voyage of the *Beagle*	1833	William Whewell popularizes the term "scientist"
	1836	
	1844	Robert Chambers publishes *Vestiges of Creation*
		Darwin writes *Essay* on evolution, but does not publish
	1845	Royal College of Chemistry opens in London
	1848	
	1851	Great International Exhibition opens in London
		Lord Kelvin (William Thomson) publishes *On the Dynamical Theory of Heat*
Alexander von Humboldt publishes *Cosmos*	1853	Commodore Perry arrives in Japan
	1855	Herbert Spencer publishes *Principles of Psychology*
	1856	William Perkin discovers aniline dye
	1858	Alfred Russel Wallace sends evolution theory to Darwin
	1859	Darwin publishes *On the Origin of Species by Means of Natural Selection*
	1860	Stanislao Cannizzaro introduces a way to measure atomic weight
		Thomas Henry Huxley debates Samuel Wilberforce about evolution
	1865	Friedrich August Kekulé introduces benzene ring
	1869	Francis Galton publishes *Hereditary Genius*
	1870	Unification of Italy
	1871	Darwin publishes *Descent of Man*
		Dmitri Ivanovitch Mendeleev introduces his periodic table of elements
		Unification of Germany
	1882	
	1895	

SCIENCE AND EMPIRE

7

The nineteenth century was the great age of European empires. Although European exploration and colonization started much earlier, European nations, especially those on the Atlantic coast, now dominated all parts of the globe economically, militarily, and politically. The steam engine, the telegraph, and the factory conquered time, space, and material desires. No part of the globe was beyond the reach of empire, and the flags of those empires were carried to the most remote and challenging places. As Western Europe underwent rapid industrialization, it turned more and more to colonial holdings for natural resources and captive markets. Both industrialization and colonialism helped to spur the development of science, which offered the ability to know the world better and revealed ways to turn the natural resources of the colonial holdings into wealth. This was not strictly a one-way exchange, as non-European countries started to adopt ideas and practices, including the study of nature, from their contacts with European thought. At the same time, for those European nations that had limited colonial holdings, science offered a way to deal with economic disadvantage by creating new tools and techniques to solve problems that resulted from a lack of cheap natural resources.

The big winner of the colonial game was Britain. After the defeat of Napoleon in 1815 the strongest rival to British power was subdued, largely leaving the world open to unfettered British exploitation. France had been defeated not just on the battlefield but also in the shops and factories. British industry, backed by its colonial strength, out-produced France, so that every bullet, tent, and naval vessel cost less and was produced faster and in greater numbers than France could match. Although the American colonies had broken away from British rule in 1776, they represented only a minor part of British holdings when compared to its control of India and parts of Africa, the Middle East, and Asia. And, despite a century of suspicion and conflict between Britain and the United States, heightened during the War of 1812 and British trade with the Confederacy during the American Civil War, Britain remained the United States' largest trading partner throughout the era.

Paradoxically, Britain's power helped pave the way for a series of political realignments on the continent that came to challenge British supremacy. The Franco-Prussian War of 1870–71 led to the unification of the German states by Otto von Bismarck, while the Italian states were unified by Garibaldi and came under the political control of Victor Emmanuel II by 1870. The wars of the twentieth century were rooted in the power blocks created in the nineteenth century.

As important as the "great game" of empire was, equally remarkable was the fact that those who stayed back home were reading reports and seeing images of those places within days or even hours of the events. In 1804 Richard Trevithick (1771–1833) and Matthew Murray (1765–1826) built the first steam engine tramway locomotives, starting a transportation revolution that saw hundreds of thousands of kilometres of rail laid by the end of the century. The first trans-Atlantic voyage by a steam-powered ship was made by the *Savannah* in 1819, and by the middle of the century steamships were replacing sailing ships. As swift as the locomotives and the steamships became, they could not beat the speed of the telegraph for moving information around the globe. Between rail, steamship, and telegraph, the information flow needed to control global empires and international trade was now possible.

Collecting and Classifying: Biology and Empire

With the major expansion of European imperialism, both scientists and the general public became more and more interested in collecting the many new and exotic species discovered. The mania for the bizarre and unique continued, and curio cabinets and displays of exotic birds, bugs, and hunting trophies were a feature of many middle-class homes. For the serious researcher, collecting gave way to a search for order and connection. After Linnaeus, the collecting sciences of the eighteenth century became the biological sciences of the nineteenth. This was deeply connected to the notion that classification and understanding were part of the process of controlling and exploiting and, so, were bound up with the imperial project.

We can trace the transformation of the study of the living world from eighteenth-century natural history to nineteenth-century biology through the career of one man, Alexander von Humboldt (1769–1859). The son of a Prussian military officer, Humboldt chose the career of a scientific traveller and naturalist rather than the diplomatic life his father had envisaged. He was influenced by the naturalists' voyages with Captain Cook and, in 1799, set out on his own prolonged travels through Spanish America. The result was a series of travel and natural history books that were bestsellers. But Humboldt's descriptions of flora and fauna differed markedly from those of Banks or other eighteenth-century naturalists. He combined exact measurement using precise instruments with intensive fieldwork and an overall concept of the interconnectedness of all living things. His greatest articulation of this came in his final multi-volume work *Cosmos* (1848–59), which was very influential for future scientists. Humboldtian science, as some historians have called it, was instrumental in engendering a strong reliance on fieldwork among biologists, especially in the United States, and in creating a proto-ecological understanding of the living world.

The categorization of new flora and fauna continued with the discovery of additional, and potentially exploitable, species. Naturalists and explorers searched out and found uses for these new plants, much as Hans Sloane had done with chocolate. One of the most important discoveries was that of quinine, the only known treatment for malaria; in an era of greater European exposure to the disease, it helped make imperialism in the equatorial regions possible. Quinine was itself a product of colonialism, since it came from the bark of a tree found in Peru and was introduced to the Spanish by the local people in the 1630s. The botanical name given to the plant by Linnaeus was *Cinchona*, after Countess Ana of Chinchon, the wife of a Viceroy of Peru in 1638. Although the story is probably apocryphal, Countess Chinchon supposedly became ill with malaria and was successfully treated with quinine bark, thus ensuring the spread of its use. Because *Cinchona* was hard to grow and expensive, the efforts to produce artificial quinine became one of the great quests of organic chemistry.

As the empires encountered, traded with, and on many occasions subjugated other peoples, the scientific move to classify came to include human beings as well. Linnaeus had paved the way for this through his classification of humans as thinking mammals (*Homo sapiens*). Further, he claimed that there were four races

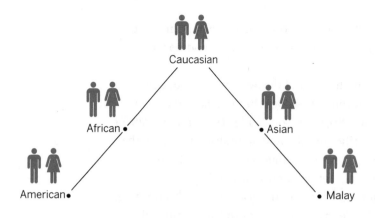

7.1 BLUMENBACH'S RACE DISTRIBUTION

of humans: Europeans, Asians, and Africans (in keeping with the Greek tripartite division of the world), plus the New World race, the Americans. Johann Friedrich Blumenbach (1752–1840), one of his protégés, furthered this categorization by developing a model of relations among these peoples. He believed that all people were of one species, but that there was a hierarchy of perfection among the races. Blumenbach used the people from the Caucus mountains, the most beautiful people he had ever seen, as the race with maximum perfection and then plotted the other races as degenerating from the norm as they moved away to the east and west. In order to have a symmetrical model, he added a fifth race, the Malays. (See figure 7.1.)

Although Blumenbach had not intended his categorization to indicate a hierarchy of development, or to imply that somehow the races below the Caucasians were closer to animals, nineteenth-century racial theorists soon took this pyramid as indicative of biological and cultural achievement and as an excuse and opportunity for exploitation. Perhaps the most egregious example of this is the story of Saartjie (Sarah) Baartmann, a Khoi native of South Africa, who was taken to Britain in 1810 by William Dunlop, a British naval surgeon, to be displayed in side-shows as the "Hottentot Venus." This resulted in a scandal, since slavery had been outlawed in Britain, and eventually Sarah's show moved to France. Scientists such as Cuvier were very keen to examine her, especially because of her unique anatomical features, but Baartmann refused. She eventually died alone as an impoverished prostitute in the streets of Paris in 1816. After her death, she was dissected by a French surgeon, and various of her body parts were on display in the Natural History Museum in Paris until 1949 and then at the Museum of Mankind in Paris until 1974.[1]

----

1. Baartmann's remains were taken off display and repatriated in 2002. She was buried near the Gamtoos River at Hankey in the Eastern Cape, South Africa. Her return to Africa was seen as both an indictment of the racism of the colonial era and the end of a terrible episode in human exploitation.

Catastrophe or Uniformity: The Geological Record

Although the study of people, plants, and animals from far-off locations attracted a great deal of popular and scientific attention, one of the most important areas of research developing in Europe was geology. It flourished under imperial competition because of its link to industrial development associated with coal, iron, and other exploitable mineral resources. Britain, for example, embarked on major field projects to understand, map, and name the strata; and imperial geological surveys were undertaken in Canada, Australia, New Zealand, and India.

Just as they had in Enlightenment Europe, geologists continued to delve into the history of geological change. The debate over causes—the Neptunism and Vulcanism of the eighteenth century—was transformed into a more quantitative discussion of the rate and type of change. The two schools of thought— Catastrophism and Uniformitarianism—that emerged from this debate relied on the idea of geological formations created by either sudden or gradual change. The crux of the problem was this: had the Earth once been hotter, had volcanic activity once been greater, or had the Earth always operated in the way it now did? Hutton had said that forces now in operation explained all geological change, but he had been labelled a dangerous radical by the end of the century, and most early-nineteenth-century geologists disagreed with him. Not until Charles Lyell's work in the 1830s did someone again bring forward the idea that forces currently in operation were responsible for changes in the Earth's configuration historically.

Georges Cuvier (1769–1832) was the most prominent of the catastrophists. He was a professor of anatomy at the newly created National Museum of Natural History in Paris. As a state employee, through his position and work he demon-strated the prominence of France in imperial competition, even after Napoleon was defeated. In addition to his geological ideas Cuvier studied comparative anatomy, establishing that the parts of individual animals must work together and, therefore, that every conceivable permutation or combination of animal parts was not possible. For example, animals with carnivorous teeth had to have suitable stomachs for digesting meat. He became famous for his ability to reconstruct an entire animal from a relatively small part of its skeleton.

When Cuvier was sent some odd fossils from Paraguay, he determined that these most closely resembled the modern sloth, although the remains suggested a giant and now extinct species. He quickly realized that the only way to prove extinction was by using this sort of giant remains, since earlier debates concerning marine fossils always remained inconclusive, given the possibility that the sea

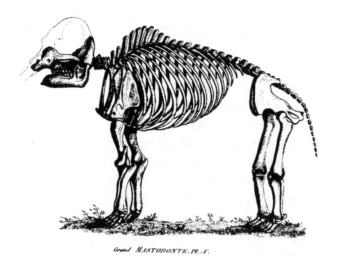

Grand *MASTODONTE. PL. V.*

7.2 CUVIER'S MASTODON FROM 1806

creatures might still be living in unexplored ocean depths. Cuvier looked at various elephant-like remains and, using comparative anatomy, established first that African and Indian elephants were different species and, second, that the remains found near the Ohio River in the United States indicated a different species altogether. He called this a mastodon, while he named a different but related specimen from Siberia a mammoth. (See figure 7.2.) Eventually, frozen carcasses of this Siberian species were found, complete with woolly coats, indicating that they were not tropical animals who had strayed far from their natural habitat, since they were clearly creatures from a cold climate.

So, Cuvier asked, why had they died out? Most of the remains of giant mammals found near Paris, particularly ancient hippos and rhinos, were in gravel pits. Therefore, he reasoned that there must have been a sudden "revolution," probably a flood, that had killed them. He and his fellow researchers discovered that the local gypsum quarries showed evidence of seven progressive floods, with alternations between freshwater and salt-water fossils, and from this determined that there must have been a series of catastrophes. Cuvier used recent work on the Alps, which demonstrated through an analysis of the strata column that they must be of recent origin, to suggest that their emergence might have been a cause of one of the catastrophes. He stressed, however, that these catastrophes were local, not universal, and therefore were not tied to the biblical flood. Although he could not identify the causes of all the revolutions, he was convinced that they were ultimately knowable.

Cuvier stressed the progressive aspect of the history of life on the planet, with extinction as part of the package. As more fossils were discovered, by Cuvier in France and William Smith (1769–1839) in Britain, they discovered that in the gypsum layer under the gravel pits were fossil mammals even more different from present animals than the hippos and mastodons had been. For example, Cuvier found the remains of an animal that looked like a combination of a tapir, pig, and rhino. The remains in the Tertiary rocks were largely mammals, while the Secondary rocks contained mostly lizards. In fact, Cuvier found a number of lizards occupying diverse parts of the ecological scene: a flying one, which he named "ptero-dactyle"; a swimming one, which he called "ichthyosaurus"; and a walking one, which he named "iguanodon." And in the Primary rocks, nothing. Here was a clear progression from one sort of life on Earth to the next.

Some geologists continued to equate the catastrophe evident at the end of the Pleistocene Age with the biblical flood, but most followed Cuvier's lead in ignoring this religious context. They concentrated instead on the evidence that life forms had progressed through different stages, and it seemed more and more likely that the forces responsible for the mass extinctions must have been much more powerful and of a different kind than were evident today. This was confirmed by physicists who were looking at theories of heat radiation and who began to argue that the Earth had originally been much hotter and was gradually cooling down.

English geologist Charles Lyell (1797–1875) disagreed with this idea that the past was of a different kind than the present and therefore unknowable. He sought to maintain the rationality of science by developing a theory that was totally consistent and that enabled scientists to understand nature through observation in the present. In *Principles of Geology* (1830–33), a three-volume work whose title consciously referred back to Newton's *Principia*, Lyell expounded his theory of uniformitarianism. He argued that gradual cumulative geological change could account for extinction and the progress of species in the fossil record. He used his own research on Mount Etna in Sicily and the work of George Poulett Scrope (1797–1876) on the French mountains and the gradual creation of valleys through lava flow to show that gradual geological change corresponded to the gradual extinction of species in the same area. He also argued that the discovery of large mammals that had escaped extinction at the last supposed catastrophe, such as the giant Irish "elk" found in peat-bogs above the Pleistocene level, showed that catastrophes were not the only cause of extinction and, perhaps, had not occurred. (See figure 7.3.)

Lyell's theory had three separate facets. The first, *actualism*, stated that forces of the kind now in action had created the world as we see it. This

7.3 1846 RECONSTRUCTION OF AN IRISH ELK
This extinct deer, sometimes called an elk, was over three metres (ten feet) tall.

was reminiscent of Newton's claim for the universality of forces now in action to explain the structure of the universe. The second, *uniformitarianism*, established that forces of the same degree as today were at work in the past. That is, Lyell refuted the claim that things had been hotter or more violent in the past. The result of these two axioms was the third facet: that the world was in a *steady state*. Just as Hutton had said earlier, Lyell claimed that there was no progress, no direction, to the changes in the world. Unlike Hutton, however, Lyell had a much richer set of paleontological data, and therefore his was a much more difficult stance to support than it had been for his predecessor. Lyell was forced to stress the imperfection of the geological record and to claim that dinosaurs might be found at some other level in future investigations.

While the actualist and uniformitarian aspects of Lyell's theory were very attractive to geologists and biologists, the non-progressionist stance was a hurdle few could overcome. Lyell had to argue for the stability of species; that is, that there had been no evolution. But the work of many scientists, from Cuvier on, seemed irrefutable. As well, Lyell seemed to have brought back the idea of a "system," an idea that had been discredited by the work of people like Humboldt in favour of fieldwork.

Despite problems with his approach, Lyell's work was widely influential because of the attraction of his methodology of comparing the past with the present. The argument that this methodology was sufficient to understand nature guaranteed the autonomy of science from philosophy or religion. More significantly, Lyell was already an influential member of the scientific community. He was well known in the British scientific establishment, as a member and eventually president of the Geological Society of London, as a fellow and Royal Medal winner of the Royal Society, and finally as president of the British Association for the Advancement of Science. Politics also favoured Lyell's views. Uniformitarianism could be seen as bolstering a conservative politics, keeping working classes in their place, as opposed to catastrophism, which fit more with the French view of the possibility of social and political improvement through revolution. Therefore, Lyell's theory did not disturb middle-class gentlemen, like Charles Darwin, who made up the scientific establishment. By contrast, on the continent scientists continued to follow Cuvier's theory, especially since it had the backing of physicists.

While these large theoretical debates raged, most nineteenth-century geologists spent their time on slow, steady fieldwork, sorting out the stratigraphy of the geological column. Field geologists, such as Roderick I. Murchison (1792–1871) and Adam Sedgwick (1785–1873) in Britain, spent many summers tramping through the countryside looking for exposed strata. This work turned out to be complex, since specific layers could overlay at one place, underlay at another, and be completely absent at a third. The competition to classify the various geological layers became intertwined with issues of political and professional power, especially seen in competing nomenclatures. In the end the strata were named for the British counties in which they were found (Cambrian for Wales, Silurian for the British tribe who had fought the Romans near the Welsh border, and Devonian for Devon). Murchison announced his identification of the Silurian system at the inaugural meeting of the British Association for the Advancement of Science in 1831.

The Question of the Origin of Species

One of the most perplexing questions arising from the new earth sciences and spurred by imperial expansion and exploitation was that of species. Where did new species come from? They seemed to appear without prior warning in the fossil record and to disappear just as rapidly. Was there a series of new creations? Did those species still exist somewhere undetected in the world? (This seemed unlikely in the case of dinosaurs or large mammals.) And what were species in any case? One of the earliest attempts to answer some of these questions came from a French scientist, Jean-Baptiste de Monet de Lamarck (1744–1829), who worked at the Jardin du Roi, as did Buffon and later Cuvier. Lamarck denied the possibility of extinction, arguing instead that one species transformed into another through evolution. In 1809 he published *Philosophie Zoologique*, in which he articulated the evolutionary theory that has come to be called Lamarckianism. For Lamarck, the environment influenced the development of various characteristics and, thus, was of prime importance in evolutionary change. When the environment changed, forces internal to the individual plant or animal encouraged physical changes to take place to ensure adaptation to the new conditions. For example, if a short-necked ancestor to a giraffe came to live in an environment where all available food grew close to the top of very high trees, the internal forces within that giraffe would encourage the growth of its neck over time. Most crucial for Lamarck's theory, of course, was that these changes acquired in one generation could be inherited by the next; this was usually called the inheritance of acquired characteristics. Thus, the offspring of this stretching giraffe would be born with a slightly longer neck, the next generation with an even longer neck, and gradually the present-day giraffe would result. There was no extinction, since long-gone forms simply evolved into something else. Lamarck in this way resurrected the Great Chain of Being, without its religious

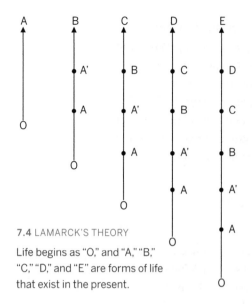

7.4 LAMARCK'S THEORY

Life begins as "O," and "A," "B," "C," "D," and "E" are forms of life that exist in the present.

connotations, since he argued that every species began at the most primitive, and the existence of so many different species today was accounted for by a series of spontaneous generations of the most primitive forms. (See figure 7.4.)

Lamarck was reviled for this theory, especially by his enemy Cuvier. While Lamarck's theory serves to illustrate the fact that people in the early nineteenth century were thinking about evolution, he had little immediate influence on the field. For example, Charles Darwin owed little to his ideas. However, the theory made its mark when many prominent American biologists in the early twentieth century adopted it and it became known as neo-Lamarckianism.

In a similar way, both Erasmus Darwin (1731–1802), Charles's grandfather and member of the Lunar Society, and Robert Chambers (1802–71), who espoused evolutionary theories in early-nineteenth-century Britain, demonstrate the prevalence of evolutionary thinking rather than serving as some linear path to Darwinian theory. Erasmus Darwin's ideas, published in a long poem, *Zoonomia* (1794–96), emerged from the radical thinking of the late Enlightenment and got him into trouble as a dangerous Jacobin free-thinker. Chambers's *Vestiges of Creation*, published anonymously in 1844, and based on a LaPlacean view of an expanding universe, was wildly popular among the general public but ridiculed by scientists. Chambers was seen as an outsider, with no proper professional standing, and thus with no right to make such speculative claims. Darwin took this lesson to heart and worked to establish his credentials before publishing any "wild" theorizing.

Darwin and Evolution by Natural Selection

Charles Darwin (1809–82) was born into the ranks of the scientific elite, both socially and intellectually. He was the son of a rich doctor, the paternal grandson of an important member of the Lunar Society, and the maternal grandson of Josiah Wedgwood, the great British industrialist. Later, he married his cousin, Emma Wedgwood, thereby solidifying his tie to the British manufacturing elite. He first attended the University of Edinburgh studying medicine, but found surgery (in an era before anesthesia was common) too disturbing. He moved to

Cambridge with the idea he would become a minister in the Church of England. An indifferent student, he first developed his enthusiasm for biology, and his life's work, when he spent one summer on a geology tour of Britain with his professor, Adam Sedgwick. He turned with alacrity to natural history and worked with botanist John Stevens Henslow (1796–1861), establishing his own collection of beetles. In 1831 Henslow recommended this eager and gentlemanly student to Captain FitzRoy as a naturalist and companion on the voyage of the HMS *Beagle*. Darwin would never have been recommended, or accepted, had he not belonged to the correct gentlemanly class.

The voyage of the *Beagle* (1831–36) changed Darwin's life. As he boarded the ship, Henslow handed him the first volume of Charles Lyell's *Principles of Geology*, a book that convinced him that forces in action today were responsible for all the changes of the past. He accepted uniformitarianism and actualism, although he never believed in Lyell's steady-state hypothesis. Darwin always saw natural change as directional. When he experienced an earthquake in Concepción, Chile, he was convinced that forces in action today could be very powerful and disruptive. He encountered a plethora of new and beautiful species, including fossils of the giant armadillo; on the Galapagos Islands, off the coast of Ecuador, he noticed that the species of tortoises and finches differed from island to island (although he did not really understand their classification or significance until later, in museum settings in London).

Darwin returned full of new ideas and soon established himself as part of the scientific profession through his published papers, first in geology and later in biology. His first scientific paper presented his theory of coral-reef formation, which followed Lyell's uniformitarianism and claimed that coral reefs were formed as an oceanic island submerged. Since most coral can only live close to the surface of the ocean, it would constantly grow on top of the submerging rock in layers to form a reef. This elegant explanation established Darwin's credentials with the Royal Geological Society.

Darwin began to keep a series of notebooks in which he puzzled over the relationship of species to each other—in fact, the whole question of just what a species was. In 1838 he read *An Essay on the Principle of Population* (1798) by Thomas Malthus, and the mechanism for evolution suddenly appeared clear. Malthus had argued that food supply would at best increase arithmetically (1, 2, 3, 4, 5 …) but that the population would increase geometrically (1, 2, 4, 8, 16 …) with the ultimate conclusion a life-and-death competition for resources. While Malthus had developed this theory in order to explain the population crisis he saw looming in Britain, Darwin immediately saw its application to the plant and animal world.

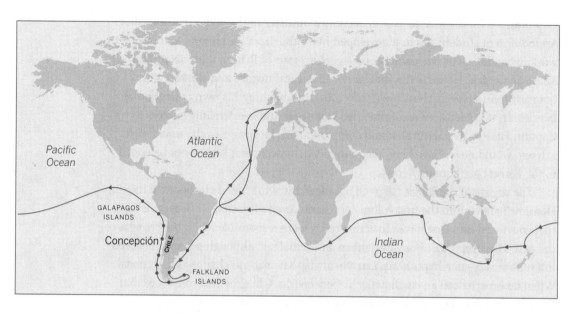

7.5 DARWIN'S VOYAGE ON THE *BEAGLE* (1831–1836)

Darwin's theory, worked out between 1842 and 1844, is often called the theory of evolution by natural selection. He began from the premise that evolution had indeed taken place; his trip to South America had convinced him of this. So how had it happened? First, he argued, variation existed within a population of individuals. You can see this by observing domestic animals, such as the pigeons he bred. These variations were random, continuous, and small. Second, certain variations turned out to be advantageous in particular environments, and nature selected those variations—this is called natural selection. Since Malthus had shown that only a tiny fraction of the number of individuals born survive, there had to be some reason why some survived and others did not. Somehow, their variation equipped them better for survival than others in the population. For example, some birds were born with sharper beaks, which allowed them to burrow for insects; this was an adaptive characteristic in an environment where there were many burrowing insects and other food was scarce. This variation was passed on to the next generation, until more and more birds had this characteristic sharp beak and blunt-beaked birds were replaced or became extinct.

Darwin described an intraspecies struggle in which each individual in a specific species competed with other individuals in that same species for scarce resources. Other potentially dangerous species simply constituted the environment in which variation occurred. Nature was "red in tooth and claw," as Alfred, Lord Tennyson, put it,[2] because each day presented a competitive struggle for existence in which winning was everything. There was no second chance, no going back, because once you were extinct, you were gone forever. The result was branching evolution, where a common ancestor might give rise to numerous different species as different variations filled different ecological niches. (See figure 7.6.) Darwin answered the question of the definition of species by including time, since the map of species, as envisaged by Linnaeus, was now a snapshot in time of a continuing process, and the connections between closely related species, genera, and so on were due to their earlier common ancestry.

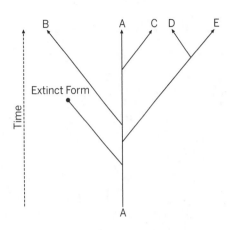

7.6 DARWIN'S SYSTEM

Darwin's system starts with an initial form that then produces related variants over time. Some variants may go extinct and are lost from the system.

This theory of struggle and competition corresponded very closely to the capitalist and imperialist struggle for resources seen in the clash of nations taking place around Darwin. He was influenced by earlier discussions of evolution and geology, but he was also a member of a rich mercantilist family, living in a nation prospering from industrialism and in major expansionist competition with other European nations. His theory was a product of a particular time and place, as much as it was the inspiration of one scientist.

Although Darwin wrote his theory of evolution by 1844 in an *Essay*, he took no steps to publish. He may have feared the kind of ridicule that Chambers endured. He was also struggling to establish his credentials as a scientist among his peers in Britain. He spent the intervening years conducting a massive research project on barnacles, which, with his earlier coral-reef theory, garnered him much respect among his fellow scientists. In 1858, however, another naturalist, Alfred Russel Wallace (1823–1913), announced in a letter to Darwin that he had developed a theory of evolution by natural selection and thus forced Darwin's hand. Darwin asked Lyell for advice, and Lyell obligingly delayed publishing Wallace's paper until Darwin had quickly written a paper of his own, so that a joint paper could be read before the Linnaean Society. Within a year, Darwin

2. Alfred, Lord Tennyson, "In Memoriam," *Alfred, Lord Tennyson: An Authoritative Text, Backgrounds and Sources of Criticism*, ed. R.H. Ross (New York: Norton, 1973) 36.

CONNECTIONS

Science and
Class: Wallace
and Collecting

Alfred Russel Wallace came from a very different family background than Charles Darwin. He was not able to pursue a scientific career as a gentleman amateur, but rather earned his living collecting for others and writing books for a popular audience. He was not a member of the elite circles in which Darwin moved, although later he was befriended by Darwin, who petitioned the government to award Wallace an annual pension, which they did in 1880. Without the time, money, connections, and social capital, Wallace was unable to take the lead in scientific discussions about evolution, which became associated almost exclusively with Darwin's name rather than his own.

Wallace was born into a precarious middle-class family (his father Thomas had studied law but never practised) and was only able to attend grammar school until he was 14. After that, he apprenticed to become a surveyor and worked in a number of surveying positions until his twenties. While working, he became interested in biology and voyages of scientific discovery. He read Malthus, Lyell, Chambers, and Darwin's account of his time on the *Beagle*. He also befriended the entomologist Henry Bates (1825–92), with whom he took his first voyage of exploration and collection to South America in 1848.

wrote a full statement of his theory in *On the Origin of Species by Means of Natural Selection* (1859).

There were a number of similarities in the paths Wallace and Darwin took. Both had travelled on collecting voyages, both had been inspired first by Lyell and then by Malthus, and both were interested in the distribution of species, that is, the question of why species occur geographically nearest to their closest relations. Moreover, both were involved in the imperial project, Darwin as a member of the industrial gentlemanly class, and Wallace as a paid collector, travelling to Amazonia and later to the Malay Archipelago. So perhaps it is not altogether surprising that both developed parallel theories on evolution by natural selection. Wallace always claimed that his was the less well-developed theory, and when he wrote his complete book on the subject, he called it *Darwinism* (1889). But the difference in their status, both scientifically and socially, also accounts for this difference in reputation. For good or ill, this theory has ever since been called Darwinian evolution.

Bates and Wallace spent two years exploring the Amazon River, and a further four years on the Rio Negro. There they found an astonishing wealth of new species of plants and animals never seen by Europeans. Wallace had hoped to pay for this trip through selling his collection on their return, but unfortunately, after 26 days at sea the ship caught fire and burned to the waterline. Wallace was only able to save a few notes and sketches; everything else perished.

From 1854–62, Wallace travelled through the Malay Archipelago, collecting specimens to sell and studying the natural history of the region. He noted that there was a pronounced difference in species distribution on different sides of the archipelago. This made a strong case for the historical division of species, based on geographical distribution (a case later made by Darwin). This distribution line in the Malay Archipelago is now called the Wallace Line. He collected over 126,000 specimens, thousands of which had been unknown to science until that time. While there, he had his insight into the mechanism of evolution (natural selection), which he outlined in a letter to Darwin in 1858, thereby forcing Darwin to get his better-articulated theory into print the next year.

On his return to England, Wallace established himself as a successful writer. His most famous book, *The Malay Archipelago* (1869), has never been out of print. He began to associate with the luminaries of natural history and he met and conversed regularly with Darwin. But his finances were always precarious. He lived from book advance to book advance. Further, his interest in social issues (inspired by Robert Owen) and his foray into spiritualism isolated him from the gentlemen fellows of the Royal Society. Wallace never achieved the fame of Darwin and largely became a sort of footnote to Darwin's story. His lack of gentlemanly connections and resources contributed to his lower status in the scientific community.

Herbert Spencer and Social Darwinism

Darwin's theory struck a chord with educated Britons because, in many ways, it corresponded to social theories that had already been articulated. Evolution by natural selection fit especially well with Herbert Spencer's (1820–1903) ideas and become a major explanation for social growth and development. Widely popular, Social Darwinism justified the belief in progress on one hand and the need to manipulate nature and society to achieve this progress on the other. This collection of theories taught people that evolution happened whether it was desired or not and, therefore, that civilized people were obligated, once conscious of it, to use this evolutionary force for good.

Herbert Spencer was a Quaker by birth, and after an early career as a railway engineer, he worked as a writer and sub-editor for *The Economist*. A small inheritance allowed him to devote his life to studying and writing, especially about the human condition. In such books as *Social Statics* (1851) and *Principles of Psychology* (1855),

Spencer argued that the development of human societies could be explained as an evolution from simplest (native tribes) to most complex (European imperialist states). Like Darwin, Spencer was influenced by Malthus and saw life as a struggle in which only the strong tended to survive. He coined the phrase "survival of the fittest" to explain this phenomenon. In order to ensure the strongest and best society, he argued that it was desirable to keep state control to a minimum. Spencer did not support a completely individualistic morality, however, but maintained that strong communities were based on understanding the natural unity of interests of individuals. Survival of the fittest was thus key to the economic and social progress of mankind. These theories, when applied to race, were used as a justification for segregation (to keep the weaker races from mixing with the stronger) and subjugation (the strong had a moral and natural imperative to control the weak).

Spencerianism and forms of Social Darwinism provided justifications for capitalism, *laissez-faire* economics, and arguments against a welfare state. It should come as no surprise that one of Spencer's biggest fans was the American industrialist Andrew Carnegie.

Other forms of social Darwinism stressed the struggle between races or nations, thus justifying the military and industrial competition taking place in Europe. These theorists believed that war was a way of winnowing out inferior nations. "Might is right" provided a justification for imperialism, since, according to its rationale, inferior races deserved to be exploited and controlled. Social Darwinism also had an individual expression in eugenics, which was presented by its proponents as the science of race. Eugenists claimed that the state had a duty to limit the multiplication of its least fit citizens and to encourage its most fit to increase. An early proponent of this was Francis Galton (1822–1911), a cousin of Charles Darwin, who wrote *Hereditary Genius* in 1869.

Darwin had not mentioned human beings in *Origin of Species*, although he was certainly aware that his theory could be so applied. At the 1860 meeting of the British Association for the Advancement of Science, Darwin's theories were attacked on this basis by Bishop Samuel Wilberforce, known as "Soapy Sam" in the press, and defended by "Darwin's Bulldog," Thomas Henry Huxley (1825–95), who replied that he would rather be descended from an ape than from an intelligent man who used his intellect to retard the growth of science! Darwin later entered the fray with the publication of *Descent of Man* (1871), in which he set out to prove that evolution of intelligence and morality were also possible through natural selection and, therefore, that humans had evolved in the same way as had animals. This book lacked the rigour of his earlier work but produced a curious research avenue in which people examined their companion animals for signs of intelligence and humour. Wallace objected to this demotion of humans to animals and spent the last years of his life investigating spiritualism, a popular Victorian pursuit, in search of the divine spark that separated humankind from the beasts and that remained even after death.

Although Darwin's books and theories were extremely popular— *Origin* sold out within days of the first printing—his theory of evolution by natural selection did not win over the biological community before the twentieth century. He certainly had his contemporary supporters, especially Huxley in Britain and Asa Gray (1810–88) in the United States. Naturalists were very favourably disposed to his theory, since it explained both why classification worked and the source of geographical distribution of species. Naturalist Henry Walter Bates (1825–92), for example, argued that natural selection could be seen in the Midlands moth. Prior to the industrialization of the Midlands, the moths were predominantly light-coloured to blend with tree bark and hide from predators. Due to pollution, trees had become

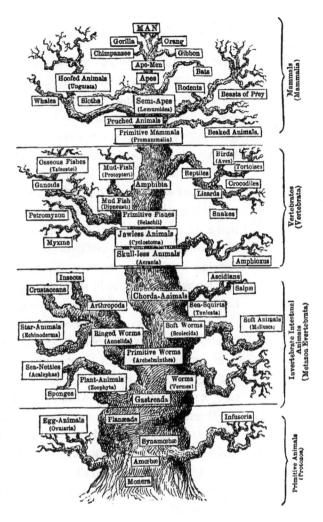

7.7 PEDIGREE OF MAN

A "family tree" based on an anthropocentric interpretation of Darwin's theory of evolution. This schema presents humans as the ultimate product of evolution.

darker, and dark moths gained camouflage advantage and multiplied, becoming the dominant form. Bates also put forward a Darwinian explanation of mimicry, the phenomenon of a harmless insect looking like a specimen of a poisonous species. As Bates pointed out, an insect that looked unpalatable was more likely to survive, and so such colouring could be a variation selected for survival.

Opposition to Darwin's Theory

Most naturalists, especially in Britain, were less concerned with evolutionary theory and more interested in collecting. A natural history collecting mania had taken hold, encouraging men, and increasingly women, to tramp the land in search of rare plants, animals, insects, and fossils. Middle- and upper-class collectors paid others, like Wallace and Bates, to travel to exotic locations and bring back specimens. Many of these amateur collectors were deeply influenced by natural theology, a doctrine articulated most clearly by William Paley (1743–1805). Paley argued that if a man were walking in the woods and found a watch, even if he had no idea what it was or how it worked, he would know that it had been made by some intelligence, that is, by the watchmaker. How much clearer it was, then, that the natural world, more intricately constructed than the watch, must have been fashioned by the Great Watchmaker. This argument, called the argument from design, was a potent one, which was later taken up by the authors of the *Bridgewater Treatises*, including Baden Powell (father of the founder of Scouting). This series of books, commissioned by a bequest from Francis Henry Egerton, Earl of Bridgewater, attempted to link various aspects of science to a proof of God's existence and a divine plan. Paley wrote the most famous of the series. Darwin himself was careful not to contradict this idea of design, although many of his detractors saw atheism in the heart of his system, since it seemed to require no final plan to create humans.

While evolution as a basic premise steadily gained adherents in the scientific world, Darwin's theory did raise questions. How could God create a world that was so violent, so wasteful, and so unlike the grand design envisaged by the natural theologians? Could the human species really be descended from apes and not created in God's image? If Darwin was right, did that mean parts of the Bible were wrong?

For biologists and other scientists, other pressing issues were also unresolved. The fossil record did not seem to contain any gradual evolutionary forms, and there were many "missing links," or unfound intermediate forms, predicted by Darwin's theory. Present-day variations seemed too small to have created evolutionary change of the magnitude of the fossil record. More damning, the socially powerful and influential physicist William Thomson (1824–1907), Baron Kelvin of Largs or more simply Lord Kelvin, suggested that the age of the planet must be between 20 million and 400 million years, too short a time for Darwinian gradual evolution.

Thomson was a prodigy, graduating from the University of Glasgow at the age of ten, then moving to Cambridge to complete his education. He was made Professor of Natural History at the University of Glasgow in 1846, holding that position until his retirement in 1895. He did ground-breaking work on electricity and magnetism and helped to lay the first Atlantic telegraph cable. For his work, he was knighted in 1866 and made a peer of the realm in 1892. He was also president of the Royal Society from 1890 to 1895. Lord Kelvin, basing his conclusions about the age of the Earth on its temperature and on the rate of cooling of the planet from a molten state, argued that its age was about 50 million years. This calculation came from his work on thermodynamics, particularly his *On the Dynamical Theory of Heat* (1851), in which he introduced the absolute (or Kelvin) scale of temperature, setting zero at the theoretical point at which molecular motion stopped. On this absolute scale, water melts at 273.16°K. Assuming that the planet was moving along a heat spectrum from molten to the temperature of space (i.e., it is a rock heated only by the energy of the sun), the time that life could have existed on Earth could be calculated. The problem for Darwinian evolution was that 50 million years, while a very long time, was a far cry from the 20,000 million years that some scientists had suggested were necessary for evolution to produce life as it now existed.

Finally, there were questions about the biological mechanism by which variations or inherited characteristics were passed on. Darwin's speculation on the mechanics of reproduction were not well received. Combined with other objections, many scientists were reluctant to accept his explanation. By 1900, while evolution was part of every biologist's credo, the mechanism was more in doubt than ever.

The Professionalization of Science and Science Education

It is not surprising that the era that shaped Darwinian theory was also the era that produced Sherlock Holmes, Sir Arthur Conan Doyle's cool, scientific, and supremely logical private detective. Scientists, even if they continued to argue the fine details, seemed to be revealing all of nature, and it seemed possible to create a complete picture of how the physical world operated. With such knowledge, no secret could be kept from the observant mind. Darwin and Sherlock Holmes were also alike in being amateurs rather than professionally trained and employed. While Holmes gave metaphorical birth to generations of private detectives, Darwin, as gifted and insightful as he was, represented the waning of a scientific style. He was the last of the great amateur gentleman scientists. For science as an occupation, the nineteenth century was a turning point. There was an increasing sense of science as a professional activity, and with this came a growing separation of the study of nature as a branch of philosophy. This was spurred by the appearance of more and increasingly specialized scientific organizations and the development of new institutions in education and research dedicated to science. While it may have been used by German academics earlier, the term "scientist" was a product of the nineteenth century, introduced by William Whewell (1794–1866) to a large audience at a meeting of the British Association for the Advancement of Science in 1833.

The reign of Napoleon offers a good starting point for the expansion of science as a profession. Although Napoleon's actual support for science was somewhat variable (for example, he abandoned a number of scientists in Egypt when he escaped the British there), he did foster a sense of the importance of technical knowledge. The Académie des Sciences was disbanded in 1793 during the Revolution, but it re-emerged in 1795 as part of the Institut National. In addition to the Académie Napoleon approved the foundation of the Société d'Arcueil in 1805, which had as members such leading thinkers as Claude Berthollet (1748–1822), Pierre-Simon Laplace, and Alexander von Humboldt.

More important than the Société d'Arcueil was the transformation of the École Polytechnique into the leading scientific and engineering school of the era. Originally founded as the École centrale des travaux publics in 1794 by the National Convention, it changed its name in 1795 and absorbed the state artillery school in 1802. First under the Ministry of the Interior, the connection with the military was completed in 1804 when Napoleon transformed it into an elite military school. The artillery had always been the most intellectual branch of the military, requiring a broad understanding of mathematics, physics, chemistry, and elements of what we would call materials science. At the same time, the artillery was not steeped in ancient traditions and offered a path to military command that was not determined solely by social rank. It had attracted many bright young men of lower birth, including Napoleon, who was commissioned as an artillery officer in 1785.

The École Polytechnique was founded by three men: Lazare Carnot (1753–1823), who organized the Republican armies and wrote on the science of fortification; Gaspard Monge (1746–1818), a mathematician and physicist, whose work on descriptive geometry laid the foundation for architectural and engineering drawing; and Adrien Marie Legendre (1752–1833), a mathematician who worked on number theory. Mathematics, chemistry, and physics were core subjects, and the material was often so advanced that there were complaints that the École Polytechnique was too theoretical to be useful as a military school. Yet its reputation was so great that it continued to function throughout the century and continues as one of France's leading educational institutions today. It has produced a stream of professional engineers and scientists for more than 200 years.

The grand paradox of Napoleon's reign was that he ended the Republic and failed to conquer Europe but set the continent on a course that brought about the very reforms that other countries had gone to war against him to defeat. With the exception of the British, who had suffered through similar reforms a century earlier, the wars forced many European countries to undergo major economic and social change. Agrarian economies and restricted middle classes could not provide the material resources to counter the Napoleonic threat. To match the power of France, the continental powers needed the support and productivity of their citizens, and in turn those citizens demanded more autonomy and a greater say in government. Pre-Napoleonic armies were based on late-Renaissance structures and commanded by the often untrained nobility, but the death toll among officers and the rapid expansion of the size of Napoleonic-era armies forced a change in military command. Napoleon had taken French peasants and middle-class burghers and turned them into a powerful army. Many new officers, who were not drawn from the nobility, led forces into battle for or against him. They were not willing to go home and return to the old system.

The freedom and broad scope of the amateur scientist had benefited British science in the years from Newton to the early nineteenth century, but with the increasing complexity of scientific knowledge, that approach was failing. The centres of scientific strength shifted to France and, even more, to German schools and research organizations. In Britain the rhetoric of scientific utility as espoused by the Royal Society rang increasingly hollow. The Royal Society was doing less and less to support science and scientists and had, in fact, become little more than a kind of exclusive social club. Most members conducted little or no scientific work, yet they sat in judgment on scientific matters and offered advice to the government. Scientists such as Charles Babbage (1792–1871) were particularly concerned about the lack of government support for science (although he received a number of grants for his work on calculating machines) and the state of the Royal Society. Babbage wrote *Reflections on the Decline of Science in England* (1830) and published a second edition with additional material and a foreword by Michael Faraday in 1831. It was a vitriolic attack on the Royal Society. While many of its points were valid, it did little to persuade the Royal Society to change its orientation.

With the Royal Society seeing no reason to change, Babbage and a number of his friends opted to create a new society dedicated to science and the promotion of science. In 1831 they founded the British Association for the Advancement of Science (BAAS), modelled on the Deutsche Naturforscher-Versammlung, which had been created in 1822 by Lorenz Oken (1777–1851). Where the Royal Society and the

Académie des Sciences were elite organizations that treated science as a superior intellectual pursuit and in which membership was tightly restricted, the new societies were organized around actual participation in science. The BAAS held meetings not only in London but all over the country and in colonial territories such as Canada. These meetings encouraged scores of naturalists by providing a venue for the presentation of local knowledge that was ignored or deemed too insignificant for the elite organization. The BAAS also developed closer ties with industry than did the Royal Society, since its membership was open to anyone interested in science including small-business owners, schoolteachers, and craftspeople, unlike the Royal Society, which continued to select members by sponsorship and election.

Babbage and his supporters were concerned about Britain losing its leading position in science, but for the most part the consequences of falling behind went unnoticed, masked as they were by the growing power of the British Empire. In 1851 the British celebrated that power with a massive display of technology and empire at the Great International Exhibition in London. The centrepiece of the Exhibition was Joseph Paxton's Crystal Palace, a display hall built of a cast iron frame and glass panels. Inside were the industrial marvels of the day, a cornucopia of machines and products brought together to demonstrate and promote technical innovation. The Exhibition was a massive undertaking, with 13,000 exhibits from around the world including industrial and commercial displays and fine arts. It also included exotic displays from the empire, with plants such as a mammoth water lily from British Guiana that had leaves large enough to support a child's weight. The Exhibition attracted over 6 million visitors. The building itself was an innovation, a forerunner of modern modular steel-frame construction. Paxton used 4,000 tons of iron for the skeleton and 83,610 m² (900,000 square feet) of glass exterior skin to enclose 71,800 m² (772,784 square feet) of space in Hyde Park. (See figure 7.8.)

Associated with the Exhibition were efforts to educate people about and to popularize fine arts, business, and science. A number of people noted that, while Britain led the world with her industrial power, she lagged behind her rivals in training and institutional

7.8 CRYSTAL PALACE

The Crystal Palace, the central exhibition hall of the Great Exhibition of 1851.

support for technical and scientific education. In the same year as the Exhibition, the Government School of Mines and of Science Applied to the Arts opened its doors, a partial response to the continuing concern about science training. The school was renamed the Royal School of Mines in 1863, and it added the "New Science School" in 1872; it was built in South Kensington with profits from the Exhibition. The New Science School contained the Departments of Natural History and Physical Science. While these efforts helped Britain, they did not match the support and scope of technical and science training in France or Germany.

Louis Pasteur

In France, the École Normale Supérieure produced generations of important scientists, including people such as mathematician Évarist Galois (1811–32), sociologist Émile Durkheim (1858–1917), and biologist Louis Pasteur (1822–95). Pasteur, who was broadly trained in chemistry, physics, and biology, began his scientific career investigating the asymmetrical crystalline structures of acids. As a vitalist, he was convinced that living organisms were intrinsically different from nonliving matter. As significant as his crystallography was, he is most famous for his bacteriological work. He discovered that the fermentation process—in wine, milk, and vinegar—was due to the activity of microscopic animals rather than a chemical reaction. Carrying on the observations of seventeenth- and eighteenth-century microscopists, and adding carefully controlled experimental work, Pasteur discovered that these micro-organisms were anaerobic (lived without oxygen) and that they could be killed by heat. The process of pasteurization, used in winemaking (and later in milk production) to kill harmful microbes, boosted the French wine industry, as well as earning Pasteur a good return on his patents.

During his career, Pasteur extended the germ theory of disease, first put forward by Robert Koch (1843–1910), to diseases in silkworms; developed vaccinations for anthrax and rabies; and helped establish Pasteur Institutes around the world to carry on his research. He was a skilled self-promoter and won an important scientific debate with Felix Pouchet (1800–72) concerning the possibility of spontaneous generation. Pouchet believed that life could be created from inanimate matter under the right conditions (such as warm moist earth, or dung). When an experiment proved inconclusive, Pasteur simply said that spontaneous generation was not possible, and his status as the most famous French scientist of the day ended the argument.

Science in Japan: The Fusion of Ideas in the Global Context

By the beginning of the nineteenth century, colonialism and global contacts had made clear to many people, often in the bloodiest terms, the power of European technology. Some non-European nations began to realize that there was a scientific foundation to this imperial power and began to seek it out for themselves. In some cases, scientific exchange was well established; for example, the Jesuits had been teaching in China since the sixteenth century. In the early days, the Jesuits learned more from the Chinese than the Chinese learned from the Europeans (a situation repeated in the McCartney mission of 1793), but by the nineteenth century the scientific ideas of Europe took on new importance for non-European powers with the rise of European imperialism.

In the case of Japan, the transformation of a small group of scholars and collectors creating an indigenous natural philosophy into a scientific community was spurred first by interest in Chinese scholarship and later by European ideas, sometimes smuggled into Japan. During the Tokugawa period (1600–1868), astronomers and teachers were employed by the shogun and received a modest salary. Physicians, who were often trained in natural philosophy as well as medicine, had a higher social standing than astronomers or philosophers. Prior to the sakoku (closed country) policy of 1633–39, Dutch and Portuguese traders brought European medical texts to Japan. Information about surgery and *materia medica* (pharmacology) were particularly sought after because of their utility. After 1639, the only foreign traders allowed in Japan were the Dutch, Korean, and Chinese, all strictly controlled. In 1650, shogun officials ordered a European anatomical textbook from the Dutch and directed a number of physicians to study Western medicine. Over time, the Japanese examined and translated a number of European medical books and growing interest in experimental physiology and anatomy following the European style led to a decline in Chinese-based medical practice.

European astronomy (primarily Ptolemy and its Aristotelian foundation) had been introduced by Jesuits prior to the sakoku policy, but the introduction of the Chinese Shou-shih calendar around 1670 offered a much more practical approach to astronomy. This was not just rote adoption, since the Chinese calendar was modified for Japan using a European world map to adjust for the different longitude and latitude. The fusion of ideas can be clearly seen in the life and work of the astronomer Goryu Asada (1734–99). Trained as a physician, he taught himself astronomy and left his clan to do astronomy in Osaka. Shigetomi Hazama (1756–1816), his patron and student, owned a copy of a Chinese astronomical text that had been edited by a Jesuit who included Kepler's three laws. According to the historian of science Takehiko Hashimoto, the Keplerian agreement between observation and theory impressed Asada so much that he began to study Western astronomy. In turn, the skills learned from European astronomy led to a mathematical survey of Japan and the creation of the first indigenous map of Japan in 1821.

One of the crucial changes in the place of science in Japan came about because of the entry of the samurai class into "technical schools," which taught subjects such as astronomy, physics, and mathematics. This increased both the interest in and status of the physical sciences. Interest in Western science was further increased by the arrival of Commodore Perry in 1853. Perry's arrival forced the Japanese to open diplomatic relations with the United States and the appearance of his ships made the Japanese realize that the technology of the Americans was far ahead of anything in Japan. By 1870, interest in Western practices prompted the Higo administration to close the Jishukan academy, a Confucian school, and open the Yogakko academy run by Captain L.L. Janes, a graduate of West Point. In addition to English instruction, he taught mathematics, chemistry, physics, and geology.

In 1868 the Tokugawa shogunate ended and the Meiji ("enlightened rule") period began. This was characterized by restoration of the power of the emperor, but also by the Charter Oath:

1. Deliberative assemblies shall be widely established and all matters decided by open discussion.
2. All classes, high and low, shall be united in vigorously carrying out the administration of affairs of state.
3. The common people, no less than the civil and military officials, shall all be allowed to pursue their own calling so that there may be no discontent.
4. Evil customs of the past shall be broken off and everything based upon the just laws of Nature.
5. Knowledge shall be sought throughout the world so as to strengthen the foundation of imperial rule.

While clause five clearly indicated Japan's desire to select the best ideas and practices, including those of science, from anywhere in the world, it is also important to note that clause four contains the same conception of natural law as was promoted by many European thinkers in the late eighteenth century.

Japan provides an excellent example of the importance of science to modernization for non-Western countries. Although this science originated in Europe, it was not a Western idea that was somehow imposed on others by colonialism or other forms of coercion. Japanese scholars and leaders had a variety of sources for natural philosophy and in the beginning used a combination of Chinese and domestic ideas as the foundation of Japanese natural philosophy. When competing ideas filtered in from Europe, they accepted or rejected them based on their utility, but they also blended and adapted those ideas. Today, Japan is one of the most science-oriented countries in the world, but it retains its own traditions and culture.

Chemistry and the State: Classification, Structure, and Utility

While French science prospered through the work of scientists such as Pasteur, Germany was creating a scientific program that proved incredibly fruitful. By 1850, the German states had promoted a strong scientific culture, transforming teaching and research at the universities and creating a network of partnerships among scientists, business, and government. At first, Britain, blinded by wealth and empire, did not recognize the extent of Germany's progress and relied on the eminence and expertise of such established scientists as Lord Kelvin. Although some British chemists and physicists were on a par with French and German researchers in terms of "pure" research, the continental scientists were supported by the money and effort poured into science and engineering as part of a larger integration of science with state and industrial needs. Chemistry had emerged as a separate category of research in the eighteenth century, but in the nineteenth century it was in a state of flux. Much industrial chemistry was based on a craft or artisan system of production, but by the middle of the nineteenth century sharply increased demands for materials led to a greater need for mass production methods. Some products such as gunpowder, dyestuffs, acids, and naval materials such as pitch were so important and in such high demand that they became issues of national security. To understand how obtaining and maintaining supplies of these materials would develop into a serious national issue, we have to head back to the laboratory and pick up some of the threads of chemistry left from the previous century.

At the turn of the century new processes were being pioneered, and more substances were being rapidly created and discovered. While this demonstrated a robust interest in chemistry and helped supply industry, it also meant there was more to be confused about. For example, Lavoisier had listed 33 elements in his work *Elements of Chemistry*, but 32 new elements were added to the list by 1860, as well as 1,000 compounds that had not existed in the previous century. Chemists were developing better laboratory techniques, made possible by new tools such as the Bunsen burner, introduced by Robert Bunsen (1811–99) and his student Henry Roscoe (1833–1915). The burner's colourless flame heated substances to a point hot enough to emit light. As a result, chemists found that each substance had a unique pattern of spectral colours, a finding that proved extremely useful for analytical work, including Bunsen's discovery of two elements: rubidium and cesium.

Although better tools and techniques provided interesting and useful materials, fundamental questions about chemical activity required an overarching theory to pull the information into a comprehensive system. Order was needed, and some felt that it could be obtained by looking for the hidden relationship of the elements. John Dalton (1766–1844) had propounded an atomic theory that distinguished elements by their atomic weights (relative masses based on the assignment of a comparison of elements with hydrogen as the lightest). This idea was published in his *New System of Chemical Philosophy* (1808) and widely influenced how matter was understood. Yet a simple linear list of elements by weight did not provide much structure. Around 1829, Johann Döbereiner (1780–1849) put elements in groups of three, called "triads," and noted similar properties in the grouped elements. In 1862 A.E. Béguyer de Chancourtois (1820–86) published what he called the "telleric helix" that placed the known elements on 45° lines arranged on a cylinder. It was a step toward organizing the elements by weight and characteristics, but it gained little notice.

In 1826, the great chemist Jöns Jacob Berzelius (1779–1848) calculated atomic weights for 48 elements and molecular weights for hundreds of associated oxides. His work was not universally accepted, however, as other researchers proposed alternative atomic weights and systems of measurement. Different scientists used different standards for atomic weight for their measurements, some using hydrogen as one on the scale, while Berzelius set oxygen as equal to 100 and measured weights against it. After several years of work, in 1860 Stanislao Cannizzaro (1826–1910) announced a new way to deal with the problem of measuring atomic weight when he demonstrated that the density of gases and vapours could be compared with the density of hydrogen in order to determine the molecular weights of elements and compounds accurately.

With a reliable method for determining atomic weight in hand, it became possible to address the question of classification of the elements. Julius Lothar Meyer created a table of elements in 1864 that placed elements with similar properties in columns in sequence with their atomic weights. These analogous elements also exhibited similar valence. Valence was initially a measure of the combining power of an element. Elements with a high valence, such as oxygen, vigorously combined with other elements to form compounds, while other elements, such as gold, only rarely formed compounds. In modern terms valence is based on the number of hydrogen atoms with which an element may combine; for example, one atom of oxygen can combine with two atoms of hydrogen or one other atom of oxygen. This system also correlated with the chemical activity of elements, not just quantity. For example, the elements with valence of 1 (such as alkali metals) were very reactive, while elements with valence of 3 (nitrogen, or arsenic) were much less reactive. The four most important elements for organic chemistry form a nice list by valence:

H (hydrogen) 1
O (oxygen)   2
N (nitrogen) 3
C (carbon)   4

In Britain, analytical chemist Alexander Reina Newlands (1837–98) independently produced a table of elements that grouped them by families and atomic weight. When he presented his work to the Chemical Society in 1866, it was criticized both for problems with the analogous groups and because it left no spaces for the apparent gaps or odd jumps in weight that seemed to exist within various groups. His paper "The Law of Octaves and the Causes of Numerical Relations between Atomic Weights" was rejected by the *Journal of the Chemical Society*. This rejection came back to haunt the Chemical Society when the periodic table, which looked very much like Newlands's work, was established not by a British chemist but by a Russian. The Chemical Society belatedly awarded Newlands its highest prize, the Davy Medal, in 1887.

Mendeleev and the Periodic Table

Julius Lothar Meyer (1830–95) and Dmitri Ivanovitch Mendeleev (1834–1907) gave order to the elements. Each independently recognized that the elements could be grouped by both atomic weight and by characteristic. In 1868 Meyer created nine columns and placed elements with similar characteristics in ascending order of atomic weight. That same year Mendeleev began to write a chemistry textbook that was to be a great synthesis of chemical knowledge. Like Lavoisier did in the *Elements of Chemistry*, he attempted to classify all known elements. He made a card for each element, listing its properties, and then arranged and rearranged the cards, looking for some pattern. A pattern emerged when the cards were sorted by

Reihen	Gruppe I. — R^2O	Gruppe II. — RO	Gruppe III. —. R^2O^3	Gruppe IV. RH^4 RO^2	Gruppe V. RH^3 R^2O^4	Gruppe VI. RH^2 RO^3	Gruppe VII. RH R^2O^7	Gruppe VIII. — RO^4
1	H=1							
2	Li=7	Be=9,4	B=11	C=12	N=14	O=16	F=19	
3	Na=23	Mg=24	Al=27,3	Si=28	P=31	S=32	Cl=35,5	
4	K=39	Ca=40	—=44	Ti=48	V=51	Cr=52	Mn=55	Fe=56, Co=59, Ni=59, Cu=63.
5	(Cu=63)	Zn=65	—=68	—=72	As=75	Se=78	Br=80	
6	Rb=85	Sr=87	?Yt=88	Zr=90	Nb=94	Mo=96	—=100	Ru=104, Rh=104, Pd=106, Ag=108.
7	(Ag=108)	Cd=112	In=113	Sn=118	Sb=122	Te=125	J=127	
8	Cs=133	Ba=137	?Di=138	?Ce=140	—	—	—	— — — —
9	(—)	—	—	—	—	—	—	
10	—	—	?Er=178	?La=180	Ta=182	W=184	—	Os=195, Ir=197, Pt=198, Au=199.
11	(Au=199)	Hg=200	Tl=204	Pb=207	Bi=208	—	—	
12	—	—	—	Th=231	—	U=240	—	— — — —

7.9 MENDELEEV'S PERIODIC TABLE FROM *ANNALEN DER CHEMIE* (1871)

atomic weight. He produced a table of elements in 1869 and an improved version in 1871. (See figure 7.9.) While there continued to be problems, especially with some of the more recently discovered elements whose characteristics were not completely certain, his arrangement made clear both the progression of atomic weight and the collection of series by characteristics and valence.

One of Mendeleev's great insights was to leave gaps representing undiscovered elements, whose particular characteristics he then predicted. Mendeleev beat Meyer to press in 1869, but Meyer included some of Mendeleev's work when he produced a new table in 1870 that was almost identical to the modern periodic table of elements.

Mendeleev's system pulled together a vast array of chemical ideas and information. Just as Lavoisier's nomenclature contained an underlying philosophy within it, so too did the periodic table present a particular philosophy of matter. It treated matter as a collection of distinct and indivisible particles. The particles were capable of combining to form more complex compounds, but they could do so only in fixed proportions. Atoms of an element were identical and behaved in the same manner. The table also codified certain conceptions of the physical world such as the quantity of atoms in a given mass of material, valences, and the definition of elements. While the modern periodic table of elements tends to be treated as a kind of index, it has behind it a long history of debate over all these questions about the structure and condition of matter. When a scientist uses the periodic table, he or she accepts the philosophic ideas encoded in it. After long use and demonstrated reliability, the acceptance of the periodic table has become unconscious, so deeply imbedded in the practice of science that it is considered axiomatic and thus very hard to challenge.

7.10 MENDELEEV'S PREDICTIONS AND
BOISBAUDRAN'S ANALYTICAL RESULTS

PROPERTY	EKA ALUMINUM	GALLIUM
Atomic weight	≈68	69.9
Density	5.9	5.93
Melting point	Low	30.1°C
Formula of oxide	Ea_2O_3	Ga_2O_3

The utility of Mendeleev's system was further supported by the discovery of the elements represented by the blank spaces in his table. Among others, Mendeleev predicted the properties of what he called *eka aluminum* (*eka* meaning first in Sanskrit). (See figure 7.10.) In 1875 Paul Émile Lecoq de Boisbaudran (1838–1912) discovered gallium, which had the basic characteristics Mendeleev predicted for eka aluminum. When his "ekaboron" was discovered in 1879 by L.F. Nilson (1840–99) and called scandium, Mendeleev's fame was confirmed.

For his achievements, Mendeleev won the Davy Medal with Meyer in 1882, was elected a Foreign Member of the Royal Society in 1890, and was awarded the Copley Medal in 1905.

The Structure of Organic Compounds

While the periodic table provided a powerful tool for understanding the material world, it also made clear a troubling problem about the nature of matter. With over 60 verified elements discovered by 1875, questions arose: Why were there so many different elements? Could small changes in mass really account for their vastly different characteristics? Once the pattern of valence was established, it raised the further question of how atoms stuck together. Affinity theory increasingly seemed inadequate, for it offered no physical explanation for why a carbon atom would have twice the affinity or combining power of oxygen.

Although elements were pure and relatively limited in number, the number of molecules that could be built up out of the building blocks was immense, especially in organic chemistry. Modern organic chemistry looks at any compound that has carbon as its central component, but many chemists in the eighteenth century thought that organic compounds could be produced only by living organisms. According to this theory, called "vitalism," there was a special force of life in plants and animals. This idea was based on the simple observation that plants and animals were alive and, although composed of chemical elements, clearly were different from inorganic matter. While many believed in vitalism, researchers did not articulate it clearly. Some saw vitalism as a special property of life that could be studied, while others saw it as the divine spark and, therefore, not subjectable to scientific investigation. Friedrich Wöhler (1800–82) struck a blow against the theory of vitalism when he produced urea from inorganic compounds in 1832, although it would take more than 100 years for vitalism to disappear entirely. By treating lead cyanate with ammonium hydroxide and removing lead oxide, Wöhler was left with urea—in modern nomenclature $CO(NH_2)_2$. This was a major step toward the manipulation and synthesis of organic compounds, one that had a fundamental role to play in the European race to create high-technology industry.

By the 1840s there had been a flurry of work on the synthesis of organic compounds and, more importantly, experiments on how the component parts combined and could be substituted. Much of this work was based on the theories of Berzelius, who introduced ideas about electro-chemical combination and radicals. Radicals were groups of atoms that were a consistent base. They could have other atoms attached to them (usually from a small list of possible substitutes), which varied the properties of the compound but created a kind of family of related compounds.

Jean-Baptiste-André Dumas (1800–84) introduced his theory of types as a way of understanding the composition and activity of organic compounds by grouping them according to properties and reactions. In turn, his student, Auguste Laurent (1808–53) proposed the "nucleus theory": since the structure of the compound determined potential chemical reactions, the place and relationship of the atoms in a substance was vital to understanding chemical behaviour. Each of these ideas, strongly argued for and against by their supporters and detractors, contributed to an understanding of the complex structure and behaviour of organic compounds. What held back a generally accepted system of organic chemistry was partly a lack of coherent information, since, for example, different chemists used different atomic weights for the same atoms in a compound. This produced apparently different results from the same analysis and, hence, did little to verify the theories being used.

The problem was bonding, or the way various atoms joined together. Since differing quantities of atoms could be made to create molecules under different circumstances, it was unclear how the rules governing molecule creation actually worked. Friedrich August Kekulé (1829–96) helped to clarify this problem. Rather than seeing radicals as the functional component, each atom should be treated as an equivalent part that could be combined following fixed rules. At the heart was carbon, which could combine with four other atoms (carbon was "tetratomic" or "tetrabasic" as Kekulé called it) but need not be limited by radical or family groups. In 1857 Kekulé represented the relationship with his "sausage formulae." (See figure 7.11.)

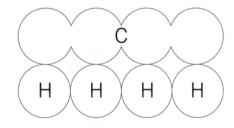

7.11 KEKULÉ'S SAUSAGE FORMULA FOR CH_4
One carbon atom is linked to four hydrogen atoms.

7.12 KEKULÉ'S ACETIC ACID TABLE

Formula	Theory
$C_4H_4O_4$	Empirical formula
$C_4H_3O_3 + HO$	Dualistic
$C_3H_3O_4H$	Hydrogen acid theory
$C_4H_4 + O_4$	Nucleus theory
$C_4H_3O_2 + HO_2$	Longechamp
$C_4H + H_3O_4$	Graham
$C_4H_3O_2O + HO$	Radical theory
$C_4H_3O_2 \atop H]O_2$	Gerhardt's type
$C_4H_3 \atop H]O_4$	Schischkoff's type
$C_2O_3 + C_2H_3 + HO$	Berzelius's copula
$HO(C_2H_3)C_2,O_3$	Kolbe
$HO(C_2H_3)C_2,O,O_2$	Kolbe
$C_2(C_2H_3)O_2 \atop H]O_2$	Wurtz
$C_2H_3(C_2O_2) \atop H]O_2$	Mendius
$C_2H_2HO \atop HO]C_2O_2$	Geuther
$C_2H_3 \atop O]O + HO \atop O$	Rochleder

The emphasis on structural ideas was critical for organic chemistry, as Kekulé made clear in 1864 when he published a list of 20 different formulae for acetic acid from various chemists to demonstrate the extent of confusion. (See figure 7.12.) While there was general agreement on the empirical formula, $C_4H_4O_4$ (representing four carbon, four hydrogen, and four oxygen atoms), there was complete disarray about how to represe nt the components and what the various functional parts might be.

Following Kekulé's ideas, a graphic method of presenting the complex structures of organic molecules based on the number of bonds that each atom could form was worked out.[3] In 1866 the system of representing links by lines appeared and continues to be the standard method to the present day. Although models of molecules had been built prior to the rise of structural theories, the clarity of the new system encouraged model building, which had important consequences for a number of research programs, most spectacularly with the race for the structure of DNA. Although the psychological shift was subtle, the change from the old alchemical concept of transforming matter to the new chemical concept of constructing molecules made chemistry more amenable to research on commercially useful compounds.

The triumph of structuralist ideas was based not only on rationalizing the nomenclature of organic compounds, however. Structural illustrations made clear that compounds with the same empirical formula (that is, with the same number and kinds of atoms) could have very different structures, and further, that certain classes of molecules differed only by small shifts of location of atoms in a larger structure.

One of the greatest breakthroughs that came out of the structural approach was Kekulé's discovery of the benzene ring. Benzene and related substances, collectively known as aromatics because of their distinctive and sometimes pungent odours, were first noted by Michael Faraday in 1825. They were found in

3. The fundamental organic components are carbon, oxygen, hydrogen, and nitrogen, with a host of others such as phosphorous, sulphur, and iron playing important roles in living cells. The mnemonic HONC lists the atoms by number of bonds (H-1, O-2, N-3, C-4).

coal tar, a by-product of coke production, and had interesting properties such as the ability to combine with a wide
range of other organic compounds. Benzene also presented a mystery, since its empirical formula C_6H_6 seemed
to leave two leftover links that were not filled by hydrogen
atoms. If the valence rules were correct, there were either
two missing hydrogen atoms, or four carbon atoms in the
molecule had valences of five. Either possibility would
require a rewriting of the definitions of atomic bonding
and elements.

7.13 BENZENE RING

According to Kekulé's own account, he found a
solution to the problem while relaxing by the fire. He imagined atoms in long rows
and moving like snakes. One of the snakes seized its own tail in its mouth. Kekulé
used this image to work out a hexagonal model for benzene with double bonds
for pairs of carbon atoms and single bonds between the pairs. (See figure 7.13.)
Although this story may have been a *post hoc* explanation (some historians
have questioned Kekulé's account), it has become one of the icons of the history
of chemistry.

Chemistry, Industry, and the State: The Creation of Synthetic Materials

By the beginning of the second half of the nineteenth century, German chemical
industries were turning to research like Kekulé's to improve production and create
new products, since the German states were far less well-endowed with domestic
natural resources than France or Russia and had far less access to colonial
resources than Britain. Although unification in 1871 improved some aspects of
domestic access to resources, Germany was very late entering the colonial game.
This had major consequences later, as Germany embarked on a series of wars to
rectify the situation. In the meantime lack of natural resources was a major
impetus to German interest in science education and research.

In 1837 the German chemist Justus von Liebig (1803–73), one of the most
influential scientists of the era, told the British Association that Britain was no
longer a leader in science. When he toured that country in the early 1840s lecturing
on various topics, he drove the point home by describing the new work being done
in chemistry in Germany and the increasing importance of chemical education,

especially laboratory-based training. Thus, the rise of German chemistry and the shift in the nature and locale of research did not go completely unnoticed in Britain. Liebig's own teaching laboratory at Giessen was a model of chemical education that combined intellectual training with practical laboratory work. It involved students in real research projects, giving them more than a technical overview of existing work.

Because there was no comparable chemical instruction at any institution in Britain, a group of prominent people, headed by Albert, the Prince Consort, and including Sir James Clark (the Queen's physician), Michael Faraday, and Prime Minister Sir Robert Peel, contributed to a fund to establish the Royal College of Chemistry. It opened its doors in 1845, under the leadership of August von Hofmann (1818–92), who had been recommended by Liebig and personally persuaded by the Prince Consort to come to Britain from Germany. Although the Royal College of Chemistry was a small step toward reviving science leadership in Britain, there was no British equivalent of the French École Polytechnique or École Normale Supérieure. Even British universities lagged far behind continental counterparts such as the University of Göttingen in the development of science programs.

The Royal College of Chemistry had as its principal aim the training of pure researchers, copying the high end of intellectual activity at continental schools, but did little to encourage any integration of research with application. Applied chemistry was, if not exactly discouraged, looked down upon as a rather second-class use of talent. It was certainly the case that many students and working chemists felt that any excursion into applied—or worse, commercial—chemistry could end a research career. Against this background, William Perkin (1838–1907) entered the College at the age of 15. Perkin's work opened the door to a whole new era in chemistry, but it was not the British who saw its importance.

High on the list of natural materials that many were attempting to synthesize was quinine, which, as we saw earlier in this chapter (see p. 209), was the only known treatment for malaria and was produced only from the bark of the cinchona tree, native to South America and difficult to cultivate elsewhere. By 1852 the East India Company alone was spending about £100,000 annually on quinine.

Hofmann believed that naphthalidine could be converted to quinine because it shared some of the same basic components. Naphthalidine was readily available because one of its components, naphtha, was a by-product of the production of coal gas, which was widely used for lighting and heat and was made by heating coal in the absence of air. The captured gas was about 50 per cent hydrogen and

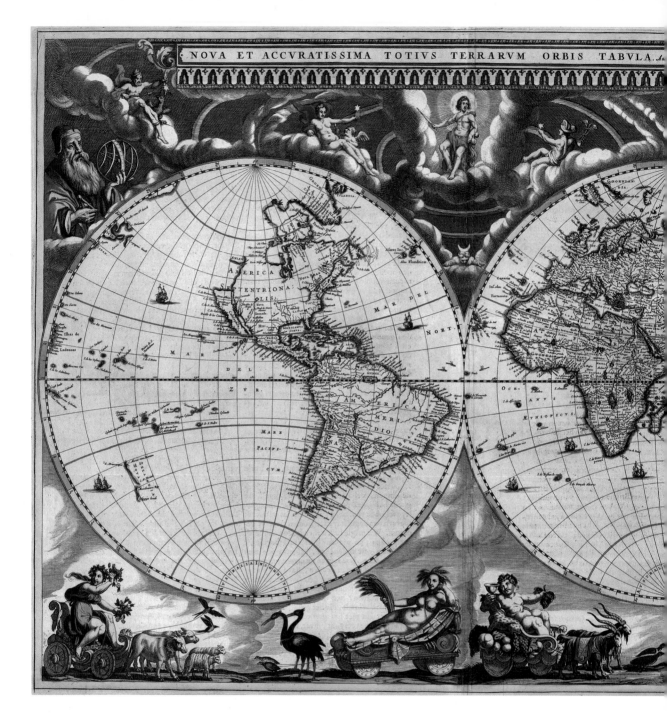

PLATE 2: JOHANNES BLAEU'S WORLD MAP, C. 1664

© Corbis

PLATE 3: HOLBEIN'S *THE AMBASSADORS* (1533)

National Gallery, London, UK / Bridgeman Images.

PLATE 4: "AN EXPERIMENT ON A BIRD IN AN AIR PUMP," JOSEPH WRIGHT OF DERBY (1768)

National Gallery, London, UK / Bridgeman Images.

PLATE 5: STEAM ENGINE
AND DYNAMO (1907)

PLATE 6: WATSON AND CRICK
AND THE DOUBLE HELIX

A. Barrington Brown /
Science Source

PLATE 7: JODRELL BANK
TELESCOPE

Andrew Barker / Shutterstock.com

PLATE 8: EAGLE NEBULA (NASA)

Appearing like a winged fairy-tale creature poised on a pedestal, this object is actually a billowing tower of cold gas and dust rising in the Eagle Nebula. The soaring pillar is 9.5 light-years or about 57 trillion miles high, about twice the distance from our sun to the nearest star. Stars in the Eagle Nebula are born in clouds of cold hydrogen that reside in chaotic neighbourhoods, where energy from young stars sculpts fantasy-like landscapes in the gas. The tower may be a giant incubator for those newborn stars. A torrent of ultraviolent light from a band of massive, hot, young stars (off the top of the image) is eroding the pillar.

Yury Dmitrienko / Shutterstock.com

35 per cent methane, with a mixture of other gases making up the remainder. What was left of the coal after the process of gas extraction was coke, which was used as a solid fuel, and a viscous brown-black tar. Coal tar was a rich source of over 200 organic compounds, including benzene, naphthalene, and toluene, which were useful in both research and commercial applications.

Perkin began his work on quinine in 1856. He created a small laboratory in his home and attempted to follow Hofmann's idea about the conversion of naphthalidine to quinine, but instead of quinine, the experiment produced sludge. However, he noticed that a rag used to clean up some spills of the failed experiment was stained an intense purple. Rather than seeing his experiment as a failure, he proceeded to purify and test this unknown substance, which turned out to be artificial aniline, the basic colouring component of the indigo plant. The intense colour led Perkin to believe that he had created a product with commercial potential. With funding from his family, he established the first artificial dye plant, producing a line of mauve and purples. His endeavour was not an immediate success, however. The dyeing industries, which used materials from natural sources—plants such as indigo, madder, and woad, as well as insects and molluscs—and methods that in some cases could be traced back to Greek or even Egyptian times, were not particularly interested in innovation.

There were, however, many problems with natural-source dyes. They varied enormously in quality, and the plant-based colours changed each year depending on growing conditions. Many of the sources of dyestuffs were also produced outside Europe, increasing cost and decreasing reliable access. These problems were felt most directly in Britain, whose dominance of international trade was based in large part on the textile industry. Perkin's aniline dye came from domestic sources and was much more uniform in quality. Perkin had fortunate timing, too, because, shortly after he started production of dye, Queen Victoria chose mauve as her gown colour for the marriage in 1858 of her daughter, the Princess Royal Victoria, to Prince Frederick William of Prussia. Overnight, mauve was the most desirable colour in fashion. Although the colours in the Queen's outfits were not aniline, the initial reluctance of dyers to buy the new product was overcome by the demand that suddenly outstripped the supply of natural-source dyes.

If Perkin's work had been concerned only with bringing more intense, uniform, and lasting colour to the world, it would have been a significant, if minor, scientific achievement. Our world is full of colours that are the chemical descendants of his work. Everything from colour photography to clothing to the finish on a new automobile can be traced back to Perkin's discovery and commercial development. However, the Perkin Medal, named after and first awarded to him in 1906 by the

7.14 CHEMICAL DESCENDANTS AND RELATIVES
OF ANILINE DYES (A BRIEF SAMPLE)

DERIVATIVES OF COAL TAR	
Dyes	Mauve (anilines), Picric, Alizarin
Pharmacology	Aspirin, Codeine, Quinine
Chemical weapons	Chlorpicrin, Brombenzylcyanide
Artificial flavours and scents	Vanillin, Coumarin
Explosives	Toluenes such as Trinitrotoluene (TNT)
Plastics	Bakelite
Biology	Methylene Blue Cell Stain

Society of Chemical Industry (American Section), was not dedicated to him because he was a commercial success. It was created in recognition of his effect on modern chemistry, especially organic chemistry.

At a monetary level, the commercial success of aniline, and later alizarin, dyes was a dramatic demonstration that research could have real applications. A chemist could get rich by taking a discovery out of the laboratory and into the marketplace. This fact was not lost on many young scientists. A number of the world's leading chemical companies were formed because of artificial dyes. Farbenfabrik vormals Friedrich Bayer (now known simply as Bayer) was founded by Friedrich Bayer (1825–80) and J. Weskott (1821–76) in 1863 to produce fuchsine and other dye stuffs. The Aktiengesellschaft für Anilinfabrikation, better known today as AGFA, was organized by Paul Mendelssohn-Bartholdy and Carl Alexander Martius in 1867. Martius had worked for Hofmann in London at the Royal College of Chemistry; he helped to entice Hofmann to return to Germany in 1865 with the promise of more funding and new laboratories.

As economically successful as the dye industry might be, Perkin's work opened the door to organic synthesis. The tools that made aniline dye possible were the very tools that made almost all modern organic products possible. Thus, the economic impact of the exploitation of coal-tar research was enormous.

By the time Perkin retired from the dye industry at the very young age of 36, the relationship of chemical research and commercial exploitation had been completely altered. Although it would be unfair to say that this was solely the result of his discovery of mauve dye, his work was both crucial and iconic.

It is also somewhat ironic that Hofmann, who had been Perkin's teacher, could not be kept in Britain to continue to foster the next generation of chemists but moved back to Germany. By 1878 the value of coal-tar production in Britain was worth £450,000, but in Germany its value was the equivalent of over £2 million annually. Seventeen artificial-dye factories were operating in Germany compared to only six in Britain. This created a positive feedback loop. Because there were many more jobs for chemists in Germany, chemistry was an attractive educational and career path. In turn, the number of chemists extended the range of research being carried out, which led to new products and a growing demand for chemists to manage the expanding lines of chemical production. Chemical production in turn meant industrial plants that needed engineers, builders, and technicians. The new chemical processes called for new technology in steel production and parts manufacturing, which could be made available for other industries. The best and brightest chemists were guided into pure research, with the next tier filling out the ranks of instructors, and the rest being absorbed by industry. By 1897 Germany had more than 4,000 chemists in non-academic positions, while British industry employed fewer than 1,000.

Conclusion

By the end of the century, Britain was still the most powerful nation on Earth, but its position was increasingly being challenged. In the great game of the colonial era Britain had the best colonies and controlled the seas, but Germany created a scientific and industrial powerhouse and was getting ready to use it. In the conflict that was to come, Germany turned to its scientists, especially its chemists, to overcome its disadvantages. The world of polite amateur gentleman scientists and upper-class academicians was ripped apart. Science was moving from understanding the world to mastering it, and few scientists working in the nineteenth century had any idea just how brutal scientific utility could be.

Essay Questions

1. How did the catastrophists and uniformitarians explain the origins of the Earth?

2. What was Darwin's theory of evolution by natural selection?

3. What were the roots of social Darwinism?

4. How and why did science become professionalized in the nineteenth century?

5. How did Mendeleev bring organization to chemistry?

CHAPTER TIMELINE

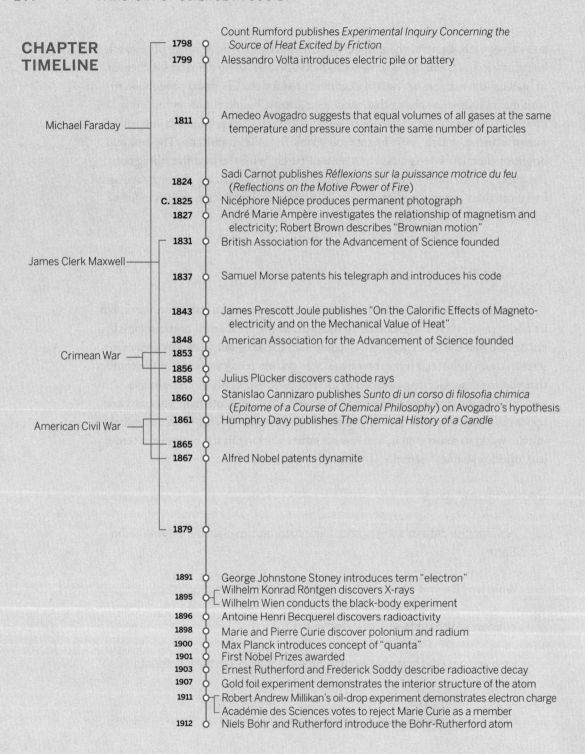

1798 Count Rumford publishes *Experimental Inquiry Concerning the Source of Heat Excited by Friction*

1799 Alessandro Volta introduces electric pile or battery

Michael Faraday — **1811** Amedeo Avogadro suggests that equal volumes of all gases at the same temperature and pressure contain the same number of particles

1824 Sadi Carnot publishes *Réflexions sur la puissance motrice du feu* (*Reflections on the Motive Power of Fire*)

C. 1825 Nicéphore Niépce produces permanent photograph

1827 André Marie Ampère investigates the relationship of magnetism and electricity; Robert Brown describes "Brownian motion"

1831 British Association for the Advancement of Science founded

James Clerk Maxwell — **1837** Samuel Morse patents his telegraph and introduces his code

1843 James Prescott Joule publishes "On the Calorific Effects of Magneto-electricity and on the Mechanical Value of Heat"

1848 American Association for the Advancement of Science founded

Crimean War — **1853**

1856

1858 Julius Plücker discovers cathode rays

1860 Stanislao Cannizaro publishes *Sunto di un corso di filosofia chimica* (*Epitome of a Course of Chemical Philosophy*) on Avogadro's hypothesis

American Civil War — **1861** Humphry Davy publishes *The Chemical History of a Candle*

1865

1867 Alfred Nobel patents dynamite

1879

1891 George Johnstone Stoney introduces term "electron"

1895 Wilhelm Konrad Röntgen discovers X-rays
Wilhelm Wien conducts the black-body experiment

1896 Antoine Henri Becquerel discovers radioactivity

1898 Marie and Pierre Curie discover polonium and radium

1900 Max Planck introduces concept of "quanta"

1901 First Nobel Prizes awarded

1903 Ernest Rutherford and Frederick Soddy describe radioactive decay

1907 Gold foil experiment demonstrates the interior structure of the atom

1911 Robert Andrew Millikan's oil-drop experiment demonstrates electron charge
Académie des Sciences votes to reject Marie Curie as a member

1912 Niels Bohr and Rutherford introduce the Bohr-Rutherford atom

ENTERING THE
ATOMIC AGE

8

Although we often associate the term "atomic" with the building of the first nuclear weapons in the middle of the twentieth century, it was actually during the nineteenth century that the atom became the focus of concentrated research. In the process, the practice of science and the relationship between science and the larger society was irrevocably changed.

In 1815 the world was Newtonian. Revolution, conquest, and war might disturb the realm of human life, but the universe swept on, a serene and inevitable clockwork governed by the laws of mass, motion, and gravity. The impact of Newtonianism stretched far beyond physics, as the Enlightenment thinkers had demonstrated, and Newton's physics itself survived over a century almost completely unaltered. Work in the physical sciences at the start of the nineteenth century tended to examine either those aspects of nature that Newton had conquered, in effect fine-tuning the Master's work, or applying Newtonian principles to those subjects he had not investigated, such as thermodynamics and electricity. The objective of most researchers was to add the new material to the cohesive system of the Newtonian world view. As this work progressed, new tools became available, and with the new information they provided, new theoretical constructs were needed to account for the discoveries being made, thus challenging Newton's dominance. Two powerful lines of inquiry coming from these new investigations were the study of energy (characterized by a view of the universe based on waves and fields of force) and the study of matter (built on a corpuscular view of nature). These two approaches seemed to propose alternative pictures of the universe. As scientists looked ever more closely at the two subjects, what had seemed like incompatible views began to intersect, eventually coming together when Einstein's work revealed the interconnection of matter and energy.

Before that insight could be reached, however, science had to be transformed. Personal patronage was increasingly replaced by institutional funding. Science itself became more specialized, with disciplines splitting into subdisciplines; for example, chemistry was divided into organic and inorganic branches. In this era the role of the scientist became more professional, and a profusion of educational and research institutions were created, particularly in Britain, France, and Germany. The nineteenth century was also a turning point for the speed at which scientific discoveries were turned to utilitarian ends. Electricity and magnetism, for example, went from scientific objects of study in the eighteenth century to industrial applications with telegraphy and commercial electrical generation by the end of the nineteenth century.

Mastering Electricity

Although it is certain that Newton was aware of electrical phenomena, if only from reading William Gilbert, it was one of the topics untouched in his works. Investigations of electricity in the eighteenth century had been hampered by the difficulty of controlling and generating electricity. The invention in 1799 of the electric pile or battery by Alessandro Volta (1745–1827) gave scientists a consistent, controllable, and quantifiable flow of electricity and made it both an object of study and a new laboratory tool. In the spring of 1820 Hans Christian Oersted (1777–1851) put this tool to work, performing a demonstration in his home on electrical heating in a wire carrying a current. He also planned to do a demonstration on magnetism and had a compass set up nearby. During the heating experiment he noticed that the compass needle moved when current was applied to the wire. Although earlier natural philosophers such as Gilbert had seen both phenomena as similar unseen forces acting at a distance, this observation was the first experimental indication that electricity and magnetism were related. When the effect was demonstrated at the Académie des Sciences in Paris in September, it was observed by André Marie Ampère (1775–1836). Ampère had wide interests in chemistry, physics, psychology, and mathematics and had been elected to the Académie in 1814 in the mathematical section. When he saw Oersted's demonstration, he recognized its importance and turned to the study of the interaction of electricity and magnetic fields. The product was the 1827 paper "Mémoire sur la théorie mathématique des phénomènes électrodynamique uniquement déduite de l'expérience," which presented both experimental and mathematical demonstrations of electrical and magnetic behaviour, including the inverse square law for magnetic action. Like gravity, the attraction of a magnet decreased by an amount proportional to the square of the distance from the magnet. Ampère's work put the study of electricity and magnetism on a solid mathematical foundation. In the same year Georg Simon Ohm (1789–1854) added resistance to the explanation. In *Die galvanische Kette, mathematisch bearbeitet (The Galvanic Circuit Investigated)*, Ohm presented his law, stating that the electromotive force was equal to the current multiplied by the total of the resistance of the circuit. In this formulation, Ohm offered a way to quantify the use of electricity by components (devices such as electromagnets and even wire) in a circuit. Ohm's original formulation was presented as:

$$x = \frac{a}{b+l'} \quad \text{which we now write as} \quad I = \frac{V}{R}$$

where I' is the current through the conductor in amperes, V is the potential difference across the conductor in volts, and R is the resistance of the conductor in ohms.

The relationship between electricity and magnetism led a number of researchers to construct electromagnets by wrapping wire around an iron core. When a current was run through the wire, the iron was polarized and produced a magnetic field. Since the wire had to be insulated from the core and from itself (or it would short circuit), and since wire did not come with an insulating layer, most early electromagnets contained few loops so that the wire would not touch itself, and varnish was used to insulate the wire from the iron core. Joseph Henry (1797–1878), working in Albany, New York, created a more powerful magnet in 1827 by covering wire with silk thread and winding 35 feet of this silk-covered wire around a horseshoe-shaped iron core. This was a very successful experiment, and Henry went on to create more and more powerful magnets, one of which could be used to lift over 2,000 pounds. While these magnets clearly had utility, his other work on electromagnets would have more far-reaching implications. In 1831 he created an oscillating device that used two magnetic coils to attract and repel a bar pivoted like a seesaw. (See figure 8.1.) From this came the telegraph, a crucial tool for the maintenance of both national and business empires and the first step toward the electrical and electronics industries.

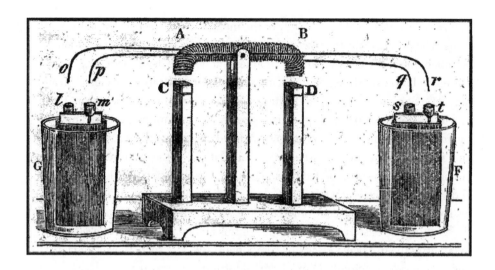

8.1 HENRY'S ELECTRIC MOTOR FROM *AMERICAN JOURNAL OF SCIENCE* (1831)

Henry went on to become Professor of Natural Philosophy at the College of New Jersey (which later became Princeton University). In 1846 he became the first Secretary of the newly founded Smithsonian Institution and later became President of the American Association for the Advancement of Science (1849–50) and President of the National Academy of Science (1868–78).

Henry also discovered that the relationship between magnetism and electricity worked both ways through electromagnetic induction; from this finding he created the first dynamo in 1830. By moving a wire through a magnetic field, a current was created in the wire. Although he was likely the first to create a dynamo, Henry did not publish his results, and so the scientific credit for the device went to Michael Faraday (1791–1867) instead.

Faraday's path to science was an unusual one. The son of a blacksmith, his early education was limited, but he was apprenticed to a bookbinder and read voraciously. After reading the *Encyclopaedia Britannica* article on electricity, he devoted himself to science. Faraday attended several public lectures given by Sir Humphry Davy (1778–1829) and corresponded with him. In 1812 Davy was temporarily blinded in a laboratory accident and hired Faraday as his assistant, the beginning of the younger man's scientific work in chemistry. Faraday had some notable successes as a chemist, including the discovery of what he called bicarburet of hydrogen, known today as benzene. His most popular chemical work was *The Chemical History of a Candle*, a series of six lectures for young people that was later published as a book in 1861.

Despite his success as a chemist, around 1821 he shifted his work more and more away from matter and toward forces. He was convinced that electricity, magnetism, light, heat, and chemical affinity were all aspects of the same phenomenon and that this phenomenon, rather than being based on some kind of fluid movement, was really a form of vibration. Just like a plucked violin string will cause a matching string to vibrate at a distance, Faraday reasoned that an electric "vibration" should be detectable. He took an iron ring and wrapped one coil of wire on one side to be connected to a voltaic pile and a matching coil of wire on the opposite side leading to a compass that would show the presence of a magnetic field. When the current was applied, the compass needle moved, demonstrating the principle of induction. (See figure 8.2.)

Faraday then demonstrated that the relationship between electricity and magnetism was dynamic by spinning a copper disc between the ends of a horseshoe magnet. When the disc was in motion, a current was created, but when it was at rest, no electricity was generated. It was only by passing through the field of the magnet that anything happened. In other words, the real energy of the magnet was in the space around it, not within the magnet itself, which only concentrated the forces. Faraday argued from these experiments that electricity and magnetism could be understood as a kind of strain (somewhat analogous to squeezing a spring) that affected the structure of matter. He found confirmation for this idea in an experiment in which he passed polarized light through a strong magnetic field; the plane of polarization was rotated, indicating that the light had been moved by the field. As he had thought, light was one element of the electromagnetic phenomena. Yet his explanation of fields of force was not accepted by his contemporaries, partly because in the increasingly quantitative orientation of physics, Faraday's work was not elucidated in clear mathematical terms.

That task was taken up by James Clerk Maxwell (1831–79). Maxwell was not an early prodigy like Newton, but his insight and mathematical skills led to recognition of his talents by the time he graduated in 1854 from Trinity

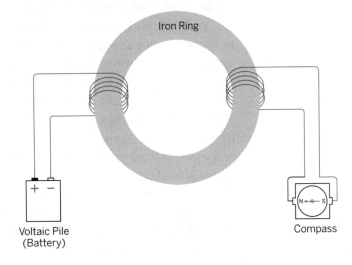

Iron Ring

Voltaic Pile (Battery)

Compass

8.2 FARADAY'S IRON RING

College, Cambridge, the same college Newton had attended. He went on to a series of senior academic appointments at Cambridge, ultimately becoming the first Cavendish Professor of Experimental Physics in 1871, when he was responsible for the design of the new Cavendish Laboratory. Maxwell was not the first choice for the position, but both Lord Kelvin and Lord Rayleigh (John William Strutt, 1842–1919) turned it down. The Cavendish Laboratory was named for the Cavendish family, who largely financed the original construction. Henry Cavendish (1731–1810) had worked on a range of topics such as the nature of airs (among other things determining that water was a compound) and measuring the density of the Earth. He had also worked on electricity, but he did not publish. The Cavendish, a teaching laboratory in physics, was a key component in the development of a new course of studies known as the Natural Sciences Tripos, begun in 1851 at Cambridge to train more scientists to compete with the French and Germans. The Cavendish continues to operate today as one of the great centres of teaching and research.

Maxwell had the profound ability to create mathematical models of physical phenomena. He took Faraday's field theory and its elegant experiments and set them on a sound mathematical footing. Because he argued that a disturbance in a field was propagated through it at a particular velocity, he was able to calculate the velocity that an electromagnetic wave travelled, which was 3.1×10^{10} centimetres per second or about 300,000 kilometres per second, equal to the experimentally determined speed of light. Maxwell concluded that this could not be a coincidence. He predicted, but did not live to see it proven, that the electromagnetic spectrum would be observed at frequencies far below and far above those of light.

Maxwell's work set the course for many aspects of physics by tying several problems together. It not only linked electricity, magnetism, and light but linked them as fields and waves in fields. This undermined the Newtonian view of light as particles but neatly accorded with experimental evidence for the wave nature of light that had been building since the work of Huygens and Thomas Young. Thus, the big questions of physics seemed to be solved in a tidy fashion. Because of field theory, the universe could be viewed as a great cauldron filled with waves that propagated through a celestial fluid known as ether in which particles existed and floated around under the influence of gravity. If the system was a bit more complex than Newton had laid out, it seemed to work smoothly and was still philosophically compact; that is, the universe was governed by a small number of fixed laws that could be expressed in absolute mathematical terms.

By the time of Maxwell's death in 1879 both the scientific and the industrialized worlds were very different than at the end of the Napoleonic age. It was now the era of the inventor. Steam power, telegraphy, photography, printing, and smelting all introduced new technologies and transformed the social, economic, and political lives of people around the world. The material world in the industrialized countries changed from handmade or small industry products to mass production and global distribution of consumer and industrial goods. (See plate 5 for a photograph of a steam engine and dynamo from 1907.)

Electricity was one of the most important factors in the transformation of European and North American society. It moved from being a curiosity for the intellectuals in the French and American salons to an industrial tool. One of its first commercial applications was the telegraph. Wheatstone and Cooke's first telegraph in 1831 moved an arrow to point to letters, although the patent and their priority were challenged. In 1833 the mathematician Karl Gauss (1777–1855) and Wilhelm Weber (1804–91) built a model telegraph system that sent a signal over a distance of two kilometres. Telegraphy was made commercially viable when the first commercial electrical generator was marketed in 1834. In 1837 Samuel Morse (1791–1872) patented his version of the telegraph and his code system of dots and dashes. In 1844 Washington was linked to Baltimore by telegraph; London and Paris were linked in 1854; and the first Atlantic cable was laid in 1858. Telegraph lines followed the railways across the globe, and New York was connected to San Francisco in 1861. When the American Civil War began, news of the horrific events were circulated wherever the telegraph went. Telegrams sent by reporters were printed by newspapers using steam-driven presses and offered coverage of events around the world in hours rather than days or weeks.

The Nature of Electricity and the Science of Thermodynamics

The speed of development of electrical devices by inventors all over the world was astounding, but even as electricity was being turned into a commercial product, its nature continued to be a mystery. Faraday and Maxwell had treated electricity as a vibration or field of force, and the generation of electricity by the interaction of conductive wire and magnetic fields corroborated this view. On the other hand, the generation of electricity from chemical sources (as in batteries) suggested that there was a material basis to electricity. Chemists were concerned about electrolysis (including chemical reactions via application of electrical currents), and that led to theories about electro-positive and electro-negative activity. For example, positive sodium reacted with negative chlorine to form sodium chloride. Although ideas about electrolysis helped scientists to understand new aspects of atomic and molecular interaction and focused the zone of electrical activity, it did not solve the underlying questions about the origin and nature of electricity.

Part of the solution was far from direct and was one of the most debated hypotheses in modern science, being rejected and revived several times. In 1811 Amedeo Avogadro (1776–1856) proposed in the *Journal de physique* the hypothesis that equal volumes of all gases at the same temperature and pressure contained the same number of particles. This was a resolution of problems presented by Joseph Gay-Lussac's (1778–1850) work on laws of gases and by Dalton's work concerning the combination of gases. Scientists knew that one volume of oxygen combined with two volumes of hydrogen to produce two volumes of water vapour. Avogadro reasoned that this could not be unless the oxygen was actually a molecule that broke up into two atoms (or "half-molecules" as he called them) producing two water molecules for every original oxygen molecule (what is now often called "diatomic" oxygen or O_2). In its simplest form, Avogadro's hypothesis posited a fixed number of atoms or molecules in a special quantity. This quantity is now referred to as a "mole" and is the amount of a substance with a mass in grams equal to its atomic or molecular weight. Determining this basic bit of information was essential to understanding the composition of materials, but it was also a key to understanding how materials formed. For the material world to function consistently, each molecule of a particular substance had to be formed of the same elements in the same proportions, but it was impossible to understand how the elements linked together to make molecules if the precise number of each element in a molecule was unknown. Figuring out the proportions provided a tool for figuring out why some elements such as oxygen and carbon combined to make many compounds, while others such as gold and silver combined with very few other elements to form molecules.

Avogadro's hypothesis was not favourably received, because it depended on particles of the same kind being attracted to each other, which ran against the affinity theory of compounds based on the attraction of dissimilar particles (like the North and South poles of magnets attracting, while North to North repelled). Ampère attempted to revive Avogadro's idea in 1814, as did August Laurent (1807–53) and Charles Frédéric Gerhardt (1816–56) in organic chemistry in the 1840s, but it was not until 1860 when Stanislao Cannizzaro (1826–1910) circulated a pamphlet, *Sunto di un corso di filosofia chimica* (*Epitome of a Course of Chemical Philosophy*), at the Karlsruhe Congress that Avogadro's hypothesis began to affect thinking about the solution to the problem of determining and comparing molecular and atomic weight.[1] The Karlsruhe Congress was the first international chemistry conference, and it was attended by some of the most important scientists of the day. The meeting had been called to try to develop international standards for nomenclature and to do something about the problem of atomic weights. Avogadro's hypothesis offered a way to deal with the problem of atomic weights that everyone would eventually accept.

What at first seemed to be a matter problem became linked to electricity because matter can have an electrical charge and that charge appears to have something to do with keeping different elements together in compounds. Avogadro's hypothesis was much better understood by the time Svante August Arrhenius (1859–1927) arrived in Stockholm in 1881 to work on solutions and electrolytes under Eric Edlund (1818–88). According to Arrhenius, when a current was passed through molten sodium chloride (NaCl), the molecule broke apart or dissociated. Its parts were not atoms, but what he called ions; the sodium had a positive charge (Na^+) and the chlorine a negative charge (Cl^-). These ions migrated to the electrical poles, sodium to the cathode and chlorine to the anode, where they lost their charge: the result was atomic (or elemental) sodium and atomic chlorine.

This suggested that atoms and groups of atoms might themselves carry an electrical charge rather than merely being affected by electrical activity. This raised as many questions as it answered, because it seemed to require the combination of two incompatible objects—Daltonian atoms and Maxwellian electric waves. One solution was to picture electricity as a kind of particle, or indivisible unit. The Irish physicist George Johnstone Stoney (1826–1911) calculated the magnitude of what he initially called the "atom of electricity," and in 1891 he proposed the term "electron" for this unit of electrical charge.

While some scientists were picturing electricity as a kind of particle, others tried to solve the problem by recreating atoms. Cartesian particles, Newtonian corpuscles, or Daltonian atoms were all small, discrete, and indivisible particles that interacted in specific, if somewhat unclear, ways. While this was a useful concept for those studying gross matter, it presented problems for those studying magnetism, light waves, heat, and so on. A different model of the atom based on a vortex was developed by William Macquorn Rankine (1820–72) in 1849 and also by Lord Kelvin in 1867. Kelvin argued that the vortex atom was perfectly elastic and that a number of important conditions, such as thermal expansion and spectral lines, could be derived from this kinetic model.

---

1. The term weight is actually incorrect, since what was being measured was the mass of the particles, but by convention molecular and atomic mass continue to be called molecular weight and atomic weight.

Kinetic Theory of Gases and Thermodynamics

These new ideas originated in work on heat. Heat, or thermodynamics as it was later called, was another one of the areas that Newton had not addressed. In the Aristotelian system heat was an element. Through the eighteenth century heat was considered to be the essence of fire or phlogiston, until Lavoisier disproved the phlogiston theory with his careful experiments. With the help of Laplace, he introduced caloric, which was an "imponderable fluid." The idea of caloric did help to explain the apparent movement of heat and the difference between heat and temperature. Caloric was thought to flow from warm areas (full of caloric) to cool areas (empty of caloric). Heat represented the total volume of caloric in a given body, while temperature measured the concentration of caloric, so that, for example, a lake could have a cooler temperature than a boiling kettle, but it contained far more caloric.

As ingenious as the idea of caloric was, it did not accord well with Newtonian physics, and, by creating a new class of matter, it also went against the trend in the nineteenth century to locate physical properties within objects rather than as additional types of things that affected objects. As early as 1738, Daniel Bernoulli (1700–82) of the famous Swiss mathematics and physics family, had presented a theory of pressure based on Newtonian principles about the movement of atoms. The faster the particles moved, the greater the pressure. The theory was seen as improbable and was mostly ignored. That heat might have something to do with the state of atoms had to be demonstrated another way before it gained attention. The attack on caloric came from Benjamin Thompson, Count Rumford (1753–1814), a man with a wildly varied career that included spying for Britain, service to the Elector of Bavaria (who created him Count Rumford of the Holy Roman Empire), building a better fireplace, marrying Lavoisier's widow (the marriage lasted only one year), and overseeing the production of cannons. Rumford had initially supported the caloric theory of heat and in fact had added "frigorific radiation" (cold rays) to the theory, but it was his observation of cannon-boring equipment that led him to reject caloric. He observed that a dull drill bit, although spinning indefinitely, could not cut through the iron of a cannon but would continue to produce heat. If caloric was a substance, it should eventually be emptied out of the iron, but it was not. In 1798 he published *Experimental Inquiry Concerning the Source of Heat Excited by Friction*. It became a classic paper in physics. Caloric theory, although proven to be wrong, had led to the quantification of heat, and Rumford's work advanced the study of heat by directly linking work (kinetic action) to heat.

In 1824, Nicolas Léonard Sadi Carnot (1796–1832) produced his *Réflexions sur la puissance motrice du feu* (*Reflections on the Motive Power of Fire*), in which he scientifically analyzed the theory of heat engines or devices in which work was done by the movement of heat from one area to another, such as the heating and condensing of steam in a steam engine. Although based on the caloric theory of heat, he demonstrated that work from heat was analogous to water falling from a high point to a lower point. From this, it became possible to calculate the efficiency of heat engines, which was remarkably low. In a steam engine that cooled steam by 100° Centigrade (that is, the steam "fell" from 150° to 50° as it passed through the pistons and condensed as water), the efficiency was only 0.236, which means that less than a quarter of the heat could be converted into work. Carnot's work was brilliant, but he died of cholera at the age of 36, and his work was not widely known until revived by Joule, Kelvin, and Clausius.

The relationship between work and heat was formalized by James Prescott Joule (1818–89) in a paper he read before the British Association for the Advancement of Science in 1843. Entitled "On the Calorific Effects of Magneto-electricity and on the Mechanical Value of Heat," it offered the first quantification of the mechanical equivalent of heat corresponding to a rise in temperature. He argued that to raise the temperature of one pound of water by one degree Fahrenheit required the expenditure of 838 foot/pounds of work. In other words, the amount of work necessary to move a one-pound weight 838 feet would, if converted into heat, raise the temperature of one pound of water by one degree. The paper met with silence from the scientific community, but he persevered and refined his experiments. In 1845 he presented "On the Mechanical Equivalent of Heat" to the BAAS, with his new calculation of 819 foot/pounds. By using a paddlewheel (which produced friction as it moved through a fixed volume of water) driven by a falling weight, he refined his figures in 1850 to 772.692 foot/pounds. (See figure 8.3.) By this time, his work was receiving more favourable attention.

Joule went on to work with Lord Kelvin on the study of heat and helped develop the kinetic theory of gases. That theory would link heat and pressure to the motion of particles in a gas with the claim that the average speed of the particles related directly to the temperature of the gas. Independently, Rudolf Clausius (1822–88) arrived at the

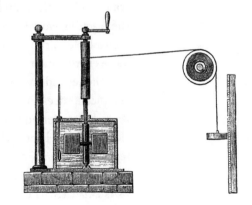

8.3 JOULE'S DIAGRAM OF THE MECHANICAL EQUIVALENT TO HEAT

Harpers New Monthly Magazine No. 231, 1869.

same conclusions about the motion of atoms. He also introduced the term "entropy" in 1865 to describe the dissipation of energy, which always increased over time. With Clausius's work, the kinetic theory of gases—that is, the relationship between the motion of the particles and the heat of the system—was established. Heat was not a mysterious fluid but the motion of atoms and molecules. Further, the motions of all those atoms moving around could be measured, at least statistically, and that meant heat could be used as a key tool for explaining states of matter.

The kinetic theory of gases was based on the existence of atoms and molecules, so it became an integral part of matter theory. In 1865, Joseph Loschmidt (1821–95) applied the theory to calculate the diameter of an atom and used this value to determine the number of molecules in a unit volume of gas. Clausius envisioned the "energy" of a molecule partitioned into contributions from its translational motion (movement from place to place), vibrational motion, and rotational motion. Based on this, he showed that the heat capacities of hydrogen, oxygen, and other elements in the gaseous state were consistent with the theory that their molecules were diatomic, lending support to the vapour density method for determining atomic weights.

The modern presentation of all this complex work can be encapsulated in two general laws that emerged in the 1840s and 1850s. The first law defined energy as a new fundamental concept in the physical sciences. It states that heat is a form of energy and that, in any closed system, the total amount of energy is constant. Thus, it would be impossible to create a machine in the real world that worked continually without an external source of energy, since the energy used to work the machine would be dissipated (by friction, for example) and thus become unavailable to run the machine. Another way of saying this is that you can't get more energy out of a system than you put into it.[2]

The second law of thermodynamics, the entropy law, says that in any physico-chemical process, it is impossible to convert all the energy into work. Some energy is always converted into heat and thus is not available for work. Further, in any closed system, heat will transfer only in one direction, from a warmer to a cooler region. Over time, entropy will make everything the same temperature if there is no external energy source. This is why a thermos bottle both works and doesn't work perfectly. The thermos, by reducing the rate of transfer of heat from the hot

----

2. An example of this problem is trying to cool your kitchen by leaving the refrigerator door open. Since the refrigerator works by taking the heat out of the interior of the refrigerator and radiating it to the surrounding air of the kitchen, you would heat up the air you were trying to cool down. And because the motor that runs the refrigerator would generate more heat from the friction of the moving parts, you would gain even more heat in the room.

interior to the cooler exterior keeps the hot beverage warm longer, but because the barrier is not perfect (nor can it be), the contents of the thermos will eventually cool to the temperature of the surrounding world. Whether it is coffee in a thermos, or the whole universe, entropy applies. It is just a matter of time.

This idea had wider implications than just physics. Kelvin used this theory to estimate the age of the Earth based on the rate of cooling seen in a heated sphere of nickel-iron. This suggested an Earth too young for Darwin's theory of evolution and became a major objection to Darwinian evolution.

Physical Chemistry, Cathode Rays, and X-Rays

Although the kinetic theory of gases bolstered belief in atoms by relating heat to the motion of particles, some scientists were coming to believe that energy—not matter—was the genuine physical reality and the appropriate basis of physical science. An increasing number of scientists argued that atoms did not exist as material objects at all. In addition to Kelvin's vortex model, Maxwell claimed that the actual existence of atoms need not be assumed for mathematical models to work, and Willard Gibbs (1839–1903) worked out his theory of thermodynamics without particles. One of the most outspoken opponents of the idea of atoms was Ernst Mach (1838–1916), who argued not only that atoms as described by atomism were hypothetical and need not exist but that theorizing about them might be misleading. Mach was one of the leading proponents of positivism, a philosophy that argued that the only reliable form of knowledge was scientific, arising from observable or empirical evidence. Since atoms could not be observed and were unnecessary for many aspects of physics, they fell into the category of metaphysics, a branch of philosophy that speculated about the origin and purpose of nature rather than looking at how nature functioned. According to Mach, nothing in science could be based on such speculation.

The anti-materialist idea even had supporters among chemists, with the influential physical chemist Wilhelm Ostwald (1853–1932) in Germany and such notable French chemists as Marcellin Berthelot (1827–1907) and Henri Le Chatelier (1850–1936) also opposed to atomism. They supported the Energetik system, according to which matter had to form a continuum, as all "matter" had to be part of the spectrum of energy. Energetik theory was based on thermodynamics and reinforced by the wave theory of light.

Many supporters of the Energetik theory called themselves physical chemists. In 1887 Ostwald, along with Arrhenius and Jacobus Henricus van 't Hoff (1852–1911), founded the *Zeitschrift für physikalische chemie* (*Journal of Physical Chemistry*), which became the leading journal in this new field. Chemistry and physics were beginning to separate both institutionally and in terms of areas of research. Physical chemistry bridged the gap by attempting to use the concepts of thermodynamics and energy to account for chemical reactivities and various physical behaviours of substances. Although physical chemistry provided some profound insights into the structure of matter and the role of forces, it was not embraced by the whole chemistry community. Some chemists objected because it depended so heavily on quantitative and mathematical analysis that seemed either superfluous or uncertain. Others, particularly those working in organic chemistry, found that the questions asked by physical chemists were simply unnecessary for their work.

The relationship between matter and energy was creating new puzzles. For example, water broke down into oxygen and hydrogen when a current was passed through it, but it was also known that these two elements would dissociate at very high temperatures. Electricity passing through a solid seemed to have no effect on the conductor, while it caused decomposition in liquids. Experiments that passed a current through gases at normal pressure either did not work or required very high voltage, which then resulted in violent sparking or even explosions. If matter was an illusion, then physicists would have to explain how energy, fields of force, and waves interacted to give what seemed to be a material universe. If matter and energy were discrete, how was an electrical charge generated from matter? What was needed to resolve these issues was a new way to study the component parts, and that meant new laboratory tools.

At very low pressure electricity would pass through gases, but it was not until the glassblower Heinrich Geissler (1814–79) devised a method of producing a good vacuum tube that research could be conducted on the phenomenon. In 1858 Julius Plücker (1801–68) noticed that when a current was passed between electrodes inside a vacuum tube, a greenish glow appeared on the cathode (negative terminal). This glow was the same regardless of the metal used for the terminal, so he concluded it was an electrical phenomenon rather than a property of the material of the electrode. Plücker also demonstrated that the luminescence was affected by a magnet placed near it.

Plücker was followed by Johann Wilhelm Hittorf (1824–1914) in Germany and William Crookes (1823–1919) in Britain. In 1869 Hittorf (who had been Plücker's student) demonstrated that whatever was being projected through the vacuum travelled in straight lines and could cast a shadow if there was an obstacle in its path. Hittorf's findings were confirmed by Eugen Goldstein (1850–1930), who named the radiation *Kathodenstrahlen* or cathode rays. Crookes independently

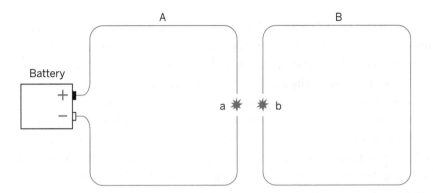

A B

Battery

a ✳ ✳ b

A loop of wire "A" is connected to a battery. A small gap "a" produces a spark as the electricity completes the circuit.

A similar loop of wire "B" with an identical gap "b" is set near loop "A." When a spark is produced at "a," it induces a current in loop "B" and a spark at "b."

8.4 HERTZ'S SPARK GAP EXPERIMENT

found the same results. He also attempted to see if the rays exerted a force by placing a paddlewheel inside the tube. The cathode rays rolled the paddlewheel away from the cathode and toward the anode. From this he argued that the rays were really particles with a definite mass, although it was later shown that it was, in fact, the trace gas molecules that caused the paddlewheel to rotate.

Work with cathode rays did not resolve the problem of atomic composition, however. For physical chemists (particularly in Germany), the behaviour of the rays seemed to confirm the wave nature of the universe, while in Britain Crookes's work was seen as confirming the particle nature of matter.

In Germany, Heinrich Rudolf Hertz (1857–94) continued to explore Maxwell's reasoning about waves. He created an elegant experiment to demonstrate that electromagnetic radiation could be generated, broadcast, and detected at a distance by making electricity jump a small gap in the air, thus producing an electric spark. To detect the waves, he used a rectangle of wire with the two ends almost touching. If electromagnetic radiation created by the original spark passed over the rectangle of receiving wire, it should produce a current and cause a spark to cross the small gap, which it did. (See figure 8.4.) These "Hertzian waves," or radio waves as they became known, validated Maxwell's ideas and extended both the spectrum of electromagnetic radiation and the ability to examine the phenomenon experimentally.

The radiation spectrum was further expanded by the discovery of X-rays by Wilhelm Konrad Röntgen (1845–1923). In 1895 Röntgen, investigating ultraviolet light, used a cathode ray tube and paper coated with barium platinocyanide that fluoresced if exposed to ultraviolet. When he covered the discharge tube in black paper, the coated paper still glowed. Even when placed in an adjoining room, the coated paper glowed when the current ran in the cathode tube. The rays seemed to originate from the glass at the end of the tube where the cathode rays struck. A range of materials were examined to determine the action of the mysterious rays. The denser the material, the less penetration was found. When Röntgen exposed photographic plates to the rays, he discovered that they recorded images. One of the first "röntgenograms" was of his wife's hand and clearly showed the bones and a faint outline of the flesh. The potential medical use for the discovery was immediately apparent, and the first medical use of X-rays took place only a few months after their discovery.

Röntgen's work and the medical application of X-rays would not have been possible without the invention of photography. Nicéphore Niépce (1765–1833) is credited with creating the first permanent photograph around 1825. His system was improved by Louis Daguerre (1787–1851) and others. By 1884 George Eastman (1854–1932) had introduced film photography, which remained the foundation of modern photography until the introduction of digital imaging. Photography is perhaps the best example of a commercial invention that was then brought into the laboratory, where it became a tool for everything from astronomy to cell biology. The use of photography raises an interesting philosophical issue for investigators about the limits of experimental reliability and objectivity, since they did not have access to the chemical composition of the film (a trade secret) and therefore could not control all the variables for their experiments.

Later work by Max von Laue (1879–1960) demonstrated that X-rays had the same fundamental electromagnetic properties as light, but at a much higher frequency, and that crystals would diffract X-rays and produce a consistent pattern. This idea was then used by William Henry Bragg (1862–1942) and his son William Lawrence Bragg (1890–1971) to create the study of crystallography, which mapped the interior structure of crystals by the analysis of X-ray diffraction images. This technique had major consequences in a wide range of fields, particularly as the clue to the structure of DNA. The other development from von Laue's work was the creation by Karl Manne Georg Siegbahn (1886–1978) of X-ray spectroscopy, which allowed a better understanding of the elements and a method of determining the composition of materials by the examination of the electromagnetic spectrum emitted by matter when heated. If a substance was heated, it produced a consistent and characteristic pattern of light. This was useful for identifying elements and would even be used to determine the composition of stars.

Röntgen was showered with awards for his discovery of X-rays, including the Nobel Prize in Physics in 1901, the first year the prizes were awarded. Named after Alfred Bernhard Nobel (1833–96) and his family, the prizes became the most prestigious award in science. Nobel was a self-taught scientist, who was particularly interested in the branch of chemistry dealing with explosives. The foundation of the Nobel fortune was dynamite, patented in 1867, and a range of other explosive products including smokeless powder. Industrial explosives were in high demand in the age of big engineering projects such as canal building, rail construction, and large-scale mining, but they were extremely dangerous. Nobel's products were stable and far more predictable. Military applications for explosives were also on the rise as European military forces expanded rapidly in the age of colonial empires.

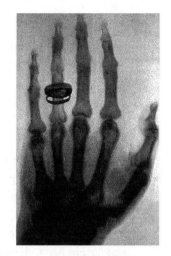

8.5 RÖNTGEN'S X-RAY OF ALFRED VON KOLLIKER'S HAND January 23, 1896.

Nobel left much of his vast fortune from the explosives industry for the creation of a series of prizes in science, peace, and literature. He wanted to reward those whose work conferred the "greatest benefit on mankind." People have speculated that he created the prizes in response to the criticism that his fortune was gained by death and destruction. The Nobel Prizes in physics, chemistry, physiology or medicine, peace, and literature were to be awarded annually and selected by special committees. The Swedish Academy oversaw the literature prize, while the Royal Swedish Academy of Sciences was responsible for physics, chemistry, and physiology/medicine. The peace prize was selected by a committee established by the Norwegian Parliament.[3] Because the prize system required the coordination of several academies and two governments, it took several years after Nobel's death in 1895 to organize.

Although some have argued that the Nobel Prizes are not always given to the best candidates, the science prizes have generally been awarded to people whose discoveries or body of work have had a major effect on the discipline and often on the whole field of science. They also represent a conduit for public exposure to science, since the awards were, and are, a major news event. Winning Nobel Prizes has become a matter of national pride. In addition to the prestige and a gold

3. It is not clear why Nobel gave the responsibility for the peace prize to the Norwegians, although they were (and continue to be) active in the international peace and disarmament movement. The economics prize was created in 1968 by the Bank of Sweden and was not part of Nobel's original plan.

medal, the winner also receives a substantial cash prize. Since 2001, the prize has been set at 10 million kronor or about $970,000 US.

Resolving the Wave/Particle Dilemma

While X-rays were making news, the battle to understand just what was going on with particles and the electromagnetic spectrum continued. In an attempt to better understand the strange particles of the cathode tube, Hertz had shown, by constructing a cathode tube that sent a beam of rays between two electrically charged metal plates, that cathode rays were not deflected from their course when they passed through an electric field. J.J. Thomson (1856–1940) thought that Hertz's conclusion had to be wrong and proved that the rays *were* deflected when passing through an electrical field. He did this by creating a better vacuum and removing gas particles that had interfered with the electrical charge on the cathode rays. The discovery had many consequences: it raised further questions about the nature of the rays, and it also formed the technical foundation for modern consumer electronics as the principle behind CRT televisions and computer monitors.

Although J.J. Thomson seemed to have transformed Hertz's ray into a particle, it was a strange particle that he found. Its path could be deflected by both a magnetic field and an electrical field. The path of the deflection suggested a negative charge on the particle, unlike X-rays which were unaffected by electrical fields. Because the velocity and path of the cathode ray could be controlled by magnetic and electrical fields, and the strength of those fields was known, it was possible to estimate the mass of the strange particle by calculating its ratio of mass to charge. The electron's mass was 1/1836 that of a hydrogen ion, or in modern terms 9.1091×10^{-28} grams. This particle appeared to be Stoney's electron.

While knowing the mass of the electron was a breakthrough, it did not automatically determine the charge of the particle. This was done experimentally in 1911 by Robert Andrew Millikan (1868–1953), who used droplets of oil floating between two oppositely charged plates to measure the electron's charge. By observing the motion of a tiny oil droplet that absorbed ions from the air (he used X-rays to ionize the atmosphere), he could calculate the charge. He reasoned that the charge could have only one size and that was the unit charge of the electron, which was the same for all electrons in all atoms.

While the mass and charge were being figured out, there still remained a troubling question: Where did the particle go after it hit the end of the cathode tube? The fluorescence associated with cathode rays linked the particle to light, but that just brought the question around full circle. Was light a particle or a wave?

Most discoveries about the electromagnetic spectrum seemed to confirm the wave nature of the universe, but there were problems with the wave theory as well. Waves, by definition, had to be propagated through a medium. Just as there is no sound to be heard from a bell rung in a vacuum, if light were a wave it had to have a medium to travel through, or there would be no electromagnetic spectrum to observe. The characteristics of the electromagnetic medium, or the ether as it was called, would have to be very particular, and for some scientists the electromagnetic ether seemed like a throwback to the Cartesian idea of vortex and plenum.

If the supporters of the Energetik theory were right, that all "matter" was simply a manifestation of waves, there remained the question of how waves could manifest themselves as a solid, since waves had the property of passing through each other. In other words, why couldn't a person walk through a wall, if both the person and the wall were not material but composed of waves? Two lines of work not only led to a resolution of the wave/particle problem but also, ironically, unravelled the very foundation of the Newtonian system. The first line came from the discovery of radioactivity and the second from a logical conundrum that was created by the impossibility of both Newton's system of mechanics and wave theory being correct.

In 1895 Antoine Henri Becquerel (1852–1908) began a detailed study of fluorescence and phosphorescence, an area he had been interested in for some time, but which was made more interesting by the work of Röntgen on X-rays. Becquerel wondered if fluorescing materials could in turn produce X-rays or cathode rays. He chose to work with uranium salts because they were known to fluoresce strongly. First, he exposed the crystals to strong sunlight, and then placed them on a sealed photographic plate. When the plate was developed, there was a darkening where the crystals had been, indicating the passage of rays from the uranium salts through the covering paper.

On February 26 and 27, 1896, overcast skies prevented Becquerel from continuing his study, so he left the crystals and photographic plate in a light-proof drawer. Curious to see if there was any residual trace of the rays from the crystals, he developed the plate anyway and discovered that the dark image was just as strong as the image from the salts exposed to light. Whatever had darkened the plate came from the sample, not from something absorbed and later emitted. Closer examination of the uranium showed that the radiation could not be reflected like ordinary light but could affect the electrical charge of objects on which it fell.

At about the same time, Marie Sklodowska Curie (1867–1934) began her work on radioactivity. She used a piezo-electric quartz electrometer that had been invented by her husband Pierre Curie (1859–1906) and his brother Jacques (1856–1914). This tool could measure very small levels of electrical charge, and she used it to identify substances that demonstrated the effects noted by Becquerel. Only uranium and thorium gave off the ionizing radiation, but when she tested a sample of pitchblende, she found that its ionizing power was greater than pure uranium. Pitchblende was for the Curies something like coal tar had been for organic chemists. It was composed largely of an oxide of uranium, U_3O_8, but also contained other components in very small quantities. Uranium was a rare material whose main use was as a colouring agent for glass-making, where it produced a beautiful blue glass. After months of work refining the raw ore or pitchblende, the Curies isolated a new radioactive element that they named *polonium*, after Poland, Marie's homeland. Further refining work produced a second radioactive element, *radium*, which they announced in December 1898. They had processed eight tons of pitchblende in order to produce one gram of radium compound. It is likely that the long-term exposure to radioactive substances led to Marie Curie's death from cancer in 1934.

Becquerel and the Curies were awarded the Nobel Prize for Physics for their work on radioactive substances. Tragedy befell when Pierre was killed in a traffic accident in 1906, but Marie continued on with her work. One of the few women allowed into the male-dominated world of science, she was the first female Nobel winner and the first female professor at the Sorbonne. In 1911 her name was put forward for membership in the Académie des Sciences, but in a highly public and acrimonious fight, she was rejected. Many still believed that women were unsuited for scientific work or that she was riding on her husband's coattails. Not only misogyny but other factors affected the rejection—anti-Semitism, the split in France between conservatives and liberals, and nationalism. Although Marie Curie was Catholic, her name suggested Jewish heritage, and she was caught in the animosity arising from the Dreyfus Affair.[4] Because she was Polish, she also became the target for radical nationalists, and, as a liberal, she was seen as a threat by the conservatives both inside and outside the Académie. When she won a second Nobel Prize that same year in chemistry, her opponents decried it as a political

4. Alfred Dreyfus, a Jewish officer in the French Army, was falsely accused of sending secret information to the Germans and was imprisoned for life. A protest, spearheaded by the novelist Émile Zola, led to a retrial and an indictment of the army's practices and verdict, although Dreyfus accepted a pardon to end his torment. The Affair caused deep fissures in French society and politics.

gesture. Marie Curie remained the only person to receive two Nobel Prizes until 1962 when Linus Pauling won his second (Chemistry 1954, Peace 1962).

Despite the detractors, Marie Curie's work not only was instrumental in understanding radioactivity and discovering new elements but also opened up understanding of the very structure of matter. Becquerel and others recognized that the ionizing radiation was not of a single type. Some radiation could penetrate only a thin layer of metal foil, while some penetrated much deeper. In addition, some of the radiation from radioactive material could be deflected by an electrical field. These different characteristics were identified as different types of rays, labelled alpha and beta rays around 1900, with gamma rays, which had more penetrating power, being identified and named around 1903.

The Atom Deciphered

While the hottest topics in physics seemed to be X-rays, radiation, and radio waves, there was a quiet experimental program studying colloids as a method of demonstrating the physical reality of atoms and matter. There were two reasons why this line of research was overshadowed by other discoveries. The first was that from the beginning colloids were associated with completely mundane things such as glue, paint, and a host of industrial and household products. The second reason was that colloids were more closely associated with organic and biological studies than with the fundamental nature of matter.

Around 1900 the prevailing theory about cell material in plants and animals was based on colloid chemistry. Colloids were identified by Thomas Graham (1805–69), a Scottish chemist and the first president of the Chemical Society of London. He classified a group of materials that would not diffuse through a membrane as colloids (as opposed to crystalloids, which did pass through the membrane). He thought these substances—like mastic, fats, paints, and parts of blood—had certain unique properties such as the ability to form gels. Cell chemistry was based on a belief that these colloids were largely undifferentiated masses with the chemically active parts—the enzymes that controlled cell activity—embedded or attached to the mass. This was called the "träger theory" from the German word for "carrier."

It became significantly easier to study colloids when Richard Zsigmondy (1865–1929) and his assistant H. Siedentopf (1872–1940) constructed the first ultramicroscope in 1903. The ultramicroscope was a marvel of optical engineering and effectively reached the limit of optical observation. It allowed the examination of

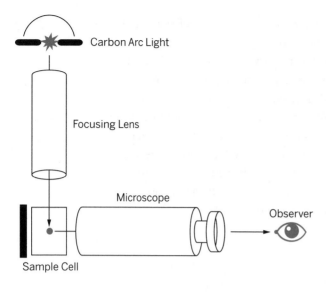

Carbon Arc Light

Focusing Lens

Microscope

Observer

Sample Cell

8.6 ULTRAMICROSCOPE

particles down to 5 millimicrons in diameter (although its general operating range was 20 to 200 millimicrons) by illuminating the object from the side against a dark background, just as one might see floating dust motes in a beam of sunlight. This allowed a quantitative evaluation of colloidal material. (See figure 8.6.)

A further step was taken when Theodor ("The") Svedberg (1884–1971) developed the ultracentrifuge. This device, which spun samples of particles in a fluid medium at up to 100,000 gravities, separated materials too small to be analyzed using other methods. Svedberg used the ultracentrifuge to study organic colloids such as hemoglobin and made a surprising discovery. Rather than finding a range of particle sizes in a sample of hemoglobin (and other such substances), he found that all the particles were the same size. This suggested that rather than being undifferentiated and random clumps of organic material, the chemical components of cells were uniform. These findings were hard to accept at first, because biochemists had trouble believing that a single molecule could be as large as the tests suggested. Hemoglobin, for example, had an atomic weight of 68,000. Biochemists had expected cellular material to be composed of small components and initially argued that the highest molecular weight for a single functional molecule could not be greater than 16,000. The hemoglobin result seemed almost absurd to some scientists, and Svedberg repeated the experiment several times to confirm the results.

Into this debate came Hermann Staudinger (1881–1965). Staudinger had a rebellious streak, having opposed German chemical warfare during World War I, and he was prepared to challenge fellow chemists regardless of their status. He was, therefore, willing to oppose the accepted opinion and champion the idea of large organic molecules. In 1924, he proposed the name "macromolecule" for the large aggregates that were showing up more and more in organic chemistry. He was roundly criticized for suggesting that these colloid compounds could be single functional molecules, and some chemists even suggested poor laboratory practice had led to the error. In 1929, as more work was done on colloids, Staudinger went to the Notgemeinschaft der Deutsche Wissenschaft (the German Research Foundation) and asked for money to purchase an ultracentrifuge. When his request was refused, he turned to other methods to determine the molecular weight of polymers, or long-chain molecules that were constructed from smaller units. He found that viscosity was related to molecular weight and that, by measuring viscosity, he could estimate the weight and size of molecular chains. With simple tools, he predicted a number of molecular weights that were confirmed by other methods. He also chemically altered a macromolecule without changing its molecular weight. While this was not the *experimentum crucis* that changed chemistry in a single moment, it did undermine the colloid theory of organic substances as being indeterminate masses. For the geneticists, the path to the ultimate control system of the cell became much clearer when macromolecules were accepted and their composition laid open to study. As long as organic molecules were thought to be indeterminate or even random aggregations, there was little way to organize or systematically study enzymes, hormones, and, ultimately, DNA.

In 1895 Jean-Baptiste Perrin (1870–1942) used the study of colloids to demonstrate that cathode rays were not waves but charged particles. He did this by showing that the rays could transfer a measurable charge to another object. Flying in the face of the popular Energetik theory, he set out to demonstrate the physical nature of atoms and molecules. Starting around 1908 he created and painstakingly observed, using the ultramicroscope, the behaviour of tiny colloid particles moving through water in a flask. The particles were too small to settle to the bottom of the container and, like dust motes dancing in air, seemed to bounce and zigzag. This random motion was called Brownian motion, since it had been first noted by the botanist Robert Brown (1773–1858) in 1827. Perrin argued that the changes in direction were due to collisions of the particles with molecules of the liquid. Since he knew the mass of the particles he had made, he could calculate the mass and kinetic energy of the water molecules. From that, he could calculate the number of molecules in a given volume of liquid and provide an experimental confirmation of Avogadro's hypothesis. From the hypothesis comes a method for calculating the specific number of particles in a given mass. This number is known as Avogadro's number and is equal to 6.022×10^{23} particles in a mole. In other words, one mole of hydrogen has the same number of atoms as one mole of gold or one mole of hemoglobin molecules. With Avogadro's number, one can figure out how many atoms exist in a given mass of an element or compound.

Perrin published his conclusions in 1909, apparently unaware that his experimental work neatly confirmed the theoretical argument for the physical existence

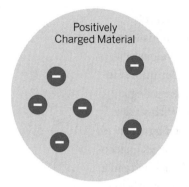

The Raisin Bun Model

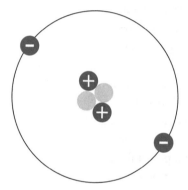

Rutherford's Orbital Model

8.7 RAISIN BUN TO RUTHERFORD ATOM

of atoms and molecules that had been published in 1905 by Albert Einstein (see Chapter 9).

While the work of Perrin and Einstein supported the actual existence of atoms and molecules, there was still the question of what the structure of the atom looked like. Around this time, the general conception of the atom was the "plum pudding" or "raisin bun" model, which pictured the atom as a single positive mass, with negative charges (the "raisins") distributed throughout. The model was reminiscent of the "träger theory." The model was not very elegant, but it fit a number of the requirements that had to exist because of the chemical and physical behaviour of the atom, such as the ability to gain and lose electrical charges and to form bonds. (See figure 8.7.)

Radioactivity disturbed this picture, since radioactive substances seemed to be shooting out bits of the atom. In 1903 Ernest Rutherford (1871–1937) and Frederick Soddy (1877–1956) published an article arguing that radioactive substances were in fact going through a series of transformations rather like a china figurine falling down a set of steps; the pieces that broke off were the alpha and beta rays. This has sometimes been called "modern alchemy" since it was an actual transmutation of matter, but unlike the medieval alchemy, it took rare metals and turned them into base metals—uranium into lead!

By suggesting that atoms were not permanent structures, Rutherford and Soddy undermined the certainty both of the corpuscular theory that had been the foundation of matter theory since the Greeks and of Newtonian corpuscularianism. In fact, the discovery that cathode rays, radioactivity, and X-rays had a material foundation suggested that atoms had sub-parts that existed separately from the whole. To confirm this, Rutherford beamed alpha rays into an evacuated double-walled glass container and found inside the chamber were helium atoms that had lost two electrons. Alpha rays were not really rays but streams of particles.

To study the penetrating power of alpha particles, Rutherford shot them at a thin piece of mica and noted that the resulting beam was much fuzzier than the original. Working with Hans Geiger (1882–1945) in Manchester in 1907, he created a detector that could record the passage of a single particle. Using this detector technique, one of his students, Ernest Marsden (1889–1970) conducted an experiment in which he shot a beam of alpha particles at a strip of gold foil. The foil was thin, only 1/3000 of an inch thick, and most of the alpha particles passed through in a straight line. A few were deflected, and a tiny portion, about one in 8,000, bounced straight back. This was an astonishing result. Rutherford deduced that it meant that the alpha particles were striking a solid core of an atom, while the rest of the atom was essentially empty space. The solid core, which he called the "nucleus," contained almost the entire mass of the atom. Thus an atom was a nucleus with electrons around it.

That the atom was made up of separate parts fit with the discoveries of others, but that it was mostly empty space seemed to some almost impossible. The picture became even more complex when Danish physicist Niels Bohr (1885–1962) came to work with Rutherford in 1912. He demonstrated that the nuclear model could not follow classical rules of physics; otherwise the electrons in an atom would lose energy and spiral down into the core in a fraction of a second. Rather, the electrons orbited in a series of layers, and the electron's orbital momentum was "quantized," or fixed at specific values. Energy was emitted from the atom only when the electrons jumped from one orbital level to another. Further, the amount of energy (the quanta) was fixed—a single "packet" of energy was needed for a jump. (See figure 8.8.)

Here we see the study of atomic structure converging with another line of scientific research, thermodynamics. Bohr could apply this insight to the structure of the atom because Max Planck (1858–1947) had already developed quantum theory out of the study of thermodynamics. A central question in thermodynamics in the late nineteenth century was about the relationship between heat and radiation. Gustav Kirchhoff (1824–87) in 1859 had devised a thought experiment to visualize this relationship. He imagined a "black body" that would absorb all radiation that fell on it from infrared (heat) through visible light and on to higher energy levels such as ultraviolet. This theoretical body also worked in the other direction. If it were heated, it should radiate at all frequencies, glowing through the spectrum. Consider a brick. At 50°C it is hot to the touch, but invisible in a dark room to the human eye. At around 700°C it is just hot enough to glow, while it emits a bright light at about 6,000°C, which is about the temperature of the surface of the sun.

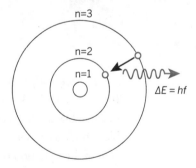

8.8 THE RUTHFORD-BOHR ATOM

A model of the hydrogen atom. Each circle represents the possible orbital zone for an electron, called the atomic shell. If an electron moves from one orbit to another, it gains or loses a packet of energy.

CONNECTIONS

Scientist and Empire: Ernest Rutherford from the Colonies

When the Right Honourable the Lord Rutherford of Nelson, OM FRS, died in 1937, he was one of the most famous and influential scientists in the world. He had won the Nobel Prize for Chemistry in 1908, been knighted in 1914, and won a number of important medals and honours in the years that followed. He served as the President of the Royal Society from 1925 to 1930, and was made a lord in 1931. What makes this doubly remarkable was that Rutherford had a colonial past. He had been born in the small town of Brightwater, New Zealand, and had done his important early work in Canada, far from the centres of intellectual power in Europe. In Rutherford's day, few people with his background were accepted into the first ranks of the British scientific elite.

Rutherford earned his BSc at the University of New Zealand. He conducted research on magnetism and developed a new form of radio receiver. In 1895, he was awarded a fellowship from the Royal Commission for the Exhibition of 1851 to do postdoctoral study at the Cavendish Laboratory at Cambridge University under the supervision of J.J. Thomson, although his doctorate would be from the University of New Zealand. He was one of the first "aliens" (those without a Cambridge degree) to be allowed to do research at the university. Rutherford was greeted with some hostility by scientists already in the lab, but he was encouraged by Thomson, who in 1900

A black body would be a useful experimental tool, but it could exist only in theory. In 1895, Wilhelm Wien (1864–1928) came up with a great dodge around the problem of a real black body. He reasoned that a hole in a graphite furnace could replicate a black body as closely as experimentally possible. Graphite can absorb about 97 per cent of the radiation that falls on it, which is good but not nearly good enough. By putting a hole in the graphite, radiation not absorbed would not radiate away but bounce around until it was absorbed on another surface inside the furnace. Some might escape out the hole, but most would be absorbed. Similarly, the system would work in reverse; as the furnace was heated, energy would radiate from the hole.

Wien found that, as he raised the temperature, the energy radiated at a range of frequencies but had a peak range, so that most of the radiation given off was at a particular frequency. This ran against what physicists had expected to find. Hot recommended him for a position at McGill University in Montreal, Canada.

In the comparatively isolated colonial world of Canadian science, Rutherford began to work on atomic structure. While in Montreal he conducted his early work on radioactivity with Frederick Soddy, creating a "Theory of Atomic Disintegration" that argued radioactive substances were transforming from a heavier substance into a lighter substance as they emitted radiation. Until that time, it had been assumed atoms were indivisible. He also coined the terms for three different types of radiation: alpha, beta, and gamma rays. This work would form the basis for his Nobel Prize of 1908. An advantage of his time at McGill was that he was free to explore what he thought was interesting. The disadvantage was that resources were limited and there were few colleagues with any knowledge of advanced physics.

In 1907, Rutherford returned to England to take up the chair of physics at the University of Manchester. Manchester was an important university, but did not have the status of Cambridge. At Manchester,

Rutherford did his famous work with Hans Geiger and Ernest Marsden, leading to the Rutherford model of the atom. He also became the first person to deliberately transmute one element into another. In 1919, he succeeded J.J. Thomson as Director of the Cavendish Laboratory and became the leading physicist in Britain. Under his direction, three separate teams would produce Nobel Prizes: James Chadwick for discovering the neutron; John Cockcroft and Ernest Walton for splitting the atom in a particle accelerator; and Edward Appleton for demonstrating the existence of the ionosphere.

Ernest Rutherford became one of the most influential scientists in mid-century Britain and the larger scientific world. His work led to a completely new understanding of the atom and led to splitting the atom. Without his discoveries, nuclear power and the atomic bomb would not have been possible. He supervised a generation of important scientists at Cambridge and was buried in Westminster Abbey. Not bad for a boy from the South Island.

bodies were expected to radiate at all frequencies with equal probability, so there should have been far more high frequency radiation (such as violet and ultraviolet), because there were more high frequency possibilities than low ones. Just the opposite happened. There was a lot of low frequency radiation and very little high frequency.

In 1900, Max Planck offered a solution to the problem. He reasoned that energy was not given off continuously but in packets. He called these packets *quanta* from the Latin for "how much." Under this theory, violet light, with a high frequency, required a quantum twice as big as the quantum for red light, which has a much lower frequency. No violet light could be radiated until there was enough energy to make up the right size packet for it, but some of the energy would have already been used up in the smaller packets required for red light. Thus, red light would appear far more often than violet light, which might never be radiated at low temperatures at all. From Planck's work on the black body problem came Planck's constant h, the smallest unit of action, which is about 6.6×10^{-27} erg seconds. When he first introduced his idea about quantum theory, it did not have a huge impact, but its importance grew as other scientists applied it to a wider and wider range of physical phenomena, including Bohr's model of atomic structure.

One of the most important applications of the quantum theory was Einstein's explanation of the photoelectric effect. Hertz had discovered in 1887 that light beamed at a metal plate would eject electrons. In 1902 Philipp Lenard (1862–1947) found that there was a threshold frequency for the effect. If light was beamed at a plate below the required frequency and regardless of its intensity, no electrons would be ejected. Einstein took the idea of a packet or particle of energy (the minimum needed to break the electron bond to the metal) and in effect said that light was a kind of particle. This atom of light, or *photon*, overturned a century of theory about the wave nature of light.

Conclusion

As the century drew to a close, the two great threads of physical science, the study of matter and the study of energy, were being knitted together. The Newtonian impulse to synthesize all of nature into simple and universal laws seemed to have come triumphantly through a period of testing that arose from the problem of adding thermodynamics, electricity, and radiation to the Newtonian universe. The structure of the atom was a surprise, but for most chemists the atom's inner structure was of less importance than the clarification of atomic weights and the practical tools that were created by electrochemistry. There were a few problem areas such as the odd concept of the quanta, but the useful applications of the new knowledge far outshone any of the irksome questions still unanswered. Many scientists, in all branches of study, felt a sense of satisfaction that science had successfully answered the big questions, and some physicists actually predicted the end of physics, as all known phenomena were effectively understood.

The end of the nineteenth century was for many a time of great excitement and accomplishment. New inventions were appearing so quickly it was hard to keep up. Railways crossed the continents, steamships ruled the oceans, and telegraph wires connected the civilized world. Photography gave us images, and the phonograph recorded sounds that seemed to offer the chance to transcend old barriers of space, time, and class.

As astonishing as the new age was, there were dark clouds on the horizon for society and for science. The Crimean War (1853–56) and the American Civil War (1861–65) had given a hint of what warfare on an industrial scale could be. The rapid urbanization of the industrial countries created massive slums and rampant social problems. Although historians have argued that the urban poor were better off than the agrarian poor, especially over the long term, the "proletariat" were both more visible and more aware of their situation, in part because of the very inventions that were transforming the industrial world. Add to this the struggles over natural resources in Europe and colonial territories in Africa and the stage was set for conflict.

In science, there were a host of questions—such as how did Darwin's macro-biological system express itself on the individual level? and what did light propagate through in the vacuum of space?—that still needed good answers, but only a few scientists had any sense that the answers would not only require new insights but would transform the very foundation of scientific knowledge. To complicate matters, these changes would take place at a time when many scientists were being asked (or in some cases ordered) to turn their work to warfare and mass destruction. Although thinkers from Archimedes to Galileo had been asked by their patrons for help in times of war, the scale of the new efforts would be vast. It would also help introduce a new model of scientific research called "Big Science." The gentleman scientist working with a few dedicated assistants in a shed or university basement would be replaced by large laboratories, teams of scientists, and massive budgets. While the individual genius would still have a revered place in science, turning ideas into research would become a much bigger process.

Essay Questions

1.   What led Ernest Rutherford to create a new model of the atom?

2.   Explain how the kinetic theory of gases helped solve the problem of the nature of heat?

3.   What experiments demonstrated the relationship between electricity and magnetism?

4.   How did Röntgen discover X-rays?

**CHAPTER
TIMELINE**

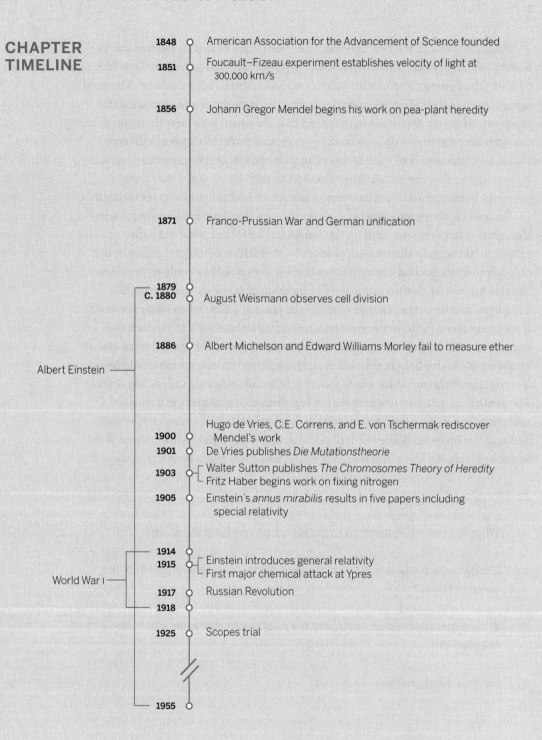

1848 American Association for the Advancement of Science founded

1851 Foucault–Fizeau experiment establishes velocity of light at
300,000 km/s

1856 Johann Gregor Mendel begins his work on pea-plant heredity

1871 Franco-Prussian War and German unification

1879
C. 1880 August Weismann observes cell division

1886 Albert Michelson and Edward Williams Morley fail to measure ether

Albert Einstein

1900 Hugo de Vries, C.E. Correns, and E. von Tschermak rediscover
Mendel's work
1901 De Vries publishes *Die Mutationstheorie*
1903 Walter Sutton publishes *The Chromosomes Theory of Heredity*
Fritz Haber begins work on fixing nitrogen
1905 Einstein's *annus mirabilis* results in five papers including
special relativity

1914
1915 Einstein introduces general relativity
First major chemical attack at Ypres
World War I **1917** Russian Revolution
1918

1925 Scopes trial

1955

SCIENCE
AND WAR

n many ways, early-nineteenth-century physicists and chemists resembled the
medieval scholars who had fit their work into the Aristotelian schema even as
they addressed totally new ideas. In the nineteenth century, the accepted
model was Newtonian. Unlike the collapse of the Aristotelian system,
however, when old observations, methods, and philosophical foundations
were all rejected, the new physics did not destroy all of Newtonianism or "classical
physics" but rather absorbed it into a larger system. While in a strict sense
Newton's laws were shown to be incomplete, they worked so well at the human
level that we still live in an essentially Newtonian world. It was only in the realm
of the very small and the very big that the seams of Newtonian physics and
philosophy did not quite mesh. The problem for science was that the universe
must be built up from the very small to the very big, so if the two ends of the
scale did not follow the rules that seemed to apply in the middle, something was
very wrong.

By the middle of the twentieth century science had torn apart the stable,
certain, and comfortable world of Newtonian physics as it replaced axioms with
probabilities. Absolute time and space gave way to observations based on the
relative position and physical state of the observer, while faith in Newtonian
certainty was abandoned. In biology, statistical understanding of populations
replaced individual field observations as the major interpretive explanation. At
the same time that science dealt these blows to Newtonianism, scientists were
also aiding in the destruction of European society as, for the first time, science

was directly used to support mass warfare. Science became a huge enterprise, involving substantial infrastructure and many people's labour. This new scientific and global warfare, in turn, caused many scientists and philosophers to lose their optimism and search for relativistic rather than absolute answers to human problems.

The Unfinished Business of Light

One of the most important sources of change to science came from the ongoing debate over the nature of light. Was light a wave or a particle? The supporters of the wave side were some of the few scientists to actively challenge a Newtonian idea, and their work in the nineteenth century had contributed to the rise of the Energetik position and the theory that all of nature was composed of waves and fields of force. The atomists responded by demonstrating the physical reality of atoms and even concluded that the electron, the basic unit of electricity, was a particle. To understand what was happening with rays and particles, a further tool was necessary. This was an evaluation of the velocity of light. Knowing the velocity of an object meant that a range of other properties could be calculated. From Galileo on, a number of natural philosophers had examined the velocity of light. In 1676 the Danish astronomer Olaus Roemer (1644–1710) had used careful observations of Jupiter's satellites for his calculations, but his work was not widely accepted. Armand Fizeau (1819–96) and Jean Bernard Foucault (1819–68) made detailed tests of the velocity of light, finding that it travelled more slowly in water than air, and measured its velocity in a vacuum to be about 300,000 kilometres per second (or 186,000 miles per second).

The nature of light waves and the method of propagation then became important subjects for investigation. Many scientists assumed that a "luminiferous ether" must exist through which light could travel. The term ether was borrowed from the Aristotelian name for the special matter that made up the supralunar sphere, although it was a different sort of substance than either the Aristotelian ether or the Cartesian plenum, and was similar to the concept of ether used by scientists such as Maxwell to explain electromagnetism. In fact, the problem was to

9.1 VELOCITY OF SOUND

MEDIUM	VELOCITY IN METRES PER SECOND AT STP*
Air	343
Water	1,484
Iron	5,120
Steel	~ 5,930

* STP is Standard Temperature Pressure. In this case, it is based on the National Institute of Standards and Technology STP of 20°C and 1 atmosphere pressure.

identify its characteristics. The rate at which waves travel depends on the rate at which the medium can deform and return to its original state. For example, the velocity with which ripples (waves in water) move away from the point where a pebble hits the surface of a pond is limited by gravity and the molecular attraction of the water molecules. No matter how fast the pebble is travelling when it hits the water, the velocity of the waves remains constant. In other words, the speed of propagation depends on the medium. If light waves followed the same rules as sound waves, then the ether should have certain characteristics similar to those found for the propagation of sound waves in different media.

Because the velocity of light was very high, the rigidity of the ether must also be high, since the more rigid the material, the faster it can deform and snap back. That meant that the rigidity of the ether had to be higher than that of steel. There was, however, a big problem. The ether also had to be almost immaterial, or it would slow the motion of the planets, collapsing the universe in just the way Newton claimed Descartes's vortex model of the solar system would collapse. This seemed to be an attack on common sense. And yet the characteristics of the ether appeared to be the inevitable consequence of wave behaviour, and the wave behaviour of light had been established by the experiments of nineteenth-century physicists.

In 1881 Albert Michelson (1852–1931) created the interferometer to test the characteristics of the ether. The device used a carefully arranged set of mirrors, glass plates, and a half-silvered mirror (half the light would be reflected and half passed through) to divide a beam of light and then recombine it. The interferometer could be used to establish precisely the wavelengths of light, but Michelson had a bigger objective: he would use it to measure "ether wind."

Even though the ether was assumed to be at absolute rest, the motion of the Earth through it should create a differential rate of "flow" in the direction of the motion of the Earth around the sun or perpendicular to it. From a mathematical point of view it was exactly the same to say that the Earth was moving through a still ether or that the ether was flowing around a still Earth. Thus, a beam of light going in one direction should not have the same velocity as a beam of light sent out at right angles, just as a boat going across a fast-flowing river does not move at the same velocity as a boat rowing up the river, even if they are being rowed at the same rate. Therefore, by constructing an experiment analogous to the boat example, Michelson hoped to measure the rate of the "current" of the ether.

In 1887 Michelson joined forces with chemist Edward Williams Morley (1838–1923) to perfect the conditions for the interferometer tests (see figure 9.2).

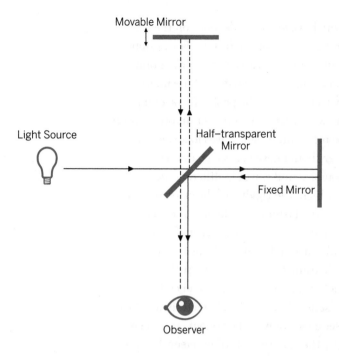

Movable Mirror

Light Source

Half–transparent
Mirror

Fixed Mirror

Observer

9.2 ETHER DRIFT APPARATUS FROM MICHELSON
AND MORLEY'S EXPERIMENT (1887)

They reasoned that if the divided beam of light travelled at different velocities because of the ether "wind," when the light was recombined it would be out of phase and thus would produce a noticeable pattern of fringes or areas of light and dark. They went to great lengths to eliminate all sources of error, but after many tests over many days and months they could observe no difference in the velocity of the light beams. Their experiment seemed a failure. Hendrik Lorentz (1853–1928) advanced the Lorentz-FitzGerald contraction theory to explain the failure by arguing that the equipment actually got physically shorter along the line of motion and that this contraction was just enough to put the light beams back in phase. While this was an ingenious solution, it was not a particularly satisfying one, since it suggested that it was not possible to study the ether.

Einstein's Theory of Relativity

There the matter of the ether stood until a new view of how light functioned was presented in 1905. When Albert Einstein (1879–1955) first published his work on light and relativity, he seemed an unlikely source for a series of brilliant insights in physics. His early life had been disrupted as his family moved a number of times. He did well in school, but because he was Jewish and a bit of an outsider, he did not initially aim for a research career, hoping instead to become a school teacher. After he graduated from the Eidgenössische Technische Hochschule in Zurich, he was employed in the Swiss patent office as a technical examiner; this afforded him the time to think about questions that interested him in physics since he could frequently do his day's work before lunch. Einstein was a very good mathematician (the story that he once failed mathematics in school is untrue), but his greatest strength was in conceptualizing problems in such a way that new approaches could be created to tackle them.

In 1905 Einstein had something of an *annus mirabilis* comparable to Newton's of 1666. In that year he published five papers: two on Brownian motion, two on special relativity, and one on light quantum in the photoelectric effect. His paper, "On the Electrodynamics of Moving Bodies," looked at the motion of light as investigated by the Michelson-Morley experiment. Einstein asked himself what one would see if one were travelling on a beam of light. He made two assumptions about this motion: first, that the motion of an object must be considered relative to the motion of some other point or object. This assumption was necessary to establish the frame of reference, which was arbitrary rather than absolute as it had been in Newtonian physics.

Second, Einstein assumed that the velocity of light in a vacuum would always be the same, regardless of the motion of the light source or the motion of the observer. Picture yourself standing at the front of a speeding train. If you threw a rock forward, the rock would have the velocity of rock plus train. But if you were to shine a flashlight from the front of the train, the light would have the same velocity as light from a flashlight switched on by someone at the train station. One of the curious things about this assumption was that it did not by itself rule out objects going faster than the speed of light in a vacuum (or in more practical terms, what would happen if two beams of light were projected in opposite directions), but it explained that such objects could not be sensed by us and therefore did not really "exist" in our frame of reference.

Here Einstein extended an observation made by Galileo, who had described a situation in which a person on a moving ship dropped a rock from the top of the mast. To the people on the ship, the rock would appear to fall straight down, landing at the base of the mast. To an observer on the shore, the rock would appear to fall in a parabola. Which observer saw the "true" motion of the rock? The answer was that they both saw the true motion, but because of their frames of reference, what they saw was different, each being relative to their own motion. The concept of relativity shattered Newtonian certainty, which had been based on a single true view of action in the universe. It also undermined ether theory, which had required the ether to be stationary for all frames of reference; that is, the ether remained still while all objects (the Earth, for example) moved through it.

While the philosophical implications of Einstein's work were radical, his argument about light accorded with observation. Yet this led to some strange implications about the relationship between mass and energy, not just about the problem of frames of reference. In his second paper on relativity, "Does the Inertia of a Body Depend on Its Energy Content?," Einstein explored the effect motion would have on objects. According to the Newtonian system—$a = F/m$, where F was force, m was mass, and a was acceleration—it followed that if mass was constant (it was unchangeable by definition in the Newtonian system) and the applied force was constant, then an object could accelerate indefinitely to any velocity. However, if the velocity of light was a maximum velocity, and there was a constant force pushing the object, the acceleration of the body had to *decrease* as the velocity increased, until eventually the acceleration became zero when the object was travelling at the velocity of light. Further, if Einstein was right, then changing the Newtonian equation around to examine the mass, $m = F/a$, produced an increase in mass as acceleration decreased toward zero but the force remained constant, which violated the conservation of mass assumed in Newton's physics. Another way of saying this is that as an object travelled faster, its mass increased, but that seemed logically impossible in the Newtonian system.

The way around the problem of increasing or decreasing mass was to realize that mass was not being created or destroyed as the object moved at different speeds, but that mass and energy were different aspects of a single entity. As Joule had shown for thermodynamics, energy could not be created or destroyed, but it could be converted from one form to another. The equivalence of mass and energy had immediate application, since it helped explain the problem of radioactivity, where energy seemed to come from nowhere. What was actually happening was the conversion of mass to other forms of energy; by careful observation it was found that radioactive materials lost mass. Einstein worked out the relation of energy to mass and produced the most famous equation in history: $E = mc^2$, where E is energy, m is mass, and c is the velocity of light in a vacuum. The use of c comes from the Latin *celeritas* meaning "velocity."

If destroying the Newtonian concept of absolute motion and constant mass was not enough, Einstein's theory of relativity also destroyed absolute time. As an object went faster relative to a stationary observer, the time elapsed for the object would be different than that elapsed for the observer. In fact, the occupant of a rocket ship travelling at 260,000 kilometres per second (close to the speed of light at 300,000 kilometres per second) would experience only half the time elapsed that a stationary observer would experience. After one hour by the stationary observer's clock, the rocket's clock would show that only 30 minutes had gone by.

Einstein's 1905 work was called "special relativity" because he was examining a special case or class of relationships in which things experienced uniform motion. While a large number of objects in the universe were covered by this, special relativity did not cover all relationships. Einstein continued to work on relativity, introducing in 1915 general relativity, which included accelerated systems, such as those represented by acceleration due to gravity. Since the universe is constructed by gravitational forces, relativity covers the biggest of big pictures. In it, space-time is considered as a four-dimensional geometric construct that is uneven. The bumps, lumps, and hollows represent the gravitational distortion of the continuum. Imagine that a sheet of rubber with a bowling ball in the centre is pulled down into a cone shape. If you try to roll marbles from one side of the sheet to the other, they will curve around the cone, some of them going down to join the bowling ball rather than making it to the far side. The others will follow a curved trajectory rather than a straight line, even if they make it to the far side. Rather than saying that the bowling ball attracts the marbles, we see the marbles and the bowling ball as linked by the geometry of the rubber sheet.

General relativity linked energy, mass, time, motion, and gravity. Many of Einstein's theoretical ideas were later demonstrated experimentally, such as time dilation, in which synchronized atomic clocks went out of sync when they were moved at different velocities. Other experiments showed that gamma rays gained energy when falling into a gravity field and that the gravity of the sun could cause light to curve. Einstein's work overturned much of the Newtonian world view, but unlike the destruction of Aristotelian physics by Galilean and Newtonian physics, much of Newton's utility remained unchanged. In the terrestrial frame of reference, the world continued to be basically Newtonian, but Newtonianism became a special case within a larger system. What relativity did undermine was the necessity of an ether.

Relativity was not, by itself, the complete death of classical or Newtonian physics, however. Consider the form of $E = mc^2$ and $F = ma$. Although the simplicity of Einstein's equation hides a world with relative rather than absolute reference points, it also suggests a degree of certainty about the universe. There might be radically different frames of reference, clocks that moved at different times and other strange phenomena, but within a given frame of reference, you could find definite answers to physical questions. Indeed, Einstein was strongly opposed to proposals that the universe behaved in uncertain ways, famously saying, "I can't believe that God would choose to play dice with the world."[1]

---

1. Albert Einstein, *London Observer* (15 April 1964); also in Ronald W. Clark, *Einstein: The Life and Times* (New York: World Publishing Co., 1971) 19.

Einstein's work opened the doors to the highest levels of academic research, and he held a number of posts until in 1914 he was appointed the director of the Kaiser Wilhelm Institute for Physics in Berlin and also became a professor at the Humboldt University. In 1921, Einstein won the Nobel Prize in Physics, ironically not for his work on relativity, but rather for his work on the photoelectric effect. Although Einstein was well known in the physics world, he was not yet the international star he would become. Such considerations were overshadowed by the calamitous events of the Great War.

Mendel and the Mechanism for Evolution

While theoretical physicists explored the limits of observation, the ideas, techniques, and tools of advanced chemistry and physics had a significant impact on another area of science. Biology throughout the nineteenth century had consisted largely of field observation, classification, and anatomy. With the introduction of laboratory-based research, some biologists adopted a more experimental approach, made possible also because of new tools. These included better microscopes and cell-staining techniques (descended from Perkin's aniline dye), the use of X-rays and crystallography, and other improved methods of organic chemistry that made dealing with the sensitive materials of cells possible. The crossover of tools and techniques from chemistry and physics created modern cellular biology. When combined with statistical methods borrowed from mathematics, biologists created the "new synthesis," which produced a powerful new model of evolution by folding the microbiology and chemistry of genetics into the macrobiology of natural selection.

At the beginning of the twentieth century, biologists were continuing the search for a fully satisfying theory of evolution. Darwin's theory had posed too many questions, and most biologists were skeptical of its conclusions. Equally, some religious leaders questioned a theory they saw as materialistic and therefore atheistic. Somewhat ironically, the most famous early-twentieth-century public dispute about evolution, the so-called Scopes Monkey Trial, occurred at a time when many biologists themselves were unsure about Darwinism. The 1925 trial in Dayton, Tennessee, must be set against the background of social turmoil in the United States, especially in the economically depressed South, and the disruption caused by the war and postwar years. Many people linked science to outsiders or foreigners, and feelings about scientists were often heavily influenced by anti-Semitism or fear of atheism.

The Tennessee House Bill 185 specifically banned the teaching of evolution in any school that received public money from the state. George Rappalyea, a geologist and mine owner, asked the American Civil Liberties Union (ACLU) to finance a court challenge; when they agreed, he asked his friend John Scopes, a young science and sports teacher, if he would be willing to be arrested for the test case. The trial was a media sensation, bringing together two famous lawyers, Clarence Darrow (for the defence) and William Jennings Bryan (for the prosecution). Reporters from across the country and around the world arrived in Dayton hoping to see a major clash. While the trial was a circus (both Darrow and Bryan were made honorary colonels in the state militia, for example), it was not quite the battle of science and religion that some had hoped for. The judge ruled that evidence for evolution could not be admitted, preventing the defence from introducing expert testimony from scientists. But in one of the strangest twists in legal history, Darrow called Bryan to testify as an expert witness on Genesis. In the end Scopes was found guilty and fined $100, which was actually the result desired by the ACLU, since they could only take the case to a higher court and have the legislation overturned if there was a conviction. Unfortunately, the ruling was overturned on appeal due to a technicality two years later, making a Supreme Court decision impossible and allowing the original legislation to stand. A number of other states enacted similar laws against the teaching of evolution, and it was not until 1968 that the Supreme Court ruled that all such specifically anti-evolution statutes were unconstitutional since they violated the Establishment Clause of the First Amendment regarding the separation of church and state.

Biologists did not wholeheartedly support Darwinian evolution by natural selection because it lacked a mechanism for inheritance—in order for variation to be selected, that variation had to be passed to the next generation, and there was no clear way for this to happen. Darwin himself had proposed a theory called pan-genesis: each parent contributed "gemmules" to the offspring, and these gemmules somehow remained in the body to be used for reproduction when the time came. Even Darwin found this explanation unsatisfactory, but it was some time before anyone took up the challenge, largely because Darwin's theory had attracted field naturalists rather than laboratory biologists. It was not until experimentation and laboratory practice became part of the biologist's tool kit that scientists could begin to investigate inheritance at a cellular level.

Mendel's Plant-Breeding Experiments

Efforts to understand the mechanism of evolution moved more and more to the examination of the cell. The major problem was how to relate cellular events to the macrobiological result. In other words, how could biologists connect what happened inside the cell with the structure, development, and behaviour of the complete organism? The first attempt to do so came before the introduction of new laboratory techniques, with a long and arduous plant-breeding experiment undertaken by an Augustinian monk in Moravia. Johann Gregor Mendel (1822–84) was the son of Silesian peasants. He got his education as many poor but bright boys did by going to a monastery school. He later joined the Augustinian Order at Brünn and went on to study at Vienna University. He began his painstaking work on plant heredity around 1856. By the time he finished his research, setting it aside when he became abbot, he had bred and examined more than 28,000 pea plants. He identified seven characteristics that had two distinct forms and that bred true from generation to generation. (See figure 9.3.)

He then cross-bred plants with alternate forms and found that the traits did not blend but remained discrete. Further, he discovered that successive generations of plants followed a clear pattern in the appearance of characteristics in the offspring. Some characteristics were dominant and others recessive. In other words, if a plant with round seeds (dominant) was crossed or bred with a plant with bumpy seeds (recessive), all the offspring in the first generation would have round seeds. But the recessive characteristic did not disappear; it was just not expressed in the shape of the plant. In the next generation, breeding two hybrid plants produced a mixture of round and bumpy seeds, as a certain percentage of offspring received dominant characteristics and a smaller number received only recessive characteristics. The ratio of dominant to recessive appearance worked out to 3:1, but in terms of the distribution of the two characteristics it represented four genetic combinations: round/round, round/bumpy, bumpy/round, and bumpy/bumpy. By crossing or hybridizing plants with different characteristics, Mendel demonstrated an algebra of inheritance to explain the transferral of characteristics. For two characteristics, the distribution was 9:3:3:1. (See figure 9.4.)

9.3 MENDEL'S SEVEN CHARACTERISTICS

1.	Form of ripe seed
2.	Colour of seed albumen
3.	Colour of seed coat
4.	Form of ripe pods
5.	Colour of unripe pods
6.	Position of flowers
7.	Length of stem

Mendel's work identified two key concepts which have sometimes been called the law of segregation and the law of independent assortment. The law of segregation identified the link between inherited characteristics and sexual reproduction. Two parts, one from each parent, were needed to create a characteristic. Those two parts had to be the same type of material, but could be different versions. In reproduction, each parent contributed exactly half of the material, which meant that whatever it was in the cell that controlled the characteristic had to segregate during gamete (sperm or egg) production. Because of the distribution of characteristic-producing material, the law of independent assortment said that variant characteristics remained independent of each other

First hybrid crossing a round seed with a bumpy seed

First generation all appear round

Second generation of hybrids produces a 3:1 ratio of round to bumpy

9.4 MENDEL'S SEED ALGEBRA

(as the algebraic illustration makes clear), and thus there were fixed ratios of characteristics in the offspring. Mendel's work explained atavism (throwbacks) and disproved the idea of blended inheritance. It had nothing to say about evolution and, indeed, seemed to make evolution less possible, since discrete units of inheritance did not seem to change.

Mendel read his paper describing this research at the Brünn Natural Scientific Society meeting in 1865, and it was published in the society's journal, but the paper had almost no impact on the biological research community. The copy of the paper that Mendel personally sent to Darwin went unread, remaining unopened in Darwin's library. The reasons for this lack of recognition are not hard to understand. Mendel was a monk in a monastery far from the main centres of scientific activity and so was not known to the biological community. The hot topic in biology was evolution, but his work looked like plant breeding, a branch of horticulture. Moreover, the results Mendel found did not seem transferable to other species. Mendel sent his results to the foremost German botanist, Karl

Wilhelm von Nägeli (1817–91), who questioned his findings and suggested he work on the plant hawkweed. Because hawkweed does not breed true, it failed to show the pattern that the pea plants had exhibited and so challenged the universality of Mendel's results. Further, there was no material basis for Mendel's statistical theory. What were these units of inheritance?

Mendel's work offered a method for predicting how characteristics were distributed in a population over multiple generations. It also directed attention to the equal importance of material communicated from each parent to the offspring. While it had been recognized for generations that reproduction was the key to understanding the likely characteristics of offspring, there was much confusion about the proportion of contribution in sexual reproduction, ranging from the female as mere vessel for the creative material of the male to male characteristics passing only to males and female characteristics only to females. Mendel's work demonstrated that in sexual reproduction each parent contributed half of the matter and information necessary to create a new organism.

During the 1880s August Weismann (1834–1915) attempted to disprove Darwin's pan-genesis theory through a laboratory demonstration of reproduction, using microscopic observation of cell division. Weismann argued that cells were immortal (since one-celled organisms reproduced indefinitely through division) and that the cell nucleus had the main role in passing information about genetic traits from parent to offspring. He observed the longitudinal split of the nucleus in mitosis (the process of division of a cell into identical pairs), which suggested that the germ plasm (as he called the material substance of inheritance) resided in the nucleus. Since sexual reproduction resulted in the combination of germ plasm from each parent, variation was produced through the many possible chromosomal combinations. The logical conclusion from these findings was that heredity was based exclusively on internal biological factors (hard heredity) rather than on any environmental influence or the inheritance of acquired characteristics.

Despite, or perhaps because of, these investigations, the link between the behaviour of cells at the microscopic level and the evidence of evolution at a macroscopic one was still unclear until Mendel's work was rediscovered in 1900. Three researchers, Hugo de Vries (1848–1935), C.E. Correns (1864–1935), and E. von Tschermak (1871–1962), all working on rates of variation in plant populations, discovered Mendel's work and recognized its importance. De Vries labelled the unit characteristics of heredity "pangens," which was soon shortened to "genes." He also recognized that although Mendel's system allowed variation through generations, it would require major shifts if evolution was to occur. He called the sudden changes "mutations" in his 1901 book *Die Mutationstheorie* and suggested that this might be a way to understand the introduction of new characteristics. In 1903, Walter Sutton's (1877–1916) book *The Chromosomes Theory of Heredity* narrowed the target area of genetic activity. He argued that genes are carried by chromosomes and that each egg or sperm cell contains only half of the chromosome pair. In effect, de Vries and Sutton discovered the cellular system predicted by Mendel's work.

The chromosome was not exactly discovered as a single moment of insight, but rather was observed by a number of people doing cell biology in the nineteenth century. Matthias Jakob Schleiden (1804–81), Rudolf Virchow (1821–1902), and Otto Bütschli (1848–1920) all noted this intracellular structure that was made observable by better microscopes and the introduction of cell stains from artificial dyes. Walther Flemming (1843–1905) observed that the nucleus of a cell split during cell division and theorized that the nucleus was passed on from generation to generation, saying "*omnis nucleus e nucleo*" ("every nucleus from a nucleus"). Flemming did not observe the chromosome split equally, missing a key point in its role in heredity. The chromosome was named by Heinrich Wilhelm von Waldeyer-Hartz (1836–1912), combining the Greek word for colour (chroma) and body (soma), since the chromosome strongly absorbed dye.

Two scientists, Theodor Boveri (1862–1915) in Germany and Walter Sutton (1877–1916) in the United States, independently demonstrated the chromosome theory of inheritance. Following the publication of Edmund Beecher Wilson's (1856–1939) *The Cell in Development and Heredity* (1902) that promoted the Boveri-Sutton chromosome theory, the chromosome became the target of intense study. In 1923, Theophilus Painter (1889–1969) published his observation that humans had 48 chromosomes. In 1956, the number of human chromosomes was reduced to 46 when better tools allowed Joe Hin Tjio (1919–2001) and Albert Levan (1905–1998) to do a more detailed examination. This in turn led to the discovery of genes, which were specific sections of chromosomes. This was first demonstrated by Thomas Hunt Morgan (1866–1945) in 1910 and led to chromosomal "maps" that linked the physical characteristics of an organism to specific locations on the chromosomes.

Although the chromosome was established as the cellular body that transmitted the information of heredity from generation to generation, it would take another 40 years to decode the chemical basis of how a gene worked.

Science and War

When the Great Powers met at the Berlin West Africa Conference in 1885, the results made war almost inevitable. In many ways the conference marked the apex of colonialism, as the major Western nations met to discuss trade and navigation and to divide up the unclaimed portions of Africa. (See figure 9.5.) For Germany, it was a test of its rising status as an economic and political superpower. Although it had only become a unified state in 1871, Germany was already challenging the industrial might of Britain and sought to establish colonial holdings in accord with its status as a European power. Germany had been late entering the race for colonies, so much was at stake. International status, access to resources, military advantage, and political benefits could be derived from colonial holdings. The conference did not produce the gains Chancellor Otto von Bismarck had hoped for, however. Although Germany did annex Tanganyika and Zanzibar, these territories were far less valuable than those held by small states, such as Belgium's control of the Congo or Portugal's colony in Angola.

Germany faced a difficult situation as a European power. It was bordered on the east by the "sleeping bear" of Russia with her huge territory, limitless natural resources, and growing industrial base. On the west was France, which was less industrialized than Germany but had better natural resources, access to two oceans, and significant colonial territory. There was much antipathy between the two countries not least because of the Franco-Prussian War, in which France was badly beaten and which led to the unification of the German states in 1871. France was forced to give up resource-rich Alsace and parts of Lorraine to Germany. Dominating the seas and holding the greatest colonial empire was the highly industrialized Britain. Despite long-standing tension with France, Britain became Germany's main opponent, and it was in opposition to British sea power that Germany embarked on a massive naval arms build-up, which resulted in the first modern arms race, as each side built more and bigger ships.

In order to successfully pursue its imperial policy, all aspects of German society had to be harnessed together, from the farmer in the field to the captain on the bridge of the dreadnought. The interest of the state and the interests of business were often synonymous, and when Bismarck introduced the first welfare programs in Europe, it was not because of socialist leanings but rather to protect the German industrial workforce and keep it as productive as possible.

To redress the lack of resources or high cost of exploiting domestic sources, Germany turned more and more to science and technology. It created an

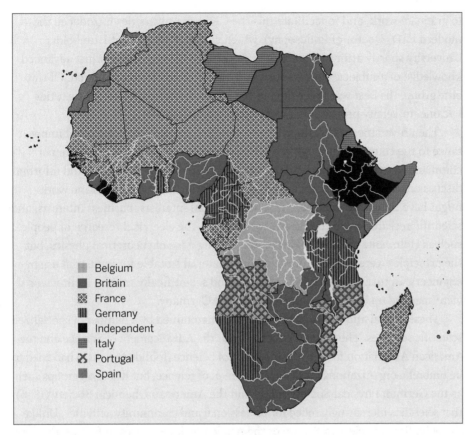

9.5 AFRICA AFTER THE BERLIN CONFERENCE 1885

Belgium
Britain
France
Germany
Independent
Italy
Portugal
Spain

integrated system that brought together schools and universities, government, and research scientists. The new approach to science had the educational system as its foundation. The government constantly promoted the improvement of education, and by the turn of the century Germany had the highest literacy rate in Europe. Technical high schools and colleges produced skilled workers for the industrial and business sectors, while advanced education was encouraged in the many universities, leading to a huge growth in the number of professionally trained scientists.

The scientific structure linked three major entities: government (including the military), business, and academic research institutions. The government funded education and created a series of elite research centres, particularly the Kaiser Wilhelm Institutes. The education system was designed to stream students, with the best science students going on to university. The best of those students went on to graduate work, and to facilitate this the German universities introduced the modern PhD (Doctor of Philosophy), which was first awarded at Humboldt University shortly after its founding in 1810. The PhD required not just advanced knowledge of a subject, but demonstrated ability to do original research. Of this elite group, the best were recruited to the research centres, while the next tier became university professors or were recruited by industry.

The universities and research centres in turn worked on problems of importance to the country. Although scientists, especially those at the top, were not commanded to work on particular problems, there were both formal and informal discussions about what topics would help the nation and what scientific work might have industrial applications. Thus, national interests, business interests, and scientific research were combined. For example, the electrical research of people such as Helmholtz and Hertz was on the cutting edge of theoretical physics, but the principles were quickly converted into material suitable for industrial use by engineers. In turn, experienced engineers and scientifically trained technicians were snapped up by the growing industries of Germany.

These formal and informal ties were also maintained by a number of specialized scientific societies. The British Association for the Advancement of Science and the American Association for the Advancement of Science (founded in 1848) had tried to be umbrella organizations for like-minded men of science, but it was on groups such as the German Physical Society (1845) and the American Chemical Society (1876) that scientists increasingly relied for professional and community activities. Unlike the older Royal Society in Britain or the Académie des Sciences in France, the new societies were composed of working scientists within a single discipline. They not only provided a conduit for scientific research through conferences and journals, but they also became a vital resource for professional development on an individual level, by helping people get jobs, and on a community level as they advised governments, lobbied for standards and laws, and promoted the discipline.

The Chemists' War

Fritz Haber (1868–1934) and Carl Bosch (1874–1940) were two scientists whose experience typifies the intersection of national interest, industry, and research. Because of the rise in German population, pressure on agriculture was increasing, but intensive agriculture exacerbated the problem of soil depletion. Germany did not have secure access to natural sources of nitrates for fertilizers, which came primarily from guano, or bird droppings. The chief source at the turn of the century was Chile, which in 1913 produced about 56 per cent of the world supply of nitrates. To make matters worse for Germany, British companies controlled most of Chile's nitrate production. In 1903 Haber began to investigate methods to "fix" atmospheric nitrogen by converting it to ammonia, which could then be used for fertilizer as well as providing a useful feedstock for other nitrogen products. The basic process created the reaction $N_2 + 3H_2 = 2NH_3$, but the contemporary method of production required large amounts of electricity, which made it uneconomical. By 1909 Haber and Robert Le Rossignol had worked out a continuous flow method to produce ammonia; it required a reaction vessel that could withstand 200 atmospheres of pressure and be heated to 500°C. Haber convinced the chemical company BASF that the process could be made commercially viable, but to put the system into place BASF had to turn to the steelmaker Krupp to manufacture the reaction vessels. Krupp, in turn, went to its scientists and engineers to pioneer a new method of forging steel to create a container that could withstand the pressure and heat.

When all the elements were put together, the Haber-Bosch system produced large quantities of inexpensive ammonia. The synthesis of ammonia earned Haber the Nobel Prize in Chemistry in 1918 and led to the creation of the Oppau and Leuna Ammonia Works, the first industrial producer of nitrate products that did not require large-scale electrical supplies to operate. By 1934, almost 64 per cent of fixed nitrogen was produced synthetically, and the leading producer was Germany, while Chile's share of world production fell to just over 7 per cent.

Although the initial concern had been to create a secure domestic source of artificial fertilizer and to supply industrial demand for ammonia, with the outbreak of war, when natural sources were cut off by Britain and her allies, the Ammonia Works provided the nitrates necessary for explosives such as cellulose nitrate. Without the synthesis of ammonia, Germany could not have continued to fight the war for more than a few months.

Germany entered World War I with expectations that it would defeat France and her allies quickly in a repeat of the Franco-Prussian War. This did not happen. French resistance was greater than expected, the British bottled up the German navy and blockaded her ports, and the war with Russia divided German forces. When both sides tried to outflank the other, the front lines stretched longer and longer until the western front ran from the Alps to the English Channel and the eastern front stretched from the Black Sea to the Baltic Ocean. Germany had created a well-trained and literate military backed by the second most powerful industrial economy in Europe, but it could not withstand a war of attrition, especially when the British and French continued to import supplies from their colonies and the United States.

Fritz Haber persuaded the German High Command to try a chemical attack. Haber arranged to transport 168 tons of chlorine gas in cylinders to the front and then waited for the wind to blow in the right direction. On April 22, 1915, the German army released the gas along a 6.5-kilometre (4-mile) section of the front line at Langemarck, near Ypres in Belgium. The yellowish-green cloud drifted across no man's land and onto the Allied position. The front line collapsed, and German forces followed the cloud through the hole in the line. Resistance behind the front lines and a lack of German reserves to carry the attack farther halted their advance. By the time the gas cleared, 5,000 soldiers were dead and 15,000 had been wounded in the attack. Haber was made the director of gas warfare after this demonstration of its effectiveness.

Chlorine was a good choice for a weapon. It was available in large quantities because of its use in chemical and textile industries, and it was slightly heavier than air so it settled in hollows and trenches. It did its damage by combining with moisture in the eyes, nose, and lungs to form hydrochloric acid. Although it was not the first use of chemicals on the battlefield (earlier in the war, the French had used tear gas, and the Germans had ineffectively used chemical-filled artillery shells against the Russians), Ypres was the first successful gas attack, and it changed the conduct of the war. It also helped change the course of science.

All sides in the conflict flung themselves into a concerted effort to create new chemicals to use on the battlefield and defensive equipment to ward off enemy attack. What had been a contest of heavy artillery, machine guns, and trench warfare now became a battle of scientific prowess. Britain, France, Canada, the United States, and Italy created national research councils. These organizations were intended to coordinate research efforts with industrial production and military needs. In short, the Allies attempted to emulate the integration of research that Germany had already pioneered. Hundreds of chemists, chemical engineers, biologists, physicians, engineers, physiologists, and others were pressed into service. Crash programs set up laboratories, and scientists began their work even before construction crews were finished. The Allies launched their own gas attack on September 25, 1915, at Loos. It was not a great success. While part of the German front line retreated, some of the gas blew back on the British and French line. The territory gained was retaken in a few days.

Chlorine was relatively easy to recognize and defend against, so Haber and his fellow chemists had to introduce more potent chemicals if they were to keep

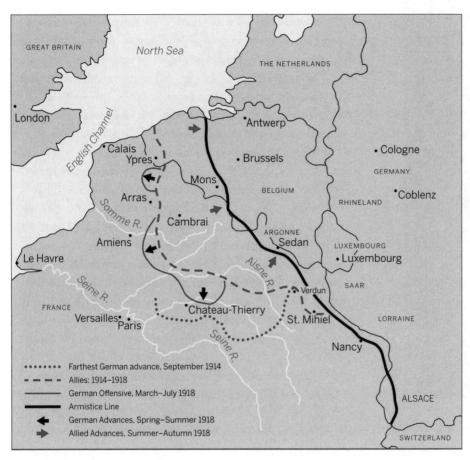

9.6 THE WESTERN FRONT, WORLD WAR I

their advantage. They turned to phosgene, which was clear, colourless, and odourless. It acted more slowly than chlorine but was more subtle, injuring or killing hours or even days after exposure, and it added a new level of terror to gas warfare. It was followed by other chemicals including chloropicrin, mustard gas, and diphenylchlorarsine. By the spring of 1916 both sides were using chemicals as standard weapons. Against an untrained or unequipped enemy, gas was deadly. It was used to great effect by the Germans against the Russians and the Italians. Against a trained and prepared force, however, it did not change the stalemate in the trenches, although it did have a profound psychological effect on both soldiers and the civilian population. The advent of chemical weapons and the development of long-range aircraft made potential targets of cities and

CONNECTIONS

Chemical War: Science and the State

World War I changed both the ideology of science and the way science was organized. Science went from being seen as one of the intellectual foundations of the Enlightenment project for reform to a morally neutral and disinterested enterprise. Although science could be used for good or evil, science itself had no moral value. The war also led to the creation of Big Science, where scientists worked in teams, operated in large laboratories, and were funded by the state or corporations at levels unheard of before the war.

Prior to 1915, most scientists believed that science was above political disputes and ideologies. Many scientists accepted the Enlightenment ideal of science as a path to truth and rationality, and thus as an active tool to create a better world for everyone. Individual scientists might be partisan, but science almost by definition could not be. It would be like saying that the force of gravity was English or the theory of acid formation belonged to France. Scientists believed participation in science was open to all and knowledge could not be constrained by legislation; anything one scientist discovered could be found by other scientists because nature was universally available for study.

This view of science began to crumble on April 22, 1915, at Ypres, Belgium, when German troops released 168 tons of chlorine gas that

vulnerable civilian populations far from the front. It also brought home to the governments and general public the power and importance of science and scientists.

In addition to chemical warfare, scientists were pressed into service working on almost every aspect of war from new forms of explosives, aerial combat, sonar, radio communication, meteorology, and medical treatment, including the use of the recently discovered X-ray.

The signing of the Armistice on November 11, 1918, which ended World War I, marked a turning point in European and world history. A generation had been devastated by the war and by the influenza pandemic that followed. A generation of young men had been slaughtered in a largely futile struggle. More than 8 million people were killed and over 20 million wounded. To add to the calamity, somewhere between 50 and 100 million people died in the swept across the Allied front line, killing at least 5,000 men and injuring more than 15,000. Haber believed that it was his duty to work for Germany during a time of war. He hoped that his chemical attack would break the stalemate of trench warfare and lead to the end of the war. His actions were not driven by a desire to kill and wound thousands of people, but to end the war as quickly as possible and stop the killing. This rationale would be applied to nearly every new weapon of war; it had been applied to the machine gun and dreadnoughts, and it would be deployed again as justification for aerial bombing and the atomic bomb.

The use of chemical warfare was the first time that science played a major role in warfare. The gas attack set off an arms race, but it also set off a brain race as each combatant nation struggled to organize its scientists for war. Before the war, only France and Germany had significant state funding of science. Because of the war Britain, the United States, Canada, Australia, and Italy established national research councils and began to fund scientific research. Scientists became a strategic resource and science was turned to the pursuit of mass destruction. The structures for scientific research that exist today were born out of that theatre of war.

The ideology of science was also changed. No longer was science associated with Enlightenment ideals of rationality, democracy, and human rights. Instead, scientists began to argue that science was morally neutral—it was neither good nor bad. Science could be turned to good or bad uses, but had no inherent morality. Haber expressed this idea by saying that a scientist should serve mankind in times of peace but the fatherland in war. Scientists around the world made a kind of Faustian bargain: in return for support for scientific research at levels far beyond those of the nineteenth century, most scientists gave up the idea that science was morally uplifting. Further, they tacitly agreed that their services could be called upon for political or military needs, rather than to uplift the whole people.

worldwide influenza pandemic that raged from 1918 to 1920, with the European nations losing around 3 per cent of their populations to the disease. European economies lay in ruins, and Russia was still in the throes of the revolution that had started in 1917. The old political systems, the old ways of living, had changed forever. France, humiliated by its defeat in the earlier Franco-Prussian War and reeling from the destruction of a four-year conflict fought largely on its soil, sought to punish the Germans. The terms of the 1919 Treaty of Versailles forced the Germans to give up colonial holdings and make huge reparation payments, imposed trade restrictions, and dismantled parts of its government and military. German society was thrown into turmoil, exacerbated by economic depression and hyperinflation. In 1923, when Germany was unable to make reparation payments, the French invaded the Ruhr district, taking over coalfields and industrial facilities in an armed occupation.

Science in the National Interest

The war had brought the utility of science to the forefront, as well as demonstrating its dark and dangerous side. Generations of philosophers and scientists had argued that science would benefit society and help nations prosper; now, its power and utility had been graphically illustrated by the "Chemists' War." Even as the victors tried to settle back into a pre-war life, the industrialized world could not ignore the system that had allowed one nation to challenge and almost defeat a coalition of larger and better supplied states. In the years following World War I, scientists worked to establish their national status and rebuild their network of international contacts. Almost in defiance of the political divisions appearing everywhere, scientists published in internationally read journals and worked both in competition and cooperation with laboratories all over Europe and the United States. The experience of big labs and big projects, which had been so important to war work, led scientists to look for projects that might fit that model. As physics emerged as the pre-eminent scientific discipline of the 1920s and 1930s, scientists in other areas turned to using methods, theories, and models from physics as they investigated evolution and inheritance at a molecular level, looking at cellular structure and DNA. At the same time, physicists themselves were delving ever further into the subatomic realm, with ultimately explosive results for the atomic age. Before the war, in Britain and the United States in particular, science as a profession had had a lower status than other academic pursuits, and with the exception of a few research centres such as the Cavendish Laboratory or Cornell University, most students seeking advanced training had travelled to Germany. When the war closed off this avenue of study, domestic programs had to fill the gap. And, as the war progressed, it became apparent to governments and educators that science education was a valuable national resource. The change helped the Unites States rise to scientific power.

In Germany, national interest, combined with the sense that science and technology could help the German situation, made research an exception to the generally dismal circumstances of the postwar years. Perhaps because money was in short supply for laboratories and equipment, a host of powerful minds in the universities and research centres flourished, doing advanced theoretical work. Max Planck, Werner Heisenberg, Erwin Schrödinger, Albert Einstein, Leo Szilard, Otto Hahn, and Lise Meitner were only a few of the people who did pioneering work in Germany after the war. International contacts were quickly re-established. Scientists on both sides of the conflict had felt a duty to work for their countries during the war, but they also felt a duty to the community of science. Many went further, arguing that science and the rationality it embodied would bring peace to the world, in an echo of the Royal Society's search for a "third way" in the seventeenth century.

Conclusion

Science, which had contributed to the marvels of the Victorian era, bringing new devices and, for many, greater comfort than any previous age, now showed a dark and dangerous side. It was not a coincidence that the certainty of the Newtonian world view crumbled at the same juncture as European society was rent by a terrible war. The infrastructure that made the esoteric research of Rutherford, the Curies, Max Planck, and Einstein possible also made chemical warfare, high explosives, and aerial bombardment the face of modern warfare.

Although governments cut funding to science after World War I, the success of scientific research for the war effort made it impossible to return to the polite Victorian era of the gentleman scientist. In the years between the two world wars scientists themselves worked diligently to promote large-scale research programs and were ready to heed the call to arms when it came again. Science had been brought to public attention. The social status of science improved, as indicated by the increase in science education, particularly at the advanced level. Governments began to pay attention to the power of scientific research, often seeking to bend it to commercial, industrial, or military ends in the years to come.

Essay Questions

1. What was the Michelson–Morley experiment supposed to demonstrate?

2. How did Einstein's theory of relativity challenge the Newtonian system?

3. Explain how Mendel's seed algebra supported the theory of evolution.

4. Why was World War I also known as the "Chemists' War"?

CHAPTER TIMELINE

1905	William Bates describes blended characteristics in heredity Clarence McClung discovers XX (female) and XY (male) chromosomes
1910	Thomas Hunt Morgan discovers sex-linked characteristics
1916	Morgan argues for a genetic theory of natural selection
1926	Erwin Schrödinger describes electrons as standing waves
1927	Werner Heisenberg describes indeterminacy Trofin Denisovich Lysenko promotes "vernalization"
1928	Fred Griffith demonstrates bacterial transfer of characteristics
1933	Leo Szilard conceives chain reaction
1935	"Schrödinger's cat" thought experiment
1939	Lise Meitner describes "fission"
1941	Szilard persuades Einstein to write to President Roosevelt about atomic weapons
1942	Enrico Fermi demonstrates nuclear chain reaction
1944	Oswald Theodore Avery, Colin MacLeod, and Maclyn McCarty isolate DNA
1945	Trinity Test; Bombing of Hiroshima and Nagasaki
1949	USSR tests its first atomic bomb
1950	Erwin Chargaff determines ratio of DNA bases
1951	Rosalind Franklin produces X-ray crystallography of DNA
1952	United States tests first hydrogen (fusion) bomb
1953	Francis Crick and James Watson present structure of DNA

World War II

Manhattan Project

THE DEATH OF CERTAINTY

10

I n the wake of World War I, long-standing social, political, and economic standards were challenged and overturned. In politics, the power of monarchies declined and new ideologies such as modern liberalism, industrial capitalism, democratic socialism, and communism gained adherents. Many democratic countries started extending voting rights to women: Canada in 1917, Britain and Germany in 1918, the Netherlands in 1919, and the United States in 1920, although French and Italian women would have to wait until 1944 and 1946, respectively, for the right to vote. The Russian Revolution changed the diplomatic map of the world.

In the arts the Impressionist movement, led by people such as painters Claude Monet and Edgar Degas, that had so shocked the cultural establishment in the 1870s and 1880s gave way to the even more radical art of Expressionists such as Wassily Kandinsky and Franz Marc and the "anti-art" art of the Dadaists such as Marcel Duchamp and Hannah Höch. In architecture the modernist movement was using cutting-edge technology and a rejection of ornamentation to create buildings for an industrial age. Charles-Édouard Jeanneret, better known as Le Corbusier, said, "A house is a machine for living in."[1] As well as the visual arts, dance, theatre, film, and literature were all marked by a period of experimentation in the first part of the twentieth century. If these disparate arts had any common ground, it was two things: an attack on the old forms and a desire to look beneath the surface of things.

---

1. Le Corbusier, *Vers une architecture* (Paris: G. Crès et C., 1923).

The postwar years were a period of decadence for some people, as they tried to forget the horrors of the war, and of depredation, especially after the collapse of international finance in 1929 and the start of the Great Depression. It seems more than a coincidence that just when the stability of government and finance was shaken by rapid technological change, war, and depression, so many artists were rejecting old ideas. It is sometimes hard to see the outside world in the world of science, where Einstein's theory of general relativity appeared in 1915 during the course of the war and Pluto was discovered in 1930 during the depths of the Great Depression, but turmoil in science was just as much a challenge to the old world view. Two areas in particular would force a reconsideration of how nature worked. The first was the continued development of nuclear physics, in particular the problem of indeterminacy, and the second was the discovery of genetics and the synthesis of evolution and biochemistry. Nuclear physics would open the door to nuclear power and nuclear weapons, while evolution and genetics would challenge social and religious ideas.

The New Physics: Indeterminacy

Although World War I was dubbed the "Chemists' War," chemistry was giving way to physics as the premier area of study. Chemistry and physics were not separate subjects at the beginning of the nineteenth century, but by the start of the twentieth century the distinction had become increasingly institutionalized and the territory more clearly defined. Chemistry focused more and more on the atomic and molecular level, while physics looked at the extreme ends of observation—the subatomic realm at one end and the structure of the universe at the other. These investigations undermined what came to be called "classical" physics, which was characterized less by its subject matter or even its experimental methodology than by its assumption of the absolute condition of nature and the resulting certainty of its laws. Prior to relativity and quantum theory, the laws of nature were considered to be simple, universal, and invariant. The location of the event or the position of the observer made no difference. The new physics removed the comfortable certainty of the old and replaced it with a more precise, but contingent system. The condition of the observer became crucial to the answers one got from observation of the physical world.

At its most basic level, indeterminacy is easy to understand. Take, for example, what seems like the important but mundane task of creating a star catalogue. The catalogue identifies stars by a set of coordinates that allows astronomers to look at the same object in the sky. Such catalogues have existed since at least the time of Ptolemy. But there is a problem. Every telescope is different, and even with the same telescope astronomers know that things such as temperature and atmospheric conditions (for example, air density and amount of water vapour) make each observation slightly different. So astronomers make multiple observations of a star and combine the results to statistically create its coordinates. Although any astronomer using the catalogue will certainly be able to find the desired object, is the star actually at those coordinates? The answer is that we don't know, and the problem is that we can't know. No physical instrument can be perfect, and no instrument can observe the universe from outside the universe and thus must interact in some way with the thing being observed. For astronomers—and essentially for anything above the atomic level—this indeterminacy has little practical effect. At the subatomic level, it created a serious scientific problem.

As in the case of the death of the phlogiston theory, rather than established physicists converting to the new view it took a new generation of scientists to develop and embrace the new ideas of quantum physics. This gulf was widened by the introduction of wave mechanics within quantum physics. Bohr's quantum picture of the atom, which seemed to work well for individual atoms, broke down in more complex arrangements, such as the simplest form of a diatomic molecule (for example, atmospheric oxygen O_2). Louis de Broglie (1892–1987) in France, Erwin Schrödinger (1887–1961) and Werner Heisenberg (1901–76) in Germany, and Paul Dirac (1902–84) in Britain all tackled the problem. De Broglie argued that waves and particles were two aspects of a single entity. Thus, electrons and photons could have particle properties if looked at one way and wave properties if looked at another. While the wave–particle duality flew in the face of about 300 years of argument about particles and waves of light, it worked neatly, particularly since anything that moves can be mathematically described as a wave.

In 1926 Schrödinger argued that electrons could be described as standing waves with modes of vibration analogous to the vibrations of violin strings. While this allowed a better mathematical description of the behaviour of the electron, it was not universally accepted, as both Heisenberg and Dirac offered alternative models that were in a sense "purer" since they were strictly mathematical. What linked the wave models was the necessity of dealing with electrons and other particles as a series of probabilities rather than as completely knowable objects. This had profound philosophical consequences, not the least of which was the rejection of aspects of quantum theory by Einstein and Schrödinger, who each felt that there must be some ultimate reality in the universe, not simply a range of possible realities. They were both concerned that limits to observation constrained what could be absolutely known about the very structure and function of the universe.

Heisenberg formalized this problem in 1927 when he reasoned that it would not be possible simultaneously to know the exact position and the momentum of an electron as it orbited the nucleus of an atom. Any method that could be used to determine the position of the electron would have to interact with the electron, and that would alter its momentum. Similarly, anything that could record the momentum would interfere with the motion of the electron and change its position. This was the principle of indeterminacy, and Heisenberg expressed it as a mathematical formula: $\Delta p \times \Delta x \sim h$, where p was the uncertainty of the momentum, x was the uncertainty of the position, and h was Planck's constant.[2] Planck's constant was a measurement of quantum energy, or the smallest "packet" of energy. Another way to think of this was to say that if Δp for an electron was completely known, then Δx would be completely unknowable, but the whole system would have a value equal to the smallest energy packet for the system.

The principle of indeterminacy has also been called the "uncertainty principle," although this is not an accurate label. While it might not be possible to determine the position and momentum of a single electron, the behaviour of the electron is consistent. In other words, the electron's behaviour is not uncertain in the sense of being random, but we cannot determine certain things about it because we are part of the same physical universe. Further, since we can be as completely certain about the behaviour of electrons as a class of objects as it is possible to be certain about anything that can be examined in the physical world, the indeterminacy principle does not make physics (or science in a wider sense) impossible or unreliable. In fact, just the opposite. The probabilistic approach to physics is more precise than classical physics since it describes a system that can be in more than one state. In other words, it models the real world better than the ideal world assumed by Newtonian or classical physics.

Indeterminacy was disturbing because it seemed to open physics to a kind of strange metaphysics. Indeterminacy can be read in such a way that all possible conditions exist simultaneously. In 1935, Schrödinger tried to illustrate the difficulty of taking indeterminacy to its logical conclusion in an essay describing

....

2. A more modern expression is $DE \times Dt \sim h/4p$.

conceptual problems in quantum mechanics. In this article he offered a thought experiment that has become known as "Schrödinger's cat":

> One can even set up quite ridiculous cases. A cat is penned up in a steel chamber, along with the following diabolical device (which must be secured against direct interference by the cat): in a Geiger counter there is a tiny bit of radioactive substance, so small that *perhaps* in the course of one hour one of the atoms decays, but also, with equal probability, perhaps none; if it happens, the counter tube discharges and through a relay releases a hammer which shatters a small flask of hydrocyanic acid. If one has left this entire system to itself for an hour, one would say that the cat still lives *if* meanwhile no atom has decayed. The first atomic decay would have poisoned it. The Psi function for the entire system would express this by having in it the living and the dead cat (pardon the expression) mixed or smeared out in equal parts.[3]

In other words, because we cannot know the state of the cat until we look, the cat is equally alive and dead *at the same time* since there is a 50/50 chance of either condition. Only by opening the diabolical device will the universe resolve into one state or the other.

The simple idea that the act of observation changes the thing observed found general acceptance since there could be no tool of observation that did not have physical properties such as mass or use waves or particles in its operation and thus interacted with the thing observed. Philosophically, the concept of the participation of the observer in the thing observed, the end of complete objectivity, had far-reaching implications, helping to produce the cultural relativism that transformed the social sciences in the twentieth century. Literature, anthropology, and history all began to take into account the reader, the anthropologist, and the archivist as part of the system rather than outside and unbiased observers. The anthropologist Franz Boas (1858–1942), who earned a doctorate in physics before turning to the study of human culture, rejected the idea that there was a hierarchy of civilizations and strongly opposed the scientific racism of people like Galton. Boas argued that there were no higher or lower races or cultures and that all people see the world through the lens of their own culture. There was no "objective" measure of culture, only contingent observations. Activities that are considered moral in one

3. E. Schrödinger, "Die gegenwartige Situation in der Quantenmechanik," *Naturwissenschaftern* 23 (1935): 807–12, 823–23, 844–49; trans. John D. Trimmer, *Proceedings of the American Philosophical Society* (1980): 124, 323–38; repr. *Quantum Theory and Measurement*, ed. J.A. Wheeler and W.H. Zureck (Princeton, NJ: Princeton University Press, 1983) 152.

society might be considered immoral in a different society, and had to be understood in context of the culture. This came to be known as "cultural relativism."

Still, for scientists to conclude that things do not have a specific state of existence unless they are observed seemed absurd, even if indeterminacy prevents us from knowing the specific state. This had an anthropocentric corollary: the universe exists only because we (or some entity) observe it. Although such an idea could be used as an argument for the existence of God (who observes all the universe at all times, thus keeping it in existence), such an idea at a human level seemed too outrageous for all but the greatest of egoists. The philosophical pitfalls of deep quantum physics seemed so great that Schrödinger later said he wished he had never met the cat.

Another way that indeterminacy was interpreted was to suggest that there were an infinite number of universes since each quantum transition state should exist. The cat in the box could then be alive in one universe and dead in another.

While scientists have speculated about the various philosophical issues associated with indeterminacy and quantum physics, they are very clear that quantum effects are unnoticeable above the subatomic realm. This has not stopped unscrupulous (or deluded) people from making wacky claims for products ranging from quack medicine to extraterrestrial travel based on "quantum physics."

Evolution, Cellular Biology, and the New Synthesis

The work of biologists in the years before World War I had revealed the existence of genes and chromosomes and had suggested that Mendel's earlier observations contained a possible mechanism for evolutionary change. However, other research showed that the genetic system was more complex. William Bateson (1861–1926) in 1905 demonstrated that some characteristics were blended, rather than segregated, while in the same year Clarence McClung (1870–1946) showed that female mammals have two X chromosomes and males have an X and a Y. This led to the concept of sex-linked characteristics, introduced by Thomas Hunt Morgan in 1910. New tools, borrowed from physics labs, allowed biologists to manipulate nature and to move decisively away from field observation to the probabilistic universe of population studies.

The research on genes and inheritance changed the basis for the continuing discussion of evolutionary theory. Mendelians argued that populations could not have continuous variation and that evolution could happen only through mutation, taking away any need for natural selection as a mechanism. Biometricians (those who followed Darwin more closely) claimed that populations varied around a mean and that the mean could be moved over time. In order to argue that new characteristics would not be swamped and that evolution could happen, biologists needed to think in population terms, which required a statistical approach that borrowed mathematical techniques from the physical sciences. Two such approaches developed, the first through population studies, especially of fruit flies, and the second through a purely mathematical analysis. The result was a rearticulation of evolutionary theory, known as the new synthesis.

T.H. Morgan best exemplifies the population approach to understanding evolution. He worked with *drosophila* or fruit flies. These small flies were a perfect experimental subject since they bred quickly, so many generations could be easily traced. In addition, they required little maintenance and were large enough to examine without special equipment. By inducing mutations using heat, chemicals, and X-rays, Morgan traced the distribution of genetic traits. At first, he was skeptical about Mendel's laws of inheritance, but his breeding program tended to confirm Mendel's ideas, although it added to the system in several ways. One change was the discovery of sex-linked characteristics; the "white eye" mutant was almost completely confined to males, thus demonstrating that some changes in the chromosomes were restricted to either the *X* or *Y* chromosome. Further work with Alfred H. Sturtevant (1891–1970) led in 1911 to the first chromosome map, which located the position of five sex-linked genes. By 1916 Morgan was using his findings to argue a genetic theory of natural selection: harmful mutations were naturally prevented from spreading (since the individuals with these harmful mutations died out), while beneficial ones gradually took over the population.

Morgan's student, Hermann Joseph Muller (1890–1967), continued the research with *drosophila* to investigate evolutionary change. Muller, who with Sturtevant had helped found a biology club at Columbia University, was fascinated by the concept of genetic change. He carefully studied mutation rates of the fruit flies and concluded that, although there was a certain rate of spontaneous mutation in the genes, the mutations were rarely viable or passed on and, in many cases, were lethal. To demonstrate the physiochemical basis for mutation, he used heat to increase mutation rates (heat increased the chances of random chemical interaction); in 1926 he turned to X-rays, which greatly increased mutation rates. Further, he demonstrated that some of the mutations were inheritable.

This might have been celebrated as a key link between Darwinian evolution and the genetic model of inheritance, but Muller's work was overshadowed by criticism of his politics. Muller was an ardent socialist and even helped publish a Communist newspaper. Because of his concern about political suppression in Depression-era United States (the FBI kept tabs on him), he left for Europe in 1932. He was invited by Nikolai Ivanovitch Vavilov (1887–1943), president of the Lenin Academy of the Agricultural Sciences and director of the All-Union Institute of Plant Breeding, to become the Senior Geneticist at the Institute of Genetics of the Academy of Sciences in Leningrad and later in Moscow. While there, Muller encouraged the Russian school of *drosophila* studies, led by Sergei Chetverikov (1880–1959) and Theodosius Dobzhansky (1900–75). These biologists combined naturalist work with Morgan's lab genetics. They studied fruit fly populations exposed to natural conditions rather than artificially created ones and developed the idea of the gene pool as a reservoir of potential genetic combinations. They maintained strong links with the Columbia research group as well; Dobzhansky travelled to the United States in 1927 to join Morgan's lab for a time.

Muller's time in the Soviet Union was short. He left after only three years, when Vavilov fell from favour around 1937 and genetics as an area of study came under attack. One of Vavilov's most vocal opponents was Trofin Denisovich Lysenko (1898–1976). Lysenko was an agronomist whose reputation was based on a system of winter planting to provide a pea crop before a cotton crop in 1927. He was pictured as a peasant scientist, a practical man who had no time for vague theories or esoteric experiments. His interest was in manipulating the maturation process of seeds, with the aim of bigger plants, higher yields, and shorter growing seasons. He called his system "vernalization," and it involved (among other things) soaking and cooling seeds to promote rapid germination. He claimed he could transform the behaviour of plant species through such environmental manipulation, following a neo-Lamarckian form of evolution. Lysenko promoted his system as the solution to all of the Soviet Union's food problems.

Lysenkoism has often been pointed to as a classic case of placing political expediency ahead of good science, but the story is more complex than simply bad science endorsed by a bad political system. Science, and Darwinism in particular, had always had an appeal to Communists, who argued that it made the rise of socialism scientifically inevitable. The Soviet Union was still an agrarian society, with a strong interest in biology, but its agricultural system was in deep trouble following the Revolution. The destruction caused by wars, natural disaster, mismanagement, and in some cases deliberate policies of oppression produced famine. The Soviet leadership turned to its leading scientists, such as Vavilov, who said that genetics would help but that it would take time and a great deal of work. Lysenko's system was available immediately and was far more tolerable to a leadership suspicious of intellectuals. The twist on the story was that vernalization did work for a limited number of crops such as peas and corn, but the change in yield was insignificant in a mass system. It did not work at all for other crops such as wheat and, in fact, made things worse in many cases by hurting yields and using up limited resources. Equally damaging to long-term Soviet science was that political support for Lysenko meant the dismantling of rival research programs, especially genetics, which were seen as decadent and Western. Vavilov lost his position and in 1940 was arrested and charged with treason and sabotage. Despite the efforts of the Western scientific community to save him (he was elected a Fellow of the Royal Society in 1942 in order to give him international status), he quickly became an "unperson," his name erased from Soviet records. He was sent to the gulag (Siberian labour camps), where he died in 1943.

What might have been a minor piece of research in agronomics was changed into a national policy by Lysenko's self-promotion and the active support of political leaders who used his work to advance their own agendas. While the politics of a highly centralized state ensured that Lysenkoism with its grave faults caused great damage to Soviet agriculture, it is false logic to conclude that politically motivated support for weak science was confined to totalitarian states, as the cases of cold fusion and missile defence research would later show.

The New Synthesis

The new synthesis gelled with the work of three men: R.A. Fisher (1890–1962), a Cambridge mathematician; J.B.S. Haldane (1892–1964), an Oxford biochemist; and Sewall Wright (1889–1988), an American biologist. These men combined Darwinism, Mendelism, and the statistics of biometry to redefine continuous and discontinuous variation in calculus-like terms, so that the problem of saltation (discontinuous variation) became part of a larger Darwinian continuum. They examined the statistical survival rates of genes and showed that variability was maintained in large populations so that favourable genes could be selected. By thinking in terms of populations rather than individuals, the new synthesis allowed for an integration of the geographical and species concerns of field naturalists with abstract mathematical population genetics. By 1940, scientists had a clear picture of the process of evolution, on both a microscopic and a macroscopic level, opening the door to new debates concerning genetic drift, punctuated equilibrium, and the structure of the gene and genetic material itself.

Studies in population genetics provided much information on the macrobiological investigation of heredity. Chromosome studies had narrowed the site of genetic activity and even mapped the physical location of the control centres or genes. The question now was: How did a gene function? This hinged on a molecular-level examination of nuclear components, and that was a difficult task. Chemical analysis of the fluid in cells showed that each cell contained a soup of different molecules including proteins, enzymes, sugars, and phosphates, among many others. Separating these was a difficult and complex task made possible by the discovery of enzymes. In a sense, enzymes had been used for generations in activities such as fermentation and making bread, but the classification of these organic catalysts as specific chemical entities was first made by Wilhelm Friedrich Kühne (1837–1900), who isolated trypsin, a chemical found in pancreatic juices that aid in digestion. He called these chemicals enzymes, from the Greek *enzumos* meaning leavened. By 1900 dozens of enzymes had been identified, and it was clear that they were the chemical engines of most cell activity. What was not clear was how they were formed and how they worked at a molecular level.

Science and the State: The Atomic Bomb

In Germany and Italy, the Fascists used genetics and evolutionary theory to impose their racist ideology. The Nazis ordered a purification of "Aryan blood." In the midst of economic depression, deprivation, and anger, Hitler and Mussolini found many willing supporters for their radical political ideals. German scientists, in particular, found themselves in a difficult position. To continue their work, they had to accept the imposition of political control and in many cases even join the Nazi Party. Some did this willingly, many out of necessity. Others, particularly after Adolf Hitler seized power in 1933, chose to leave Germany, Italy, and Austria to escape fascism. Some, like Albert Einstein who left Germany in 1932, did not have much choice because they were Jewish or had too close associations with groups rejected or outlawed by the new rulers. They could stay and give up their scientific careers and perhaps their lives, or they could leave.

Among the escaped scientists were physicists who brought with them the fear that their German colleagues were developing a super weapon. They recognized how the advances made in radioactivity research by people such as the Curies and

Rutherford and the theoretical insight into the relationship between mass and energy articulated by Einstein suggested the potential power of materials such as uranium to produce a super bomb.

The secrets of radioactivity were still being unravelled after World War I, since the process turned out to be rather difficult for a number of reasons. One of the most practical was the lack of material to work with, since radioactive elements were rare and hard to refine. The second was contamination. Radiation affected laboratory equipment (and the health of the physicists themselves), and even though the amounts of material were small and precious, they were spilled, smeared, and dissipated around the labs until the labs were so contaminated that researchers had to abandon them and move to fresh space. Finally, radioactive materials were hard to work with because they didn't sit still. As they radiated, they literally turned into new substances.

One of the most fruitful research programs involved bombarding radioactive substances with neutrons. Enrico Fermi (1901–54), following work done by Jean Joliot and Irene Joliot-Curie, demonstrated that radioactive isotopes of many elements could be produced by neutron bombardment. In turn, Lise Meitner (1878–1968) with her research partner Otto Hahn (1879–1968), who had been working on problems in radioactivity since 1907, turned to the products of uranium bombardment around 1936. Meitner was a remarkable scientist, one of the first women to graduate from an Austrian university after they were opened to women. She travelled to Berlin and, against the advice of many, studied advanced physics, earning her doctorate in 1906. She became Germany's first female physics professor in 1926 and the first woman to receive a salary as a researcher at the Kaiser Wilhelm Institute.

In 1933, when Hitler took power, Meitner was protected because she was Austrian, but in 1938 Germany

10.1 LISE MEITNER

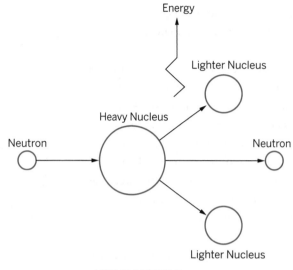

NUCLEAR FISSION

10.2 FISSION

annexed Austria, and Meitner came under German law. As a Jew, she was in danger, so she fled to Sweden. Hahn, who continued working with Fritz Strassmann (1902–80) on the problem of the creation of radioactive isotopes, corresponded regularly with her. He and Strassmann produced a series of experiments in which they bombarded uranium with neutrons, unexpectedly producing barium, a much lighter element. When Hahn sent Meitner the strange results, she followed the "liquid-drop" model of the nucleus suggested by Niels Bohr to determine that the uranium nucleus must have split apart and this explained the formation of the lighter element, barium. In Bohr's model, a neutron hitting the nucleus of a heavy element might do one of three things: it might get stuck, making the nucleus heavier by a neutron; it might knock off a bit of the nucleus, releasing a few protons and neutrons; or it might cause the nucleus to break apart, converting a small amount of matter to energy. Meitner called the splitting *fission*. She sent a letter with her conclusion to the periodical *Nature* early in 1939. (See figure 10.2.)

Bohr was at a conference in the United States when he heard about Meitner's finding. He rushed to tell other physicists about the discovery, and it caused a sensation. Several went back to their labs and replicated the discovery, confirming the work of Meitner, Hahn, and Strassmann. To Leo Szilard (1898–1964), however, the discovery opened up the horrific possibility of a nuclear bomb. Szilard, who was Hungarian, left Germany in 1933 in order to avoid persecution. That same year, while walking around London, he conceived the idea of a neutron chain reaction. The idea was a remarkably simple one. Since a neutron could strike an atomic nucleus and cause the release of one or more neutrons, these released neutrons could initiate a potentially continuous number of reactions, releasing an awe-inspiring amount of energy. (See figure 10.3.)

Szilard regarded this possible process to be so dangerous that, when he got a patent for the idea in 1936, he assigned it to the British Admiralty, the only way to both register the patent and keep it secret. When he learned about the fission of uranium, he recognized that a practical path to a neutron chain reaction was now available, and the results could be devastating.

World War II began in 1939, pitting the Axis powers of Germany, Italy, and Japan against the Allied forces of Britain (and the Commonwealth), France, and Belgium. The power of the German blitzkrieg overwhelmed Allied forces and Belgium surrendered in 1940, putting the Belgian Congo, then the greatest known source of uranium, in German hands. Werner Heisenberg sent a secret paper entitled "On the Possibility of Technical Energy Production from Uranium Splitting" to the German Army Weapons Bureau in early 1940. The threat of a German atomic bomb seemed very real. Suddenly, the ingredients for a super weapon were in Hitler's hands. Germany certainly had the industrial capability to create such a weapon, and, even though a number of the best and brightest physicists had fled Germany, there were still powerful minds such as Hahn and Strassmann available to work on such a project.

Szilard's early efforts to interest the American government in the idea of an atomic bomb had little apparent effect, so in 1941 he persuaded Einstein (now living in the United States) to write directly to President Franklin D. Roosevelt. Although this letter was very influential in the decision to proceed with the project, the military also consulted with other scientists such as Enrico Fermi, Neils Bohr, and John von Neumann about the potential of atomic power and weapons as they made their decision to go forward with the bomb.

One of the key figures who advised President Roosevelt on science policy was Vannevar Bush (1890–1974). He convinced Roosevelt to create the Office of Scientific Research and Development (OSRD), a federal research body, and he served as its head along with James Conant (1893–1978), president of Harvard University. One of the OSRD's main interests was nuclear power. Bush was an electrical engineer with experience in military research during World War I; he

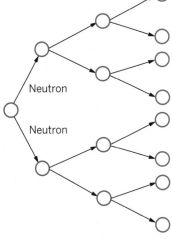

UNMODERATED CHAIN REACTION

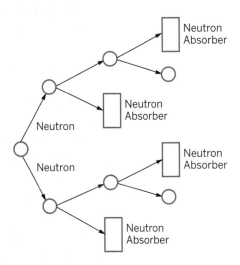

CONTROLLED FISSION CHAIN REACTION

10.3 CHAIN REACTIONS

was also a vice-president of the Massachusetts Institute of Technology (MIT) and head of the Carnegie Institution. He planned to use a network of universities to do research rather than expand or establish new federal laboratories. The universities were happy to take on the work since the funding provided was on a scale unseen since the days of chemical warfare. Under this system of federal money directed to private researchers, many of the most important American research centres were established, including the Jet Propulsion Laboratory at the California Institute of Technology (Caltech), the Radiation Laboratories under the auspices of the University of California and MIT, and the Metallurgical Laboratory at the University of Chicago. Federal money for research (not including the Manhattan Project) rose from $74 million in 1940 to $1.59 billion by the end of the war.

The Manhattan Project

In October 1941, Roosevelt was briefed on the potential of building a nuclear weapon. He authorized this research on December 6, 1941, the day before the Japanese attack on Pearl Harbor. The Army's initial part of the top secret project was called the Manhattan Engineering District so as to hide its true nature (and location). It brought together some of the most powerful minds in science in a race to beat the Germans to a super bomb. Better known simply as the Manhattan Project, it changed the course of world history.

The first steps toward a weapon were to confirm the possibility of a sustained chain reaction by fission and to evaluate the technical problems associated with the production of the necessary materials. The fission problem was undertaken by a team headed by Italian émigré physicist Enrico Fermi, who had already won a Nobel Prize for his work on radioactive elements. When he travelled to Sweden in 1938 to receive the prize, he was criticized in Italy for failing to wear a Fascist uniform or give the Fascist salute. He and his family took the opportunity of their trip abroad to escape and never returned to Italy. Fermi began his work on the Manhattan Project at the University of Chicago. He built a small reactor, called an atomic pile, under the bleachers at the west end of Stagg Field in a space that had previously been squash courts. It was constructed of blocks of graphite (a kind of carbon that absorbed neutrons), uranium, and uranium oxide. Control rods made of cadmium, designed to limit the rate of fission, were inserted in holes through the blocks. On December 2, 1942, at 3:25 in the afternoon, Fermi's team slowly pulled out the control rods and a controlled self-sustaining fission reaction began. (See figure 10.3.) This was the official start of the "atomic age."

Arthur Compton (1892–1962), a member of the committee investigating the possibility of building a nuclear bomb, was present at the reactor test. He called James B. Conant at Harvard to pass on, in code, the news of the success in a now famous telephone exchange:

"The Italian navigator has landed in the New World," said Compton.

"How were the natives?" asked Conant.

"Very friendly."[4]

With a number of the theoretical questions answered by Fermi's atomic pile, the next stage was to produce an uncontrolled chain reaction. This was a big project in terms of cost, scope, and originality. Brigadier-General Leslie Groves was selected as the military head of the project. Trained as an engineer, he had experience organizing large construction works, including the building of the Pentagon. He chose Robert Oppenheimer (1904–67) as the Scientific Director. Ultimately, they would be in charge of hundreds of scientists and thousands of military and civilian workers.

Two major hurdles had to be overcome. The first was the uranium itself. At the start of the war, the amount of refined uranium in the world could be measured in grams, but tons of uranium ore would be needed for the Manhattan Project. With the best-known sources in Europe and the Belgian Congo under the control of Germany, new mines were needed. A massive search for ore was undertaken. Sources had to be secure and in friendly territory. Fortunately, large resources were found in northern Canada, which became one of the main suppliers of raw ore. Aboriginal workers who had carried out the radioactive material reported disturbing trends of chronic illnesses and cancers in later years.

The second problem was a sticky technical issue about the kind of material needed. Neils Bohr had pointed out that the isotope U_{235} would sustain fission far better than U_{238}. While both occurred naturally, only about 0.7 per cent of uranium atoms were U_{235}. Separating the two was an enormous job because they were chemically identical and only about 1 per cent different in mass. Three methods of

---

4. Arthur Compton, *Atomic Quest* (New York: Oxford University Press, 1956) 144.

separation were tried: magnetic separation, gaseous diffusion, and gas centrifuge. Magnetic separation looked promising. Uranium tetrachloride in gas form was passed across a strong magnetic field. The heavier U_{238} was deflected less and thus separated from the desired U_{235}. This system was created by Ernest O. Lawrence (1901–58), one of the inventors of the cyclotron, at the University of California at Berkeley, but after millions of dollars were spent on the factory, it failed to produce the needed quantities. The gas centrifuge worked in experimental operation but could not be scaled up to industrial capacity. That left gaseous diffusion. As uranium hexafluoride gas was passed through porous clay filters, the lighter U_{235} passed more easily; after repeated filtration, uranium of the required purity was obtained. The massive Oak Ridge plant in Tennessee used this method to produce much of the material for the uranium bomb.

With the uranium supply issue resolved, a new material problem was created when Glenn Seaborg (1912–99) suggested that plutonium offered even better fission properties than U_{235}. The best isotope of the recently discovered element Pu_{239} was produced by placing U_{238} in a reactor and letting it pick up neutrons. Uranium was converted to plutonium, and the plutonium was enriched by having neutrons added to make fissionable material. Such a reactor was called a "breeder reactor." Plutonium was therefore added to the production system, and Seaborg was put in charge of producing it for the project.

The scientific team was assembled at Los Alamos, New Mexico, to work out the scientific and engineering difficulties involved in creating an explosive reaction. At the heart of the problem was getting the mass of material at the centre of the device to go from no sustained fission to fission at a specific moment. They achieved this by creating an implosion with conventional explosives that compressed the fissionable material and set off the uncontrolled chain reaction. The effort to bring together all the elements took several years, but finally the Trinity Test, of a bomb called Gadget, took place at Alamogordo, New Mexico, on July 16, 1945. The Trinity Test exploded at about 5:30 a.m. Enrico Fermi, an eye witness, estimated the power was about equal to 10,000 tons of TNT. Oppenheimer quoted the *Bhagavad-Gita*, saying "... now I am become Death, the destroyer of worlds...."[5]

The explosion presented both a scientific triumph and an ethical dilemma. Germany had surrendered on May 7, 1945. The race against the Germans was over.

....

5. Robert Oppenheimer, quoting the *Bhagavad-Gita*, in Ferenc Morton Szasz, *The Day the Sun Rose Twice* (Albuquerque: University of New Mexico Press, 1984) 89.

For many of the scientists, especially the Europeans displaced by the Fascists, that meant the end of the need for a super bomb. What few realized was that as early as 1943 Groves had already been looking at the use of the weapon in the Pacific theatre. The construction and delivery of the weapons were pushed ahead. In the end, three bombs were constructed: Little Boy, Fat Man, and Bomb #4.

Leo Szilard was horrified that the project was still continuing and once again went to Albert Einstein for a letter of introduction to Roosevelt. Szilard wanted to persuade the president not to use the weapons. He was concerned about both the devastation they would cause and the resulting arms race that might destroy the planet. Before the meeting could be arranged, Roosevelt died on April 12, 1945. Harry S. Truman became president, and among his briefings on assuming the office he learned about the Manhattan Project. Szilard could not meet Truman, meeting instead with James Byrnes, Truman's Secretary of State. Byrnes had already advised Truman to use the bomb as soon as possible, and he rejected Szilard's concerns.

Truman's advisors concurred with the military that the bombs should be used without warning on Japan. Further, targets were to be chosen so that the effects of the blasts could be studied. Truman faced a hard decision. Although the Japanese were losing badly, they were offering stiff resistance and publicly vowed to fight to the death. Military planners suggested that as many as 1 million casualties might be expected from an invasion of mainland Japan, whereas many fewer would die from the bomb attacks. There was still anger over the sneak attack on Pearl Harbor, and to further complicate the situation the defeat of Germany had freed the Soviet military to join the battle against Japan. Truman wanted to avoid the division of Japan following the precedent of the partitioning of Germany if the Soviet Union were to participate in the war.

Therefore, Truman authorized the use of the atomic bombs on July 21, 1945. On July 26 the Allies released the Potsdam Declaration calling on the Japanese to surrender unconditionally or face "prompt and utter destruction." Two days later the Japanese government rejected the call for surrender. Preparations were made for dropping the bombs. With complete air superiority, there was no interference on August 6 as the *Enola Gay* flew to its target and dropped Little Boy on Hiroshima. On August 9 Fat Man destroyed Nagasaki. Japan surrendered unconditionally on August 14. More than 200,000 people were killed in the two nuclear blasts, the vast majority of them civilians. Many more were injured, some from radiation burns and poisoning.

The effort to build the bombs cost about $1.8 billion. This can be compared to the $5.4 billion the United States spent on tanks and the $2.6 billion spent on

CONNECTIONS

Science and Fascism

On April 7, 1933, the German Government passed a law called the *Gesetz zur Wiederherstellung des Berufsbeamtentums*, the Law for the Restoration of the Civil Service. This law targeted "non-Aryan" civil servants and forced them out of their jobs, including those scientists who worked at German universities and research centres such as the Kaiser Wilhelm Institutes. Although Max Planck made a personal appeal to Adolf Hitler to reverse this decision, Jewish scientists were forced to leave their jobs and to flee the country. A further group of scientists fled Italy when Mussolini came to power and Austria after the Nazi Anschluss in 1938.

Fascist ideology was based on racism, but it also attacked the idea of the universality of science. Hitler explicitly attacked what he called "Jewish science," including the theory of relativity, quantum mechanics, and uncertainty theory. He placed Jewish science in the same category as jazz music and modern art as degenerate forms of culture. This policy took a terrible toll in human terms, and it also transferred the centre of modern physics research from Germany to the United States and Britain.

Albert Einstein left for Britain in 1932, before the law had been passed, and was instrumental in publicizing the hardship of scientists displaced by fascist policies. The economist William Beveridge

all other explosives. Although building the nuclear bombs was not exactly a bargain, given the results many felt it was a justifiable expense. It remains a point of debate, however, whether the race for the bomb had really been necessary at all. At the end of the war there was little evidence that nuclear research was being undertaken in Germany, although some recently declassified documents suggest that key German scientists might have been able to construct such a weapon. The documents emerged after the war during Operation "Alsos," in which the Allies debriefed ten of Germany's top scientists, including Otto Hahn and Werner Heisenberg. Their discussion of the bombing of Hiroshima and Nagasaki suggests that they already understood the technical details as well as the basic principles behind the bomb. This changes the question from "Were the Germans working on an atomic bomb?" to "Why didn't they build one?"

set up an Academic Assistance Council in the UK, aimed at rescuing Jewish and politically vulnerable academics. Many prominent academics quickly backed this organization—J.B.S. Haldane, John Maynard Keynes, Ernest Rutherford, G.M. Trevelyan, and A.E. Housman, to name a few. This organization eventually helped over 1,500 academics escape Germany and Austria, in order to continue their work in Britain and other free countries.

The list of physicists, mathematicians, chemists, and other scientists who escaped from fascist Europe was remarkable. Physicist Rudolf Peierls (1907–1995) went on to work with the British team connected to the Manhattan Project. Nuclear physicist Hans Bethe (1906–2005) was fired from his position in the University of Tübingen and joined Peierls in 1933. Leo Szilard fled to Britain in 1933 as did Max Born (1882–1970) who went on to write a best-selling science book, *The Restless Universe*. Lise Meitner escaped in 1938, travelling first to the Netherlands and then to Sweden. Enrico Fermi escaped fascist Italy in 1938 when he travelled to Stockholm to receive the Nobel Prize in Physics and never returned. Many of the scientists moved to the United States and were instrumental to the Manhattan Project. A Stanford economist, Petra Moser, has estimated that US patents increased by 31 per cent in fields common among Jewish scientists who fled Nazi Germany for America. Of the scientists who went to Britain in the years before and at the beginning of the war, many went on to win Nobel Prizes, to receive knighthoods, and over 100 became members of the Royal Society or British Academy.

Historians have speculated that German racial policies, besides being inhumane in and of themselves, used up massive resources and crippled German scientific work. Refugee scientists helped give the Allies a scientific and technological advantage over the Nazis and contributed to establishing Big Science in the Western world.

National Security and Science Policy

In some ways, the Trinity Test let the nuclear genie out of the bottle, but in the wake of the bombing of Japan, as Szilard predicted, there followed an arms race that put the safety of all life on the planet at risk. It raises, even more than the involvement of scientists in chemical warfare, the issue of what role scientists have to play in military activity. Were the scientists responsible for the bomb, or did that responsibility lie with the politicians and military leaders? To what extent should national security determine science policy?

The postwar years provided a simple answer for many. Where nuclear weapons and national security were concerned, scientists were expected to work for the state. Three things were needed to manufacture a nuclear weapon from scratch:

access to uranium, an industrial infrastructure capable of manufacturing the components of the weapon, and intellectual resources sufficient to manage the first two elements. At the end of the war, four nations had the capacity to build nuclear weapons: the United States, Canada, Britain, and the Soviet Union. The United States had the bomb already, while Canada, as one of the lesser powers, did not pursue an independent nuclear program, preferring to disarm and not spend the money. The Soviet Union began its work on a weapon shortly after the war, and on August 29, 1949, at Semipalatinsk in Kazakhstan detonated its first atomic bomb, Joe 1, with about a 10- to 20-kiloton blast. British scientists, some of whom had taken part in the Manhattan Project, were pressed into nuclear work. Although badly battered by the war, Britain put together its own nuclear program and tested its first weapon in 1952.

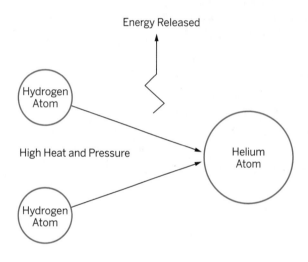

10.4 FUSION

Even before the end of the war, the alliance between the Western powers and the Soviet Union was strained. Both sides saw the other as the next enemy. General George S. Patton even secretly discussed taking "his boys" on to Moscow after he defeated the Germans. So it was not surprising that, unlike during the years following World War I, the scientific teams of World War II were not disbanded and sent back to the groves of academe. Rather, in this charged environment, which has become known as the Cold War, the next step in destructive weapons was taken. As early as 1938 Hans Albrecht Bethe (1906–2005) had studied thermonuclear reactions involving light elements in order to understand how the sun (and all stars) functioned. He concluded that under the right conditions hydrogen would undergo fusion. At the great temperature and gravitational pressure inside a star, hydrogen atoms were squeezed together (or fused) to form helium. When that happened, a small amount of mass was converted to energy. The basic principle looked simple: $_1H^2 + {}_1H^2 \rightarrow {}_2He^4$. (See figure 10.4.)

The energy of the stars comes from the loss of mass that accompanies the transformation, following Einstein's relationship of $E = mc^2$. About 0.63 per cent of the mass of the original atoms is lost by conversion to energy. While this seems tiny, it is huge compared to the energy derived from fission, in which the splitting of a uranium atom releases only 0.056 per cent of the mass as energy. In other words, fusion released more than ten times as much energy as fission. Hence the power of the sun, which converts about 650,000,000 tons of hydrogen to helium every second.

Edward Teller (1908–2003) had also been thinking about the power of fusion in the early 1940s. Following the invention of the fission bomb he saw a way to create conditions for a fusion bomb. There is still a great deal of debate over the creation of the first fusion bomb, or H-bomb. A priority fight between Stanislaw M. Ulam (1909–84) and Teller as well as the secrecy surrounding the nuclear arms race have clouded the exact history, but it is likely that the theoretical work was done by Teller and others at Los Alamos, and the initial design of the device was done by Richard L. Garwin (1928–), a young physicist on a research visit to the facility. The basic structure of the hydrogen bomb was a fission bomb surrounded by, or next to, a high hydrogen package. The fission bomb exploded in such a way that it reached the ignition point for a fusion reaction. Teller championed the development of the weapon, and the first test bomb was ready in 1952. It stood two stories high, and when it was detonated, it vapourized the Pacific island of Elugelab. Its power was equal to 10.4 million tons of high explosive, or about 700 times the power of the atomic bomb dropped on Hiroshima.

Unlike the fission bomb, there was no theoretical limit to the destructive power of a hydrogen bomb. Scientists conceived of weapons powerful enough to blow huge holes in the atmosphere, flatten entire countries, or create gigantic tidal waves. The measurement system for destructive power jumped an order of magnitude from kilotons to megatons. In addition to the raw destructive power of the fusion bomb, there was also massive radiation danger. The larger weapons released far more radioactive materials such as strontium-90 and cesium-137, which they blew high into the atmosphere from where they rained down on a wide area.

A number of scientists opposed the development of these super bombs. Robert Oppenheimer was stripped of his security clearance and effectively prevented from doing further research for the military because of his opposition to the development of bigger weapons. Beginning in 1947 the *Bulletin of Atomic Scientists* placed a doomsday clock on its cover. The first clock was set at seven minutes to midnight. In 1953 after both the United States and the Soviet Union tested fusion weapons, it was set to two minutes to midnight. In 1955 Bertrand Russell and Albert Einstein issued a manifesto calling on all governments to find peaceful means to resolve conflicts and to give up nuclear arms. It was signed by Max Born,

Percy Bridgman, Leopold Infeld, Frederic Joliot-Curie, Hermann Muller, Linus Pauling, Cecil Powell, Joseph Rotblat, and Hideki Yukawa. Although the war effort had forced scientists to choose national interests over the international community of science, the profile of these scientists indicates a countermove to recreate an international republic of science. From the end of World War II, these competing loyalties vied with one another, with national pride and security (and funding sources) pulling against a belief in the universal nature of scientific understanding. Einstein said, "I know not with what weapons World III will be fought, but World War IV will be fought with sticks and stones."[6]

Discovering DNA

One of the signatories of this manifesto, the geneticist Hermann Muller, studied mutations in genes and was particularly aware of the dangers of radiation from his biological work. In one of the ironies of modern science, the science of destruction and the science of life drew on material from the same source. X-rays and radioactive tracers played a key role in taking the work of the geneticists into the heart of the cell. Quietly, almost without notice by the public or the politicians, whose attention was on physics, the quest for the control mechanism in inheritance continued. Using many of the resources and techniques from the physics lab, genetics and cell biology gained momentum in the aftermath of World War II.

The area of investigation shifted from tracing population inheritance patterns to understanding the structure of the genetic material itself. Biologists were influenced by the understanding of nature at an atomic and subatomic level that physics provided. They increasingly sought to unravel genetics at the chromosomal and molecular level. The first step was to identify the substance within the cells that contained genetic information.

At about the same time that Mendel was working on pea plants, the German scientist Friedrich Miescher (1844–95) was analyzing pus from discarded hospital bandages, hoping to find a cure for infections that frequently occurred in bandaged wounds. He tested the material he had isolated and discovered that some of it was not a protein. This was a puzzle: What was this substance doing in a mass of proteins? He found that the material, which he called nuclein (later nucleic acid), contained a high level of phosphorous and concluded that it was part of the

6. Alice Calaprice, *The New Quotable Einstein* (Princeton, NJ: Princeton University Press, 2005) 173.

nucleus of a cell and acted as a storage space for phosphorus, which the body needs in small quantities. In the 1880s, August Weismann, while investigating chromosomes, discovered that this same nucleic acid was present in them. By the time Morgan's lab at Columbia University began its *drosophila* experiments, it was accepted that genetic information was passed on through something in the chromosomes, either the protein or this nucleic acid.

In 1928 Fred Griffith (1881–1941), a British medical officer, was studying the bacteria responsible for a pneumonia epidemic and found that the pneumococcus bacteria occurred in two forms. One form (*S*) was smooth and highly infective; the other (*R*) was rough and harmless. Because both were found in patients, he wondered what the relationship between the two might be. Griffith injected mice with harmless living *R* cells and heat-killed *S* cells. The mice died, and living cells of both types were found in their bodies. He concluded that some substance from the *S* cells was transferred to the *R* cells, changing them into virulent *S* type cells. If it could be isolated, this would be the control substance.

In 1944 the team of Oswald Theodore Avery (1877–1955), Colin MacLeod (1909–72), and Maclyn McCarty (1911–2005), at the Rockefeller Institute in New York, reported that they had isolated Griffith's transforming material and that it was a nucleic acid—specifically, deoxyribonucleic acid, or DNA. Although not everyone was convinced, the large molecule came under increased study. The Phage School (named for the phage viruses they studied) used radioactive tracers to follow molecular events in phage infection; in 1952 A.D. Hershey (1908–97) and Martha Chase (1930–2003) discovered that phages leave their protein coats behind and infect bacterial cells with their DNA. There is much debate as to whether the Avery/MacLeod or the Hershey/Chase experiments constitute the moment of the identification of DNA as the genetic material. In many ways, the answer depends on which professional subdiscipline (molecular biology or bacteriology) is considered more important, not which one supplied the definitive answer. The two findings together, however, pointed all interested geneticists toward the structure of the DNA molecule as the key to understanding inheritance.

The work of Hershey and Chase had more immediate impact than that of Avery and MacLeod because the chemical composition of DNA had been determined. In 1950 Erwin Chargaff (1929–92) made a major step in unravelling the complex nature of DNA. He established that the molecule contained four types of nitrogenous bases, which existed in a one-to-one ratio of adenine to thymine and guanine to cytosine. This held true for all the different samples from a range of organisms. These bases could follow each other in any arbitrary order on the polynucleotide chain, so there was a possibility that the order of the bases somehow affected inheritance. This discovery was one of the key components needed to create a model of DNA.

The race to discover the structure of DNA was on. It has become one of the iconic case studies in the history of science, raising questions about how scientists actually work, the way credit is awarded, and what effect cultural expectations, both within the scientific community and in the larger society, have on the practice of science. Historians have been able to ask these questions about these events because many of the people involved reported what happened. Particularly telling in this regard were the frank account of his own work written by James Watson in *The Double Helix* and an alternative view of events found in Anne Sayre's study, *Rosalind Franklin and DNA*.

A number of laboratories were working on DNA. Linus Pauling (1901–94) in the United States turned his attention to the structure, while at the University of London Rosalind Franklin (1920–58), in conjunction with Maurice Wilkins (1916–2004), used X-ray crystallography to analyze DNA. Franklin did her scientific training at Newnham College, Cambridge, obtaining her MA in physical chemistry in 1947. After a series of research positions, including the Centre national de la recherche scientifique, she obtained a fellowship at King's College, London. In 1951, Franklin produced extremely fine X-ray diffraction photographs of what was called the B form of DNA, but she did not immediately see the structure, because it was only one of a number of different images. She was also in an unpleasant situation at London, as Wilkins treated her like a technician rather than a colleague, and the male-only world of the university and the laboratory excluded her from the informal networks so important for both support and the contacts that often underlie scientific work.

Into this situation came Francis Crick (1916–2004) and James Watson (1928–). Crick had been employed in war work as a physicist but was now working on his PhD at Cambridge in biophysics, having read and been influenced by Schrödinger's book *What Is Life? The Physical Aspects of the Living Cell*. He met James Watson, an American who had already earned his PhD at the age of 22 and who had done work as a phage geneticist. Although both men were supposed to be working on other projects—Crick on his thesis and Watson on viruses—they decided to try to find the structure of DNA. They set out to build a model that would agree with the X-ray diffraction data and account both for autocatalysis (DNA splitting in half) and heterocatalysis (transferring information to create proteins and other cells, as took place in reproduction).

It was clear to Watson and Crick that the molecule was a long-chain polymer with a constant diameter. They already had the basic chemistry of bases and sugars, and they knew that any arrangement of nucleotides in the DNA molecule had to account for the regularity of the molecule's structure as well as its chemical stability. It also had to account for how the molecule could replicate itself faithfully. Three factors influenced their thinking. First, unlike bases seemed to attract each other. Second, Chargaff had shown that the ratio of bases was 1:1. Finally, Pauling, whose model building inspired theirs, introduced the idea of the helix. While an early attempt went badly wrong and they were warned by their superiors to concentrate on their own work, Watson and Crick persisted. One major breakthrough came when Maurice Wilkins showed them, without Franklin's knowledge or permission, her crystallographic photographs of DNA. The image revealed that the molecule had to be a

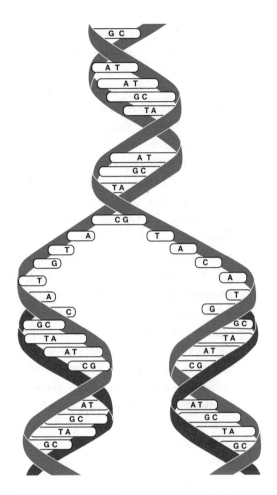

10.5 THE DOUBLE HELIX

double helix; Watson and Crick constructed a model that looked like a twisted ladder made of two spines with the base pairs (adenine, thymine, guanine, and cytosine) arranged as rungs. (See figure 10.5.) It accounted for genetic, biochemical, and structural characteristics and explained auto- and heterocatalysis. Equally, it predicted the mechanism of storage of genetic information in the order of the base pairs. On April 2, 1953, Watson and Crick published their model in the journal *Nature*. The brief paper, only a single page, was a sensation. Their model was confirmed, and the cellular control system was revealed. For this work they won the Nobel Prize for Physiology or Medicine in 1962, sharing the honour with

Wilkins. (See plate 6 of Watson and Crick with their model of DNA.) Franklin, who had spent her career working with X-rays, had already died of cancer and so was not eligible to share the prize.

There is no doubt that Watson and Crick made a brilliant discovery, but their methods seem to run contrary to what is commonly taught as good scientific practice. They did no primary research but rather sought information from others, with little to share in return. Their access to, and use of, Franklin's work seems underhanded and questionable, and they did not acknowledge her contribution in their initial publication. On the one hand, scientific information was regarded as "public" in the sense that it was (and still to a large extent is) expected that work and results would be talked about informally over dinner or beer, as well as being presented formally at conferences and published. It was also accepted that others would make use of that material, regardless of its formal or informal source. On the other hand, professional success was based in part on priority and the acknowledged importance of work. If scientists used other people's work, they were expected to acknowledge their sources. Watson and Crick ignored part of the code of gentlemanly behaviour implicit in the ideology of science dating back to the scientific revolution. The race for the structure of DNA suggests that in science "anything goes" (as philosopher Paul Feyerabend claimed) and that its ideology has come to resemble more closely some Darwinian struggle for existence than the gentlemanly witnessing espoused by Robert Boyle. Should scientific research be constrained by rules of polite conduct, or does the pursuit of knowledge override such limitations?

Conclusion

The utility and power of science was increasingly recognized by governments in the postwar years, and the organization of science was transformed as a result. With the start of the Cold War, governments did not dismiss their teams of scientists as they had after World War I but instead chose to support scientific research and direct it to politically determined objectives. All the industrial nations and many of the less technically advanced countries recognized the necessity of having and developing scientific knowledge and integrating national interests with research interests. After World War II, almost all important scientific breakthroughs came from scientific teams rather than individual scientists. These teams were increasingly funded by major research institutions or by federal governments through grants and contracts. While the emergence of "Big Science" had its roots in the Chemists' War, it was really the Physicists' War that cemented the power of large laboratories and established the massive infrastructure needed to support them.

In disciplinary terms, physics replaced chemistry as the premier science and became the focus of increasing public, governmental, educational, and industrial attention. Biology gained by its connection to physics but remained a distant second in terms of funding. While the military forces of the world powers continued to regard science and scientists with some suspicion, the effectiveness of science-backed military activity was demonstrated first by Germany's blitzkrieg and more spectacularly by the first belligerent use of nuclear weapons. The utility of science after the atomic bomb was too powerful to allow scientists to slip back into their quiet academic laboratories as they had largely done at the end of World War I. The new way to do science had been proven.

"Big Science" with big funding, big laboratories, and larger and larger teams of scientists working on research agendas established by national governments was the way of the future. The United States and the Soviet Union, staring at each other across an ideological chasm, mustered their scientific forces. This marshalling of science produced the new arms race and its corollary, the race to space. The place of science and the practice of research were permanently changed.

Essay Questions

1. What was the "new synthesis" in biology?

2. What two ideas were needed to make an atomic bomb possible?

3. Did the discovery of the structure of DNA circumvent traditional methods of scientific research?

4. Why was indeterminacy necessary for modern physics?

CHAPTER TIMELINE

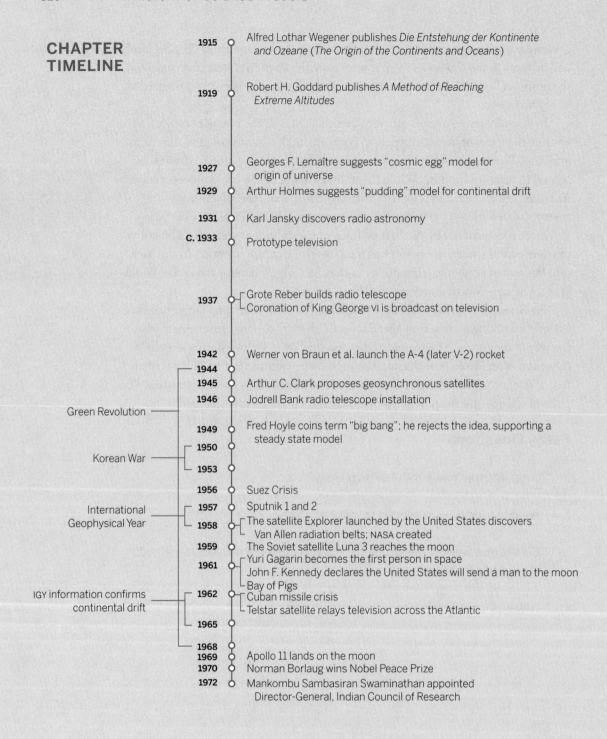

1915 — Alfred Lothar Wegener publishes *Die Entstehung der Kontinente and Ozeane* (*The Origin of the Continents and Oceans*)

1919 — Robert H. Goddard publishes *A Method of Reaching Extreme Altitudes*

1927 — Georges F. Lemaître suggests "cosmic egg" model for origin of universe

1929 — Arthur Holmes suggests "pudding" model for continental drift

1931 — Karl Jansky discovers radio astronomy

C. 1933 — Prototype television

1937 — Grote Reber builds radio telescope
Coronation of King George VI is broadcast on television

1942 — Werner von Braun et al. launch the A-4 (later V-2) rocket

1944

1945 — Arthur C. Clark proposes geosynchronous satellites

1946 — Jodrell Bank radio telescope installation

Green Revolution

1949 — Fred Hoyle coins term "big bang"; he rejects the idea, supporting a steady state model

1950

Korean War

1953

1956 — Suez Crisis

International Geophysical Year

1957 — Sputnik 1 and 2

1958 — The satellite Explorer launched by the United States discovers Van Allen radiation belts; NASA created

1959 — The Soviet satellite Luna 3 reaches the moon

1961 — Yuri Gagarin becomes the first person in space
John F. Kennedy declares the United States will send a man to the moon
Bay of Pigs

IGY information confirms continental drift

1962 — Cuban missile crisis
Telstar satellite relays television across the Atlantic

1965

1968

1969 — Apollo 11 lands on the moon

1970 — Norman Borlaug wins Nobel Peace Prize

1972 — Mankombu Sambasiran Swaminathan appointed Director-General, Indian Council of Research

1957: THE YEAR THE WORLD BECAME A PLANET

11

On October 4, 1957, Moscow radio announced that the Soviet Union had launched Sputnik and the distant stars were one step closer. Dreams of space flight went back centuries, but only with the triumph of Big Science after World War II did the necessary components of scientific organization and technological development make it possible to launch objects into space. While individual achievement was still important, especially when Nobel Prizes were increasingly touted as a benchmark for the prowess of national science, the days of the lone scientist working in a homemade laboratory were essentially over. The bigger the question, the bigger the team created to solve it.

In addition, science was now a public commodity. Sputnik, which means "fellow traveller," orbited the Earth at an altitude of 900 kilometres and flashed across the sky at 29,000 kilometres per hour. It was a beacon in the night, proudly proclaiming the scientific and technological superiority of the Soviet Union. Publicity, prestige, and national pride were now linked to scientific achievement in a much more public way than seventeenth-century natural philosophers had found in their patrons' courts. Interest in science sent ripples through popular culture, with the appearance of movies and novels about spies stealing secret formulas or smuggling researchers out of foreign countries. Nevil Shute wrote about the end of the world by atomic holocaust in *On the Beach* (1957), and his book was turned into a film of the same title in 1959. The Nobel Prizes, which had for the most part been a minor news item before the war, now made newspaper headlines around the world.

The launch of Sputnik had not only political and military implications but also a psychological impact. For the first time, something created by human hands had left the Earth. While the concept of the Earth as a planet was firmly established, in practical terms it had been the limit of human experience until 1957. With the opening of the space era, it was possible to conceive of the solar system and the galaxy as something more than images seen through a telescope. What had been the realm of fantasy and science fiction became reality, expanding the potential zone of human activity enormously. It also hastened the spread of the idea of the Earth as a single biophysical unit—a single world, not a collection of continents or politically distinct regions that could function without regard for the state of other parts of the globe.

The same year that Sputnik was launched, 1957, was proclaimed the International Geophysical Year (IGY). The International Council of Scientific Unions (ICSU), an umbrella organization of various national scientific associations, initiated a massive research project to study both the Earth and the Earth's interaction with the universe. IGY ran from July 1957 to December 1958, with some of its projects running much longer. IGY and Sputnik also embody the ideological tension facing science and the scientific community in the modern world. The science of the IGY was to be cooperative, international, creative, nonpartisan, and open. Many scientists felt that these were the true characteristics of science, and they worked to get governments and the international community to celebrate this vision. The race to space, on the other hand, was characterized by secrecy, nationalism, and partisanship; moreover its military foundation was dedicated to destruction. Realizing that to compete militarily it was necessary to compete scientifically, military leaders around the world had little trouble finding scientists to work on their projects and generally enjoyed the support of the scientific community. Thus, the race to space, the arms race, and IGY represent different kinds of utility to the state. It is not surprising that IGY and Sputnik coexisted despite their apparently different characteristics, since both offered benefits to those who sponsored the work.

The significance of IGY and Sputnik must also be seen in the context of a world on the edge. The United States tested a non-deliverable fusion bomb in 1952. The Soviet Union tested its first partial fusion bomb, Joe 4, in 1953. In the same year, the Korean War ended in a stalemate. The Warsaw Pact, the Eastern Bloc equivalent of the North Atlantic Treaty Organization (NATO), was formed in 1955, heightening fears of another European war. And 1956 was particularly tense. The Egyptians, who were establishing ties with Eastern Bloc countries, planned to nationalize the Suez

Canal, thus cutting off transportation to the East and oil supplies to Europe. To thwart this, Britain, France, and Israel planned a daring attack. Israel seized the Suez Canal and then surrendered it to the British and the French. When the American government failed to back them and the Soviet Union threatened intervention, the invading troops were forced to withdraw. Britain and France were left looking powerless and defeated. In the same year, a revolt against Soviet rule in Hungary was brutally suppressed, and the American military exploded the biggest hydrogen bomb ever at Bikini Atoll.

The launch of Sputnik 2 on November 3, 1957, confirmed the capabilities of the Soviet rocket program and clarified the military threat. Sputnik 2 was not only much larger than Sputnik 1 but on board was Laika, a Siberian dog. She lived in orbit for eight days until her oxygen

11.1 ATOMIC BOMB TEST AT BIKINI ATOLL (1956)

supply ran out. Weighing 508 kilograms, Sputnik 2 was large enough to be a weapon. Where World War II had expanded the zone of combat from the few kilometres of the trenches of World War I to the range of heavy bombers and V-2 rockets, the launch of the Sputnik satellites effectively eliminated the concept of the "front line" altogether.

And yet the launch of the first satellite fascinated the world's populations. Around the globe, people tuned into the radio beep that Sputnik broadcast and stood staring at the night sky hoping to see the satellite zoom by. Some people, particularly science fiction writers and fans, considered the launch of Sputnik the end of human childhood, a milestone that signalled the beginning of the next era of human development when we would leave the cradle of the Earth and move out to the stars.

The International Geophysical Year (IGY)

Lloyd V. Berkner (1905–67) and Sydney Chapman (1888–1970) initiated the idea of the IGY in 1950. Berkner was an electrical engineer by training but had wide interests in electronics, nuclear development, radar and radio, rocketry, and atmospheric science. In 1950 he was working on ionospheric physics in the Department of Terrestrial Magnetism of the Carnegie Institution of Washington; from 1951 to 1960 he was head of Associated Universities, Inc., which ran the Brookhaven Laboratories for the Atomic Energy Commission (AEC). Chapman, who had degrees in engineering, physics, and mathematics, had been interested for a long time in atmospheric science, having proposed a theory of ozone creation and depletion in 1930; in that same year, in association with Vincent Ferraro, he theorized that magnetic storms were caused when plasma ejected from the sun enveloped the Earth.

Since Berkner had served with Admiral Byrd on his Antarctic expedition during the second International Polar Year in 1932–33 (the first occurred in 1882–83), he used the Polar Year as a model for this new and bigger project. The IGY was initiated in 1952 by the ICSU, an arm of the United Nations created to promote international scientific cooperation. To manage the actual operations ICSU established the Comité Spécial de l'Année Géophysique Internationale, with Chapman as its president and Berkner as vice-president. In the end, some 67 nations took part. In addition to gathering information, the International Council of Scientific Unions established the World Data Centers, which archived and disseminated the information produced. The World Data Centers continue to function to this day.

The year 1957 was picked for IGY because it coincided with an expected peak in sun-spot activity, which followed an 11-year cycle. Other areas of research were aurora and air glow, cosmic rays, geomagnetism, glaciology, gravity, ionospheric physics, longitude and latitude determination, meteorology, oceanography, rocketry, and seismology. The result of this massive undertaking was one of the greatest records of the structure and function of planet Earth and its neighbourhood.

The geological and oceanographic work provided the first global map of the structure of the planet. Although huge amounts of work had been done mapping the globe and geological surveys had been undertaken by various governments, especially the British, the material was not well integrated. Therefore, one of the tasks of the IGY was to coordinate masses of previously gathered data.

One of the major topics of interest that Berkner and the geologists hoped to address with this compilation of existing and new data was the question of ocean and continent formation. Although the specific approaches to the research question came primarily from nineteenth-century geology, particularly from Lyell and his *Principles of Geology* (1830), new tools such as the seismograph were changing the way geologists and geophysicists studied the Earth.

Continental Drift

Like the shift from the Newtonian to the Einsteinian universe, the stability of geology was called into question when Alfred Lothar Wegener (1880–1930) introduced his ideas about continental drift in 1912. Trained as an astronomer and working mostly in meteorology, Wegener had noticed both the apparent "fit" of geological regions, such as South America with the west coast of Africa, and certain similarities in the distribution of animals and fossils; he combined these with paleoclimatological evidence. He was not the first to notice these relationships. Eduard Suess (1831–1914) had suggested the existence of supercontinents called Gondwanaland and Laurasia in his massive five-volume geological work *Das Antlitz der Erde* (*The Face of the Earth*), published between 1883 and 1909. Suess explained that the land bridges that connected the continents in prehistoric times had since disappeared, eroded, or collapsed into the oceans. In contrast, Wegener argued that the Earth had been molten in its earliest days and that it had contracted as it cooled, allowing lighter continental material to rise above the denser basalt of the crust, forming one supercontinent, which he called Pangaea. This continent subsequently broke up, and the parts representing the current continental masses "floated" apart. In 1915, Wegener published a detailed and expanded theory in *Die Entstehung der Kontinente und Ozeane* (*The Origin of the Continents and Oceans*). While the theory was popular (Wegener's book was translated into several languages and reprinted more than five times), it was almost completely rejected by the geological community. Although part of the rejection was territorial—what did a meteorologist know about geology?—there were good reasons to be skeptical. Wegener offered no reasonable mechanism for why the continents moved and seemed to suggest, among other things, that the various mountain ranges were about the same age if they were created during the formation of the supercontinent. It was clear to all geologists that the mountains of the world were not the same age.

In 1929, Arthur Holmes (1890–1965) attempted to revive Wegener's theory by postulating thermal convection in the Earth's mantle. Hot material was less dense than cool material, and thus it would rise. When it cooled, it would sink back again, like pudding in a pot. This cycle of heating and cooling might make the continents move. Holmes's idea received little serious attention, although continuing work in geology, revealing significantly more information about the structure of the planet, did not rule it out. The question of continent and ocean formation still remained to be resolved.

During IGY, oceanographers worked hard to map the ocean floors. The mapping revealed not only a series of mid-ocean ridges but that near them the ocean floor was not uniform. It was newer near the mid-ocean ridges, with less sediment and younger rock. This ran against expectations. If the oceans were very old, then their floors should be thick with sediment laid down over the millennia. An even more surprising discovery was geomagnetic reversals. Using equipment developed to detect submarines, scientists mapped the magnetic orientation of the rock that made up the ocean floors. To their surprise, they found that the magnetic orientation of the mineral magnetite changed across the oceans. When rock was in a liquid state, the particles in it were randomly oriented toward the Earth's magnetic field, but once it started to solidify, the particles aligned along the magnetic lines of force, like millions of little compasses. Once the rock was solid, the orientation was frozen, pointing in one direction and unaffected by changes to the fields. The oceanic bands of rock made it clear that the fields had shifted and even changed places, with the magnetic north and south switching. These reversals were rapid and resulted in bands of rock with radically different magnetic orientation, which appeared to be laid out more or less symmetrically from the ocean trenches toward the continents. (See figure 11.2.)

With this new information, Holmes's idea received more attention, but "drifters" were still largely scientific pariahs in the 1950s, unable to publish papers in scientific journals and often risking career advancement. John Tuzo Wilson (1908–93), an influential geologist, IGY committee member, and president of the International Union of Geodesy and Geophysics (1957–60), did not initially support the theory of continental drift. However, by 1962 new information from IGY research published by Harry Hess (1906–69) and others, which gave scientific evidence for the spreading of the ocean floor, convinced him otherwise. Greater understanding of the ocean floor and the discoveries of features such as mid-oceanic ridges, geomagnetic anomalies parallel to the mid-oceanic ridges, and the association of island arcs with oceanic trenches near the continental margins, suggested convection might indeed be at work. Wilson's 1965 paper, "A New Class of Faults and Their Bearing on Continental Drift," helped transform the direction of geology. Although the term "continental drift" was eventually replaced by the term "plate tectonics" because it encompassed a broader geological system, in many ways Wegener's idea was vindicated.

Another piece of evidence for continental drift came from space. Careful examination of the variation in the orbits of IGY satellites indicated gravitational anomalies, which suggested convection currents inside the Earth. This discovery was a by-product of another area of IGY research. Rockets and, more importantly, the instruments they carried were central to the study of the upper atmosphere and conditions in space. Many scientists wanted to use them as peaceful tools rather than weapons of mass destruction. James A. Van Allen (1914–2006), who was an alumnus of the Department of Terrestrial Magnetism at the Carnegie Institution and who had worked with captured German V-2 rockets at the end of World War II, was particularly keen on getting scientific satellites into orbit. Van Allen's work was the first American success to counter the Soviet Sputnik program.

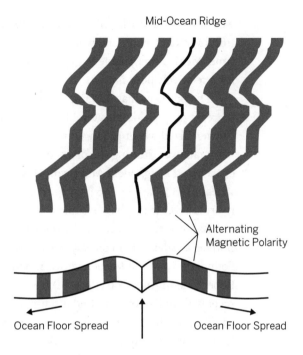

11.2 MAGNETIC OCEAN FLOOR MAP

The Green Revolution: Science and the Global Agriculture

One of the most significant applications of science to global problems in the Cold War era was the Green Revolution. The term "Green Revolution" was first used in 1968 by William Gaud (1907–77), the Director of the United States Agency for International Development, although the scientific work had been going on since the 1940s. The term referred to the use of scientific principles and technology to increase agricultural production, particularly in the developing world. This was part of a belief that all the emerging problems of the modern world could be solved by science and technology. The leading figure in this revolution in agriculture was Norman Borlaug (1914–2009), a microbiologist. In 1944 Borlaug took a job working for the Rockefeller Foundation's International Maize and Wheat Improvement Center and went to Mexico to improve wheat production. At this time, Mexico was unable to grow enough wheat for its population, a major issue for a relatively poor nation.

The problem that Borlaug faced was simple to state but difficult to solve. To increase crop yield, wheat fields had to be fertilized with nitrogen fertilizers. The fertilized wheat produced more grain, but the weight of the kernels made the plant fall over, ruining the plant and reducing the yield. Borlaug's solution was to cross breed a high yield American wheat variety with a strong-stemmed dwarf wheat from Japan. By 1963, 95 per cent of Mexico's wheat consisted of Borlaug's hybrids and Mexico had become a net exporter of the grain.

Because of his success in Mexico, the Indian Ministry of Agriculture invited Borlaug to advise them on ways to increase their grain production. The Rockefeller Foundation and the Mexican government sent Borlaug and the wheat specialist Robert Glenn Anderson (1924–81) from the Canadian Department of Agriculture to meet with Indian officials and discuss ways to increase production. India and other countries in the region were experiencing low-level famine, but there was resistance to the new techniques. In 1965, Borlaug and Anderson arranged to send almost 500 tons of hybrid wheat seed to India and Pakistan, but the outbreak of war between the two countries slowed the project. To prevent starvation (and to encourage India to have closer ties to the United States), the United States sent 20 per cent of its wheat production to India in 1966. Ironically, the starvation caused by the war overcame local resistance to the new method of farming. Working with scientists such as Mankombu Sambasiran Swaminathan (1925–), who is considered in India to be the father of the Green Revolution, the basic hybrids were adapted for regional conditions. Swaminathan had been trained as a geneticist in India and was a post-doctoral fellow at the University of Wisconsin from 1952 to 1954. He worked on the Mexican project before returning to the Indian Agricultural Research Institute, where he was instrumental in developing the strains of wheat and rice for the South Asian region. By 1968, India, Pakistan, Turkey, Mexico, and several South American countries were using Borlaug's agricultural system. Wheat production nearly doubled and India became self-sufficient in grain production in 1974.

In 1970, Norman Borlaug won the Nobel Peace Prize for his work in this area. His efforts and the efforts of those who worked with him have saved the lives of millions of people. Borlaug was proud of his work, but he was also aware that he could be creating a kind of Malthusian trap if it was not properly used. In his Nobel acceptance speech he said:

It is true that the tide of the battle against hunger has changed for the better during the past three years. But tides have a way of flowing and then ebbing again. We may be at high tide now, but ebb tide could soon set in if we become complacent and relax our efforts. For we are dealing with two opposing forces, the scientific power of food production and the biologic power of human reproduction.... Man also has acquired the means to reduce the rate of human reproduction effectively and humanely. He is using his powers for increasing the rate and amount of food production. But he is not yet using adequately his potential for decreasing the rate of human reproduction. The result is that the rate of population increase exceeds the rate of increase in food production in some areas.[1]

The Green Revolution was, like the International Geophysical Year, an international effort by scientists to use science for the benefit of all humanity. Although certain aspects of agricultural change were used to garner support for the West as part of the Cold War, in the long run Borlaug and his supporters were more interested in people than politics. Modern critics have suggested that the Green Revolution did more for US agribusiness than for the adopting countries, and that monoculture farming can have bad environmental consequences including the loss of wilderness areas and a decline in biodiversity. On the other hand, this revolution allowed many countries in the developing world and especially in Asia to achieve food security and stability.

Mapping the Universe: The Steady State vs. the Big Bang

While the IGY program focused on the Earth and the sun, some astronomers were creating a startling new image of the universe. By the time Sputnik was launched, much more was known about those stars that science fiction writers yearned to visit.

Astronomers had discovered that the universe was in motion. While this was not a new idea, no method had been developed earlier to observe the motion of objects outside the solar system, due to the difficulty of calculating the distance to stellar objects. In 1838 the astronomer Friedrich Wilhelm Bessel (1784–1846) was the first to measure the distance to a nearby star using stellar parallax. By measuring the tiny

---

1. Norman Borlaug, "Acceptance Speech for the 1970 Nobel Peace Prize," at www.nobelprize.org/nobel_prizes/peace/laureates/1970/borlaug-acceptance.html.

variation in the angle of observation from one side of Earth's orbit to the other (Bessel found 0.314 seconds of arc), he calculated that the star 61 Cygni was 10 light years from Earth. Although this method worked for close stars, it did not work for objects farther away since the light of distant objects had no measurable angle of intersection.

By the turn of the century, more than 500,000 stellar objects had been mapped. In 1917 George Ellery Hale (1868–1938) oversaw the installation of the 2.5-metre refracting telescope at the Mount Wilson Observatory. Named in Hale's honour, it remained the largest telescope ever made until the building of the 5-metre reflecting telescope at Mount Palomar in 1948. In 1926 the Bruce Proper Motion Survey began to trace the motion of stars by comparing photographic plates of the same region of space taken 25 years apart. The motion of the stars and the theory of relativity led Belgian astronomer Georges F. Lemaître (1894–1966) to argue in 1927 that the universe was created from an explosion of matter and energy, which he called the "cosmic egg."

Using data from the Wilson Observatory, in 1929 Edwin Powell Hubble (1889–1953) applied new methods based on the Doppler effect,[2] or the red shift of light, to calculate the distance to the Andromeda nebula as 930,000 light years away. He also found that most galaxies were moving away from the Earth and from each other. This fit Lemaître's idea, since it seemed possible to "rewind" cosmic history by tracing the movement of the nebulae backward and drawing everything together at some ancient time.

Fred Hoyle (1915–2001), one of the greatest astronomers of the era, objected to the cosmic egg theory, arguing along with Hermann Bondi (1919–2005) and Thomas Gold (1920–2004) that the universe was in a steady state. The Steady State model, presented in 1948, pictured the universe as uniform in existence, with no beginning or end, and looking much the same everywhere (on the large scale) and at every time. Hoyle, Bondi, and Gold dealt with the apparent expansion of the universe by arguing that matter was constantly created. Some considered the creation of matter from nothing to be an odd idea, but the amount required was very small (on the order of a few atoms per cubic light year per year) and it was no more philosophically unsettling than the question of the origin of the cosmic egg. Ironically, it was Hoyle who coined the term "Big Bang" as a derogatory label for the competing theory.

The debate between Steady State and Big Bang supporters was heated and often acrimonious. The evidence that tipped the argument in favour of the Big

2. The Doppler effect, named for Christian Doppler (1805–53), explains the apparent rise and fall in pitch of a moving sound. If an object, such as a star, is moving away from us, the distance between the light waves appears to get bigger, making it look as if the light is shifted toward the red end of the spectrum.

Bang (although there are still a few Steady Staters today) came from an unexpected source—radio waves from space. In 1931, Karl Jansky (1905–50) was researching radio interference for Bell Telephone Laboratories. He found three types of naturally occurring interference: local thunderstorms, distant thunderstorms, and an unknown but constant source, which he initially thought might come from the sun. This was a reasonable idea, since back in 1894 Sir Oliver Lodge (1851–1944) had claimed that the sun emitted radio waves, although he had been unable to detect them. However, after a year of study Jansky determined that the unknown source was actually the Milky Way. Our home galaxy was emitting radio waves! When Jansky finished his investigation, Bell assigned him other research, and he never followed up on his discovery. He died of a heart attack at the age of 44, but in his memory the unit of energy flux, or "radio brightness," is called a Jansky (Jy).

Jansky's discovery was largely ignored, but Grote Reber (1911–2002), a radio engineer, was intrigued with the idea of radio waves from space and built a 9.4-metre parabolic dish antenna in 1937 in order to capture these emissions. He set out to create a radio map of the heavens. Between Jansky's serendipitous discovery and Reber's work, the field of radio astronomy was born. Because of the interest in looking at extreme events through physics (the interior of stars, for example) and the relationship of general relativity to the structure of the universe, money for radio astronomy became available after World War II, as did stocks of left-over electronic equipment, often available as cheap military surplus. In 1946 the first major radio telescope installation—a 218-foot parabolic aerial—was built at Jodrell Bank in Britain. (See plate 7.) Radio astronomy took a leap forward when Martin Ryle (1918–84) introduced the first radio interferometer. Essentially a system of two receivers hooked together, it provided a much sharper image and more detailed information about radio sources. With simple interferometers and arrays of many radio telescopes linked together, radio astronomy became as good at resolving light sources as optical telescopes. Reber, Ryle, and others began mapping the heavens, creating a much more complex picture as things visible only by radio detection were added to centuries of optical observations.

The Space Race

The technical advances that made radio astronomy possible also made access to space, especially in local Earth orbit, increasingly attractive. The race to space, however, was more than an intellectual foot race. It was also a battle between scientific styles. It is easy to confuse the search for the most productive style of research with the political battle between the Western democracies and the totalitarian regime of the Soviet Union, but the relationship between scientific research and the demands of the state, regardless of the structure of government, remains complex. Scientific success does not depend on a free populace, but the translation of discovery into products for the larger population has usually been slower under totalitarian governments. Yet, under both democratic and totalitarian governments, research deemed to be in the national interest is developed and controlled by bureaucracies that may not even understand the science involved, that measure success or failure by objectives met rather than discoveries made, and that ensure that money rules the course of research.

The story of rocketry that leads to Sputnik is one of despots and dreamers, technocrats, technicians, politicians, and researchers. While the race to space did not create the integrated system of scientific research, it is the greatest example of the power of such integration. The development of nuclear weapons was in many ways a more complex integration of research and the demands for a "useful" final product, but it was shrouded in secrecy and was presented as so advanced and esoteric that it was accessible only to geniuses. The rocket race was, in contrast, a very public demonstration of scientific prowess that led to a revision of the place of science, and especially scientific education, in the industrialized world.

The first rockets appeared around 1150 when the Chinese used gunpowder to propel them. They were later developed into a weapon at the siege of Beijing in 1232 and made their first appearance in Europe in 1380 at the battle of Chioggia between the Genoese and the Venetians. Although small rockets continued to be made for fireworks and military purposes, they were largely overshadowed by cannons and artillery. The exception to the general trend was in Russia.

In 1881 Nikolai Kibalchich (c. 1853–81) built the bomb that killed Tsar Alexander II. He made it in the chemical laboratory of the St. Petersburg Technological Institute, founded by Alexander's father to help propel Russia into the scientific and industrial age. After his arrest, Kibalchich spent the time before his execution designing rockets, including a design for a rocket plane and a passenger rocket. Kibalchich was not an aberration; Russian interest in rocketry was long-standing. One hundred years earlier the *Raketnoe Zavedenie* (Rocket Enterprise) was created to design and manufacture rockets. The first missile unit in the Russian army was formed in 1827, and in 1867 the engineer Konstantin Konstantinov opened a rocket-manufacturing plant in St. Petersburg.

The same year that Alexander II was assassinated, Konstantin Tsiolkovsky (1857–1935), a self-taught physicist, sent a paper to the Russian Society for Physics and Chemistry in St. Petersburg that dealt with the kinetic theory of gases and the mechanics of living organisms. He was told that his ideas were already well known to scientists, but rather than being discouraged he continued to pursue his interests. His driving passion was the conquest of gravity. Throughout the 1880s he contemplated what life would be like in zero gravity, and in 1903 he published the mathematical treatise "Exploration of Cosmic Space with Reactive Devices" on orbital mechanics and the principles of rocket propulsion. He also sketched a liquid propellant rocket, which used liquid oxygen and liquid hydrogen as fuel. (See figure 11.3.)

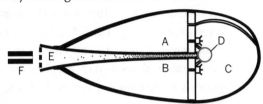

A. Liquid Oxygen Tank
B. Liquid Hydrogen Tank
C. Cargo or Crew Area
D. Combustion Chamber
E. Exhaust Nozzle
F. Control Surfaces (guidance)

11.3 LIQUID FUEL ROCKET

Illustration based on Konstantin Tsiolkovsky's 1903 sketch of a liquid fuel rocket.

Tsiolkovsky was not the only person working on advanced ideas in rocketry in this period. Robert H. Goddard (1882–1945) in the United States began serious experimental work in 1909 when he was a student at Clark University. He eventually published his PhD thesis, *A Method of Reaching Extreme Altitudes*, with financial help from the Smithsonian Institution. Goddard was a persistent and ingenious inventor, but he was also secretive and isolated. In 1926, he moved to a ranch near Roswell, New Mexico, to get away from reporters and other rocket enthusiasts and carried out foundational work on liquid-fuel rockets. While he was granted more than 200 patents, his work was little known at the time.

Rocket Science in Germany and the Soviet Union

Through the 1920s and 1930s the general public became interested in rocketry. In 1930 the legendary science fiction editors Hugo Gernsback (1884–1967) and David Lasser (1902–86) founded the American Interplanetary Society (AIS). The AIS corresponded with Robert Esnault-Pelterie (1881–1957), the leading rocket engineer in France, and with members of the German amateur rocket society Verein für Raumschiffahrt (VFR), as well as financing liquid-fuel rocket work at home. Despite this popular interest, however, most institutions and governments paid little attention to rockets. To a certain extent, this lack of interest was due to the rise of powered aviation, which was more immediate and personal since it offered human flight, but it was also because dreams of flight into space seemed both preposterous and pointless. Two governments proved the exception to this disinterest: the German government under Hitler and the Soviet government under Stalin.

The German story is better known, but was ultimately less successful. In the 1920s the reorganized German army developed strategies for the next war. They chose a mechanized, highly mobile approach that used aircraft as a form of long-range artillery. This left a tactical gap between heavy artillery, with a maximum range of about 100 kilometres but an effective range much less than that, and bombers that could fly hundreds of kilometres to attack targets but were expensive and required a great deal of logistical control. The military planners began to consider rockets, which were not prohibited under the terms of the Treaty of Versailles, and the Ordinance Ballistic Section of the German army was given the task of developing a liquid-fuel rocket to fill the gap between artillery and air power. The actual job fell to Walter Dornberger (1895–1950), who had studied ballistics at the School of Technology at Charlottenberg.

In 1929, Dornberger visited the "rocketport" used by the rocketeers of the VFR and met Wernher von Braun (1912–77). He encouraged von Braun to complete his Bachelor of Science degree in engineering at the Berlin Institute of Technology. While earning this degree, which he finished in 1932, von Braun worked with Hermann Oberth (1894–1989), the leading rocket scientist of the day, on the construction of liquid-fuel rocket engines. Oberth, who had written on the physics of space flight in the 1920s, was von Braun's original inspiration. In 1934 von Braun completed a degree in physics at the University of Berlin.

By this time, Dornberger was head of Research Station West at Kummersdorf, but German interest in rockets was waning as other aspects of military technology, backed by an industry that already existed to produce the materials, took centre stage. With von Braun, Dornberger established a test facility at Peenemünde in 1937; it opened in 1939. Their efforts were limited by budget constraints, but in 1942 they launched the first A-4 rocket.

As the war began to go against Germany, Hitler looked for weapons that would provide a technological fix to the problem of military failure. The V-1 rocket, which consisted of a jet engine on a winged bomb, was so slow it could be shot down by fighters, but the A-4 as a ballistic weapon was unstoppable. With a maximum range of about 400 kilometres and a payload of 1,000 kilograms, it was the kind of superweapon that Hitler wanted. Put into mass production as the V-2, or "Vengeance Weapon," its destructive power was significant. Although by 1945 some 300 V-2 rockets were being assembled each month by slave labour at the Mittelwerk factory near Nordhausen, they had virtually no strategic or tactical effect on the outcome of the war. The rocket designers were pleased with the success of the A-4, but they were planning much greater things. The A-9 was to be a two-stage rocket capable of intercontinental flight, the A-11 a three-stage rocket capable of lifting a pilot into space.

As the German military collapsed under attack from east and west, von Braun and the Rocket Team contemplated the end of the war. If they stayed at Peenemünde and Nordhausen, they would be in the Russian zone of control, so they decided to surrender to the American army. In February 1945 von Braun and more than 500 people from the rocket program, plus mountains of documents, travelled south across the devastated country. At the beginning of May they contacted and surrendered to the American military in Bavaria. When the American command realized what kind of prize they had in hand, they staged a special mission to clean out everything of use from the Mittelwerk factory, sending the equipment, unused V-2 rockets, and eventually many of the German scientists and engineers to the White Sands Proving Grounds in New Mexico.

The Americans scooped most of the rocket program out from under the noses of the advancing Russian forces. Only one major German scientist, Helmut Gröttrup (1916–81), and a small group of workers went to the Russians, but Gröttrup's work did not contribute directly to the next stage of rocket development. When Russian experts arrived at Nordhausen and Peenemünde, they concluded that German manufacturing efforts were more advanced but not more theoretically sophisticated than the Russian. The loss of the German rocket specialists did not appear to pose a risk until the first atomic bomb exploded over Hiroshima on August 6, 1945. The implications of a sudden boost to rocket development by the United States changed the whole complexion of the technological race.

Rocket development in the Soviet Union brings to the fore the whole issue of the relationship between science and the state. Just as tsars of an earlier era had hoped to catapult Russia into the industrial age by luring selected industrialists and founding scientific societies, the Soviet regime hoped to use rocketry and space flight to help propel Soviet society into a leading role in science and technology. There was a particular match between rocketry and the ideology of the revolutionaries who toppled the Russian monarchy. Alexander Bogdanov's (1873–1928) popular novel *Red Star* (1908), reprinted in 1917, linked communist ideology and space flight with a workers' paradise created by Martians. Lenin himself had argued for the promotion of science and technology and political tolerance for great scientists because he recognized one of the lessons of World War I: the side with the best technology wins.

The Soviet government created new research institutions, and while many were short-lived, interest in rocketry remained strong. In 1924 the Central Bureau for the Study of the Problems of Rockets (TSBIRP) was created to coordinate research and focus attention on the military development of rockets. A private group also formed that year, the All-Union Society for the Study of Interplanetary Communications (OIMS); it was a parallel to the AIS in America and the VFR in Germany. In 1927 TSBIRP and OIMS hosted the Soviet International Exhibition of Rocket Technology in Moscow.

When Stalin came to power, he was far less tolerant of independent research and purged many scientists and engineers, sending them to jail, exile in the gulags, or death. At the same time he was prepared to support science and technology in certain areas. The Academy of Sciences was expanded, and its budget rose from 3 million rubles in 1927 to 175 million in 1940. Enrollment in technical schools also rose dramatically. Stalin set up a system of *sharashkas* (prison design bureaus) to utilize the talent of the jailed scientists and engineers. One such scientist was Sergei P. Korolev (1906–66), who graduated from the Kiev Polytechnic Institute but was jailed and sent to work in the Kolmya gold mines, one of the worst of the gulags. Korolev was saved by the great aviation designer Sergei Tupolev, himself a prisoner, who arranged for him to be transferred to TSKB-39 *sharashka*, which was filled with aviation specialists who continued their work, even under appalling conditions. The war with Germany put an abrupt halt to most of this work; however, one of the prison camp inmates, Georgy Langemak (1889–1938), created a militarily useful rocket, the Katyusha, which was fired by multiple launchers carried on trucks.

After the war Soviet efforts to turn the dreams of rocketry into reality began to bear fruit, but it was not until after Stalin's death in 1953 that work began to accelerate. In 1954, A.N. Nesmeianov (1899–1969), president of the Soviet Academy of Sciences, declared that it was feasible to send a rocket to the moon or to place satellites in orbit. The Academy created a high-level Commission for Interplanetary Communications, which signalled significant support for the project of satellite launches and for the scientific foundation for intercontinental ballistic missiles (ICBMs). Korolev, now working on long-range missiles under better conditions for the government, was directed to turn his attention to ICBMs in 1953, the same year he became a Corresponding Member of the Soviet Academy of Sciences. By 1955 there was a new test facility at Tyuratam, and test firing of rockets was almost routine. The hopes of the project lay with Korolev's R-7, a stubby but massive rocket that used kerosene and liquid oxygen as the propellant. It combined 20 rocket engines in clusters and produced more than a half-million kilograms of thrust.

The first launch failed, but a second rocket launched on August 3, 1957, travelled from the launch site to land in the Pacific Ocean off the Kamchatka Peninsula. On October 4, Sputnik was put into orbit. Premier Khrushchev congratulated the team and in the following days began a propaganda campaign based on their success.

The United States Enters the Space Race

The launch of the first artificial satellite was not a complete surprise to the West; in fact, Moscow had announced the ICBM launch and even made public on October 1 the frequency on which Sputnik would broadcast. Yet many in the West, particularly in the American rocket programs, were startled. Both the Soviet success and the fragmented state of American efforts highlighted a basic problem with American research.

During the war, the Office of Scientific Research and Development (OSRD) had coordinated civilian research efforts and been the conduit for federal funding, which ended with the war. The OSRD, which had overseen significant work on a range of wartime projects, was in danger of being closed down.

Vannevar Bush, one of the heads of the ORSD instrumental in persuading President Roosevelt to mobilize American civilian scientists for war work in the first place, wanted to continue the organization in some form after the war. Although a peacetime version of the OSRD seemed desirable, it was considered to be unconstitutional since it required federal funding not controlled by elected representatives. For this reason, a number of bills proposing the establishment of a research organization failed.

A recommendation from the military led to the creation of the Research Board for National Security. It was not well received. Civilian scientists feared it was too open to interference from the military and contractors, while many politicians and bureaucrats felt that it would be too powerful an organization to operate without the oversight of the president or Congress. It was terminated in 1946. While virtually all the parties involved—the executive branch, Congress, the scientists and their universities, and the military—were interested in continued research, they could not decide on a postwar format. Scientists wanted federal money with few or no strings attached, while the government by law controlled funding and by preference wanted a high level of oversight. This left researchers with no national organization, and large-scale research funding fell to the military. By 1950, more than 60 per cent of all federal money for research came from the military.

The American rocket program—or more accurately, programs—grew slowly during this period of muddled research policy. When Congress restructured the military in 1947, creating a single Department of Defense, major missile research was divided between the Army and the Air Force. The Army, for whom von Braun was working, had control of tactical missiles, while the Air Force worked on strategic missile programs. But the Air Force, being composed of pilots, was naturally more interested in aircraft than missiles and concentrated on the development of long-range bombers and fighter planes.

When Dwight Eisenhower was elected president in 1952, he wanted to combat the rising cost of the federal budget and keep the United States out of "brush fire" wars, such as the struggle in Vietnam over French colonial rule. More than 57 per cent of the federal budget was going to the military, and military demands for bigger and more powerful weapons continued to climb. It was hard to limit military spending during the Cold War. Because the United States could not or would not match the size of Soviet ground forces, its general policy was to use air power and technical superiority to counter Soviet numbers. This meant more than bombers with nuclear weapons; it meant missiles. As Intermediate Range Ballistic Missiles (IRBMS) and ICBMs increasingly became the objective of research and development, a competition between the different branches of the military gave rise to a series of different missiles, each with different capabilities and championed by competing organizations. The Army's Redstone (a direct descendant of the V-2) and Jupiter rockets as well as the Air Force's Atlas, Titan, and Thor rockets were all rushed into development. Although they were used in the race to space, these rockets were actually designed to replace weapons such as the Minuteman and the Polaris missiles.

The "civilian" space program, Project Vanguard (actually run by the Naval Research Laboratory), was the public face of the American race to space. It did not fare well, as many scientists and engineers were absorbed by other projects, particularly in nuclear research and ballistic missile development. Regarded as a second-class project by the military and a threat to funding, it was also plagued by technical problems. When its funding requirements rose from $20 million to $63 million in 1956, it appeared to be a project out of control.

Despite all these problems, when the R-7 sent Sputnik 1 into orbit, the United States was not significantly behind the Soviet Union in development and was ahead in guidance, miniaturization, and weapon building. A crash program by the Jet Propulsion Laboratory (JPL) and the Army Ballistic Missile Agency built Explorer 1 in 84 days and launched it on January 31, 1958. It was followed by Vanguard 1 on

March 17, 1958. The military missile programs were successful, although the division of labour limited their achievements to certain specific goals. None of that mattered, however. The launch of Sputnik meant that American efforts looked second-best, the line taken by many leading media outlets such as *Life* magazine. A national outcry followed, and blame for the perceived failure of American science was cast far and wide. President Eisenhower was forced to confront a political dilemma. Should the United States collectivize its scientific effort as the Soviet Union had done and commit even more federal resources to military spending in a visible effort to surpass the Soviets? How had the Soviets managed to achieve so much in such a short time? Was there a problem with American education?

There was much soul-searching about this problem. While science and engineering training had vastly increased since World War I, many people, such as the editors of *Life* and Vannevar Bush, wanted to see a significant increase in spending on education. The problem was politically sensitive because education was a state responsibility and therefore outside the purview of the federal government. Eisenhower was reluctant both to commit large sums of money and to be seen to trample on state rights. In 1958, Congress enacted the National Defense Education Act (NDEA) as a compromise between those who wanted no federal spending on education and those who wanted a massive increase. The bill authorized an expenditure of about $1 billion over four years. It created a $295 million loan fund for students in financial need and allocated $280 million of matching federal grants aimed at purchasing equipment for science, mathematics, and language training. There was an additional $60 million fund for 5,500 graduate fellowships in areas related to national defence.

When James Van Allen and his team successfully launched Explorer on January 31, 1958, they demonstrated that American efforts were not so far behind the Soviet's. Van Allen's satellite project began back in 1955 as part of the IGY research on radiation. Although the American satellite was far smaller than either Sputnik 1 or Sputnik 2, weighing only 14 kilograms, it nonetheless established the American capability to orbit satellites. As a tool for scientific research rather than for national security, its instrument package contained a radiation detector that demonstrated the existence of bands of radiation above the Earth. Named after their discoverer, the Van Allen radiation belts range from

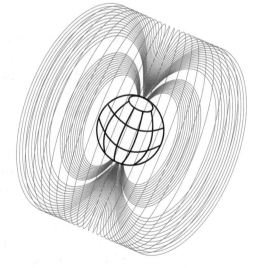

11.4 VAN ALLEN BELTS

The Van Allen belts form a protective electromagnetic field around the Earth.

650 kilometres to 65,000 kilometres and circulate along the Earth's magnetic field. (See figure 11.4.) Explorer was both a scientific success and a political statement, but it lacked the status of being first.

Despite Explorer getting into space and the successful launch in 1958 of the Atlas missile, many Americans were concerned about the technological gap between the United States and the Soviet Union in the development of ICBMs. Missiles had made long-range bombers obsolete, so the air superiority the United States had enjoyed now disappeared. The issue was further inflamed when journalist Joseph Alsop (1910–89) published articles claiming that by 1963 the Soviet Union would have 2,000 operational ICBMs compared to only 130 planned by the American military. The American Central Intelligence Agency (CIA), for various reasons, also predicted a large "missile gap" by 1963, despite little evidence. (In actuality, by 1961 the Soviet Union had deployed only 35 ICBMs.) Eisenhower was forced to increase spending, including $186 million for a spy satellite program being developed by the Advanced Research Projects Agency (ARPA) under the direction of the CIA.

The Cold War Heats Up: Communication Satellites and Television

While Eisenhower was confident that American military efforts would eventually overtake Soviet production, secret missile programs did little to quell public concern. What the United States needed was a major project to surpass the Soviet effort. Such a project had to fulfill three requirements: it had to put the United States in the lead in space; it had to establish the superiority of American science and scientific organization; and it had to be something that could excite the public. The race was not just a technological race but a fight for international prestige.

The first step was Eisenhower's creation of the National Aeronautics and Space Administration (NASA) on October 1, 1958. It brought together the National Advisory Committee for Aeronautics (with 8,000 employees and a budget of $100 million) with the Langley Aeronautical Laboratory, the Ames Aeronautical Laboratory, and the Lewis Flight Propulsion Laboratory. It went on to include the space science group at the Naval Research Laboratory (home of Project Vanguard), the Jet Propulsion Laboratory run by the California Institute of Technology for the Army, and the Army Ballistic Missile Agency, the centre where Wernher von Braun and his team were located. The first Director was T. Keith Glennan (1905–95), who was trained as an electrical engineer. He had worked at the Columbia University Division of War

Research and later at the American Navy Underwater Sound Laboratory during the war. When he was chosen to be director, he took a leave from his job as president of Case Institute of Technology. During his term, he consolidated almost all non-military space research and development under the NASA umbrella.

The national outcry over Sputnik had led to the creation of a strange hybrid of military and non-military activity in NASA. One of its reasons for existing was to do research of military significance but not under the control of any branch of the existing military. To go along with it, Congress created two new standing committees, the Senate Committee on Aeronautical and Space Sciences and the House Committee on Science and Aeronautics. These committees wrestled with the complex problems of civilian–military relations and objectives, the degree and method of cooperation with space programs in other countries, and, most of all, the money.

In 1959 the Soviet Union upped the ante once more with the announcement of its Lunik series of experiments. Although some early attempts to send a rocket to the moon had failed, Korolev and his team finally succeeded in producing a rocket with enough thrust to escape the Earth's gravity. Luna 1 was aimed at the moon but missed the target and eventually went into orbit around the sun. In good propaganda style, the Soviet press hailed the launch as a complete success as the first artificial planet and renamed it *Mechta* (Dream). Luna 2 reached the moon on September 14, while Luna 3 passed behind it and took the first photograph of the far side, which, because its rotation matches its orbit around the Earth, had never been seen.

The American response was von Braun's Pioneer 4, which was launched July 16, 1959. The effort was at best a partial success. The rocket did lift off and gained sufficient velocity to escape the Earth, but guidance problems caused it to go off course and miss the moon by 60,000 kilometres! While this was not a huge error in astronomical terms, it was an embarrassment for von Braun and his team.

In 1960 John F. Kennedy defeated Richard Nixon to become president. At 43, he was the youngest president ever elected and the first to exploit the power of television for electoral purposes. On May 25, 1961, in a speech to Congress, he declared that it was "time for a great new American enterprise—time for this nation to take a clearly leading role in space achievement, which in many ways may hold the key to our future on earth."[3] That enterprise was to send a man to the moon and bring him home safely. It would not be cheap. An early estimate for the cost of the moon project was $7 billion. When the Apollo program ended in 1973, it had cost almost $20 billion out of a total of $56.6 billion spent on NASA.

....

3. The speech is available at history.nasa.gov/moondec.html.

President Kennedy needed mass support for such a major undertaking, and, indeed, much of the point of the race to the moon was the mass exposure and approval it would garner. When he made his 1961 declaration, Kennedy was speaking not only to Congress but to the whole nation. His words were recorded and broadcast across the country. One of the significant differences about national and international politics in the second half of the twentieth century as compared to earlier eras was that they were propagated and transmitted to a huge audience. Radio broadcasts had become a key component of government activity through Hitler's rally speeches, Churchill's wartime messages, and Kennedy's declaration of American will. Radio communication had undergone significant improvement through the 1940s and early 1950s, but the problem of long-distance transmission needed to be solved before Kennedy's mass democracy became a possibility. Communication satellites, made possible by the developing science of rocketry, offered a tantalizing solution to this problem. As early as 1945 Arthur C. Clarke (1917–2008) proposed the idea of geosynchronous communications satellites. His paper, "Extra Terrestrial Relays" in *Wireless World*, argued that a satellite placed at about 22,300 miles above the surface of the Earth could have an orbit that kept it stationary over a point on the surface, making it a very useful communications link.

Proposals for military reconnaissance satellites were made as early as 1950, but the cost and problem of the international legality of satellites travelling over foreign countries raised serious concerns. As the demand for information rose steadily during the Cold War, the U-2 spy plane was developed and began over-flights of Soviet territory in June 1956. In 1957 a U-2 returned with photographs of the Tyuratam launch facility. The theory of satellites for communication and observation had suddenly become a reality.

While the words of world leaders and soap manufacturers were being broad-cast by radio, many major events of the era were televised, often transmitted live from around the globe. In tandem with the Cold War and the race for space, television came of age in the postwar era, one of the great examples of the trans-formation of a scientific object into a commonplace device. Each television is a microcosm of modern science and represents one of the unexpected products that research has given us. Like the earlier invention of printing, then of the intersec-tion of telegraphy and the newspaper, television has been used to introduce many of the products of modern science.

Television does not have a single inventor but rather hundreds of contributors. The idea of transmitting images electrically developed along with the telegraph, and working methods were devised as early as 1875. The first practical image transmitter was demonstrated by Paul Nipkow (1860–1940) in 1905, based on an idea he had in 1884. It involved a spinning disk with a spiral of holes. Light passing through the holes fell on an image and was converted to an electric signal that controlled the intensity of a lamp set behind a matching rotating disk at the receiving end. In effect, the first perforated disk scanned the image and the second synchronized the light impulses and turned them back into the image. Using a disk system, the first still picture was broadcast by radio in 1907. By 1924 a moving image could be transmitted, but the image size was tiny, about 2.5 centimetres.

The invention of the Audion amplifier in 1906 by Lee Deforest (1873–1961) opened the door to a number of electronic devices. The Audion was used to boost an electrical signal in a cathode ray tube or vacuum tube. Numerous developments in tube technology eventually allowed Philo Taylor Farnsworth (1906–71) to develop an electronic picture tube in 1927; in that same year Bell Telephone Laboratories broadcast pictures between Washington, DC and New York. In 1923, Vladimir Zworykin (1889–1982) significantly improved camera technology when he invented the "Iconoscope," a camera tube that combined lenses and used a photoelectric mosaic to capture the image. The prototype was demonstrated in 1929 and was manufactured by RCA in 1933. By 1935 television was being broadcast in Britain, Germany, and France; in 1936 the Olympics in Berlin were televised. The coronation of King George VI was broadcast in 1937. In 1939 Franklin D. Roosevelt became the first president to make a televised speech. Although broadcasts were very local, more than 20,000 television sets were sold in Britain by the end of 1939.

The demand for television was curtailed by the war, as both the people and materials needed were diverted to war work. With the end of the war and with the creation of uniform technical standards, there was a huge demand for television, which was spurred on even further with the introduction of colour broadcasting starting around 1954 in the United States and 1967 in Europe—just in time to broadcast the events of NASA's Apollo program. Almost every move of the NASA astronauts was carried out in front of television cameras. Even if the public did not always understand what they were being shown, television changed the place of scientists in society and certainly attracted young people to the subject.

The ability through the space program to put large objects in orbit made communication satellites a reality. In 1960 the United States launched Echo, the first passive communications satellite, which was little more than a silver balloon. This was followed in 1962 by Telstar, which relayed the first trans-Atlantic television broadcast.

After the launch of Telstar, equipment that had been used with Echo was no longer needed for commercial work. Arno Penzias (1933–) and Robert Wilson

CONNECTIONS

NASA as
Big Science

In 1957, the USSR beat the United States into space when the Soviets launched Sputnik 1 and Sputnik 2. This challenged the widely held belief that American technology, and technology in the West more generally, was well ahead of that being used in the Soviet Union. Western technology was in fact ahead of Soviet technology, but the democratic and somewhat chaotic nature of US research meant that it could not be as focused as the state-directed Soviet system. In the US, various parts of rocketry were being developed by many different groups, including the Air Force, the Navy, and the Army, as well as the National Advisory Committee for Aeronautics and a number of different universities. In many ways, they had forgotten the lessons of the Chemical Warfare Service and the Manhattan Project, allowing a sort of free-for-all when it came to space exploration. With the launch of Sputnik, this became a matter of national pride and security, so President Eisenhower, taking the advice of the National Science Foundation and the National Academy of Sciences, created the National Aeronautics and Space Administration (NASA) in 1958 to bring all of the major rocket programs together under a civilian administration.

With one umbrella organization and significant funding from the state ($20 billion, equal to about $210 billion today), the Apollo project was established to land an American on the moon. This was Big Science as heroic adventure and certainly placed American space exploration

(1936–) were allowed by Bell to use a microwave detector (called a "horn antenna" because it looked like a ram's horn) for radio astronomy. Working on emissions in the microwave region, they found a constant hiss, somewhat like an FM radio tuned to an empty frequency. While this had not interfered with communications, it was a problem for astronomical work. After eliminating potential local sources of interference (such as pigeons in the receiver), the noise remained. It wasn't in the equipment. It wasn't radiation from nuclear test fall-out or even from the sun. They could detect a faint bit of radiation even in empty portions of the sky.

Both frustrated and intrigued, Penzias and Wilson looked for a theoretical explanation for the persistent background noise. They contacted Robert Dicke (1916–97), who was working at Princeton University on theories about the Big Bang. He suggested that a kind of residual "noise" in the form of low-level background radiation would be left over from the original explosion. Penzias and Wilson had the ahead of the Soviets. NASA went on to orbit satellites including the Hubble telescope, create Skylab, send probes to other planets, and to participate in several other space exploration programs.

If the purpose of NASA was only to collect moon rocks and place a flag on the moon, public concerns about its large budget would be justified. However, NASA has played an incredibly important role in the development of a high-tech economy through science that is only possible in these large, well-funded teams. In addition to the space-based technologies—ranging from GPS satellites, aviation navigation, and radar systems, which were a direct development of the race to space—NASA has also developed or co-developed a huge range of products that have become consumer goods. These include infrared ear thermometers, advanced radial tires, memory foam, solar cells, and freeze dried food. The list of technologies for aircraft alone would be over 50 pages long. Equally important to the modern techno-science world, because NASA is publicly funded, it has a mandate to make public its findings. In an age where science is becoming more and more proprietary, NASA stands out as the best of public Big Science. In 2014 alone, NASA released more than 1,500 pieces of software (including project planning tools and sound modelling) to the public for free.

NASA was created as an organization that fosters invention and inquiry. NASA is the largest collection of scientists and engineers in one organization in human history. In 2015, more than 10,000 people worked in the science and engineering sectors of NASA, forming a vast pool of talent. It trains people to take on enormous projects and those skills percolate out to the rest of society. While some people might say that the value of knowing more about distant stars and planets is low, few people would say that the world would be better off if there were 10,000 fewer scientists and engineers. Far more than the Royal Society or the American Philosophical Society, NASA has produced useful knowledge. It has lived up to its vision statement: *To reach for new heights and reveal the unknown so that what we do and learn will benefit all humankind.*

evidence in hand. The universe was producing emissions like that of a black body (as investigated by Kirchhoff and Planck; see Chapter 8) at a temperature of around 3° Kelvin. The cosmic background radiation was uniform in all directions, meaning that the universe was expanding uniformly (fitting with Hubble's and Lemaître's ideas) and further that the universe had started at the same initial temperature. Although alternative reasons for this were presented by the Steady State supporters, the Big Bang eventually became the standard model for cosmology, in part because of this evidence. Penzias and Wilson went on to win the Nobel Prize in Physics in 1978, sharing the prize with Pjotr Leonidovich Kapitsa (1894–1984).

While physicists and astronomers concentrated on the big picture, NASA was still working on getting around the neighbourhood. The first practical question that had to be answered about a trip to the moon was whether a human could live in space. Some technical aspects were well understood, such as the lack of atmosphere and the general physiological impact of launches, but others, such as the effects of weightlessness and exposure to radiation, were not clear. Answers to these questions required manned flights. Both the Soviet and American space programs experimented with animals, and then moved rapidly to be first to put a man in space. The Soviets won. The first man in space was Yuri Gagarin (1934–68), who orbited the Earth on April 12, 1961. NASA sent Alan Shepard (1923–98) on a suborbital flight on May 5 of the same year.

Although space travel now seemed a possibility, the continued existence of humanity was thrown into doubt. In 1961, an American-backed invasion of Cuba failed. The Bay of Pigs fiasco was a major embarrassment and increased tensions between the United States and the Soviet Union. A year later, the world held its breath through the Cuban Missile Crisis. American spy planes recorded the construction of missile sites in Cuba that could be used to launch nuclear weapons. Since Cuba was under 150 kilometres from the American coast, there was no need for ICBMs to deliver the weapons. President Kennedy ordered the blockading of Cuba by the navy, as Nikita Khrushchev proceeded with shipment of military material to Cuba. A tense standoff followed. The Crisis, which we now know was characterized by a lack of information, military mistakes, and confusion, brought the world to the brink of war as both sides put their forces on high alert. Fortunately, secret negotiations ended the standoff. The Soviet Union agreed to remove its weapons from Cuba, and the United States quietly took its missiles out of Turkey. Just as the world started to recover, in 1963 President Kennedy was assassinated.

After so many problems, the United States was ready for some good news. Kennedy's commitment to get to the moon was endorsed by the following administration and became one of his legacies. The Kennedy Space Center was named in his honour. NASA and the Apollo program became the largest non-military project ever undertaken by the United States. At its height, it employed almost 30,000 people directly and thousands more through contractors and represented about 5 per cent of the federal budget.

One of the main accomplishments of Apollo was the production of the Saturn rocket series. These liquid-fuelled rockets, particularly the Saturn V, were the workhorses of the space program. The program was set back in 1967 when a fire on Apollo 1 killed the crew on the launch pad, but in 1968, Apollo 8 orbited the moon, essentially resolving all the technical issues about getting there and back, short of landing. Apollo 11 touched down on the surface of the moon on July 20, 1969, a moment of national pride and international sense of accomplishment. While the Soviet Union had sent more objects into space, it had not gone to the moon. The drama of the flight and the adventure was brought straight into the living rooms of millions of people, broadcast live on radio and television around the world.

For the first time, people saw the Earth as a planet as well as a global village, a distant blue-green pebble rising above an alien landscape.

Conclusion

Claude Bernard (1813–78), a medical researcher, said in a philosophical moment that "Art was I: Science is We."[4] Nowhere was this more clearly shown than in the race for space. NASA became the world's biggest supporter of scientific research in terms not only of the overall number of people directly employed or funded by grants to private and public organizations but of the scope of research and development as well. From nutritionists and home economists to theoretical physicists, from electrical engineers to librarians, from chemists to computer programmers— all were brought together in NASA. It was the biggest of Big Science. Even in 1999, when it had been reduced by almost half its peak workforce, NASA had 5,971 employees with advanced degrees (doctorates and masters) and a further 7,255 with bachelor degrees, almost all in science and engineering.

The NASA programs also helped change the image of science, especially in the United States. It was less tainted by the destructive image of the arms race and was made more American. While men speaking in foreign accents were still around, NASA's voice and image was the television news announcer Walter Cronkite and a host of boys from American coal-mining towns and prairie farms who had the "right stuff." The space program made science adventurous and glamorous rather than sedentary and esoteric. The utility of science, in its best, worst, and most Machiavellian applications, was now a part of everyday life. It had also become part of the American dream.

Essay Questions

1.   What discoveries led to the formulation of the Big Bang theory?

2.   How did Sputnik contribute to the development of modern science?

3.   Was the Apollo program worth the money? Why or why not?

4.   How was the theory of continental drift confirmed?

----

4. www.gly.uga.edu/railsback/1122sciencedefns.html.

CHAPTER TIMELINE

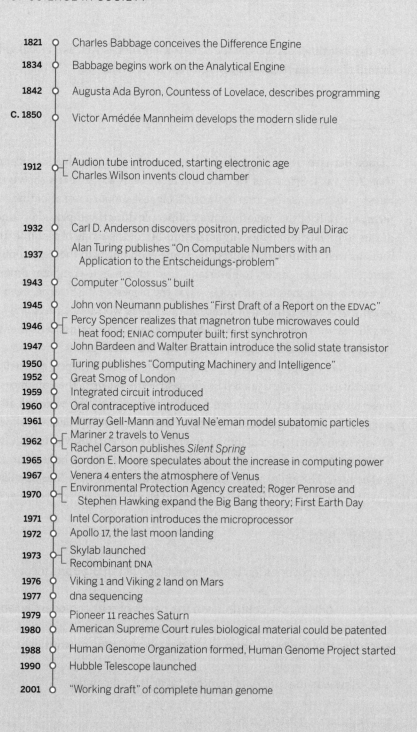

1821	Charles Babbage conceives the Difference Engine
1834	Babbage begins work on the Analytical Engine
1842	Augusta Ada Byron, Countess of Lovelace, describes programming
C. 1850	Victor Amédée Mannheim develops the modern slide rule
1912	Audion tube introduced, starting electronic age Charles Wilson invents cloud chamber
1932	Carl D. Anderson discovers positron, predicted by Paul Dirac
1937	Alan Turing publishes "On Computable Numbers with an Application to the Entscheidungs-problem"
1943	Computer "Colossus" built
1945	John von Neumann publishes "First Draft of a Report on the EDVAC"
1946	Percy Spencer realizes that magnetron tube microwaves could heat food; ENIAC computer built; first synchrotron
1947	John Bardeen and Walter Brattain introduce the solid state transistor
1950	Turing publishes "Computing Machinery and Intelligence"
1952	Great Smog of London
1959	Integrated circuit introduced
1960	Oral contraceptive introduced
1961	Murray Gell-Mann and Yuval Ne'eman model subatomic particles
1962	Mariner 2 travels to Venus Rachel Carson publishes *Silent Spring*
1965	Gordon E. Moore speculates about the increase in computing power
1967	Venera 4 enters the atmosphere of Venus
1970	Environmental Protection Agency created; Roger Penrose and Stephen Hawking expand the Big Bang theory; First Earth Day
1971	Intel Corporation introduces the microprocessor
1972	Apollo 17, the last moon landing
1973	Skylab launched Recombinant DNA
1976	Viking 1 and Viking 2 land on Mars
1977	dna sequencing
1979	Pioneer 11 reaches Saturn
1980	American Supreme Court rules biological material could be patented
1988	Human Genome Organization formed, Human Genome Project started
1990	Hubble Telescope launched
2001	"Working draft" of complete human genome

MAN ON THE MOON, MICROWAVE IN THE KITCHEN

12

When men last walked on the moon in 1972, science was so entwined with the conduct of modern society in the industrialized world that it had become an indispensable component of the world it had helped to create. Most major science projects were the work of teams and networks of teams. The information being produced was so vast that not only was it impossible for a single person to keep up with developments in science generally, but scientists also had trouble keeping up with developments in their own disciplines. There were more than 20,000 scientific journals worldwide. Disciplines were splitting into subdisciplines, which in turn often split into further specialized areas. In chemistry alone the number of types of chemists had grown more than fivefold, as the disciplinary divisions in the American Chemical Society had risen from the original 5 in 1908 to 28 in 1974.

This enormous growth meant that science was no longer the exclusive domain of a philosophical elite or even of a small group of researchers/courtiers. It was an occupation, an employment choice suggested by guidance counsellors, a column in the employment opportunities section of the newspaper. Increasingly, many of the people who could be classified as scientists by education and professional association were not doing research but were in charge of the many and varied jobs that required precision with scientific instruments and a knowledge of specialized systems. Or they were teaching the next generation those skills.

International Council of Science (ICSU)

Throughout the twentieth century, science was becoming a more international endeavour, even while state funding and war work seemed to be pulling it in the opposite direction. One of the most visible signs of the goal of international cooperation in science was the creation of international scientific unions. Although the new organizations traced their roots to the older multidisciplinary organizations like the Royal Society, the new groups focused on a single discipline and acted as a forum where representatives of national science bodies could come together to meet and share information. The international unions often undertook or coordinated projects, such as the International Geophysical Year that had participants from several unions including the International Astronomical Union, the International Union of Geodesy and Geoscience, and the International Union of Geological Sciences. The explosion of these unions after World War II and during the Cold War speaks to scientists' desire to keep scholarly lines of communication open, even as political and ideological battles heated up. (See figure 12.1.)

In 1931, the International Council of Scientific Unions (ICSU) was founded as an umbrella organization for the international unions. ICSU was the descendent of two earlier internationalist organizations, the International Research Council (1919–31) and the International Association of Academies (1899–1914). Although the scientific unions were not directly affiliated with the United Nations, they often worked on UN projects, particularly with UNESCO and the World Health Organization. In addition to representing science on the international stage, ICSU has worked to preserve the universality of science and defend scientific freedom. In 1998, ICSU changed its name to the International Council for Science, but continues to be known by its original acronym.

Women in Science

In the 1970s, women began to pursue scientific careers in large numbers. From the early modern period on, wives, sisters, and daughters of scientists had often made important and independent contributions to their husbands', brothers', or fathers' scientific work. Still, before the late nineteenth century, few women could choose an independent scientific career. Perhaps the most telling example of women's precarious position in science was the 1910 decision of the Académie des Sciences to deny Marie Curie a place among its academicians. Beginning in the late

12.1 INTERNATIONAL UNIONS BY DATE OF FOUNDING

1922	International Astronomical Union	Astronomy
1922	International Mathematical Union	Mathematics
1922	International Union of Geodesy and Geophysics	Geodesy and Geophysics
1922	International Union of Geological Sciences	Geology
1922	International Union of History and Philosophy of Science	History of science and Philosophy of science
1922	International Union of Pure and Applied Chemistry	Chemistry
1922	International Union of Pure and Applied Physics	Physics
1922	International Union of Radio Science	Radio science
1923	International Geographical Union	Geography
1925	International Union of Biological Sciences	Biology
1947	International Union of Crystallography	Crystallography
1947	International Union of Theoretical and Applied Mechanics	Mechanics
1955	International Union of Biochemistry and Molecular Biology	Biochemistry and Molecular Biology
1955	International Union of Physiological Sciences	Physiology
1966	International Union for Pure and Applied Biophysics	Biophysics
1968	International Union of Nutritional Sciences	Nutrition
1972	International Union of Basic and Clinical Pharmacology	Pharmacology
1976	International Union of Immunological Societies	Immunology
1982	International Union of Microbiological Societies	Microbiology
1982	International Union of Psychological Science	Psychology
1990	International Cartographic Association	Cartography
1993	International Brain Research Organization	Neuroscience
1993	International Union of Anthropological and Ethnological Sciences	Anthropology and Ethnology
1993	International Union of Soil Sciences	Soil science
1996	International Union of Food Science and Technology	Food science and Food technology
1996	International Union of Toxicology	Toxicology
1999	International Union for Physical and Engineering Sciences in Medicine	Medical physics
2002	International Society for Photogrammetry and Remote Sensing	Photogrammetry and Remote Sensing
2005	International Union for Quaternary Research	Quaternary period
2005	International Union of Forest Research Organizations	Forestry
2005	International Union of Materials Research Societies	Materials science
2011	International Sociological Association	Sociology and Social sciences

nineteenth century women began a concerted effort to break into the male preserve of science. This was part of the larger political and social feminist movement, linked to the struggle for suffrage, although it is a mistake to see the vote as the prime goal for many feminists. Education was a key component in their quest for equality, and one strategy that was used to gain greater education was the promotion of a good education in order to make women better mothers. Science education was central to this, since science promoted rationality and was useful. Women worked to establish a number of women's colleges in order to develop this useful science agenda. At institutions such as the "seven sisters" in the United States (Vassar, founded in 1865; Smith, 1871; Wellesley, 1875; Bryn Mawr and Baltimore, 1885; Mount Holyoke, 1888; and Barnard, 1889) women taught and learned science. The same philanthropic women who supported these colleges also began the slow campaign to allow women into the regular universities, especially to earn PhDs in science.

Women developed branches of science that were seen as particularly suited to them, especially astronomy, psychology, anthropology, and home economics. As early as the nineteenth century women were employed as calculators in astronomy, since they were better at the tedious mathematical calculations necessary to establish accurate star locations and would work for a fraction of men's wages. Women's assumed empathy with others made psychology (especially child psychology) and anthropology seem apt areas of study. Home economics, with its subject the science of the home, also seemed an obvious fit. Unfortunately, these proved unhappy choices. Women's services were less needed as more powerful astronomical instruments and computers replaced long calculations, and the other sciences were always seen as second-class compared with physics, chemistry, or biology.

During World War II more women entered the sciences to fill positions, including university science teaching, that were vacated when many male scientists were seconded to the war effort. However, most of these new positions were not permanent, since many universities froze tenure hirings during the war, and after the war most women were expected to step down to make space for returning servicemen. Increasingly, universities and colleges, in an effort to become more prestigious, insisted that their faculty have PhDs, which far fewer women had earned. Universities also began to establish serious anti-nepotism regulations, making it impossible for the wife of a faculty member to have a job in her own right.

After Sputnik, some of the federal money earmarked for improving science in the United States did go to women, but many voiced the concern that national scholarships to women were wasted money, since girls would marry and not use the training the state had provided. The Soviet Union had many more women scientists, in part because the regime there made a conscious decision to promote science as a woman's career. Americans ridiculed these female scientists as proof of the weakness of the Soviet state.

It was not until second-wave feminists of the late 1960s and 1970s began to protest loudly the treatment of women in science that changes began to happen. Equal Pay legislation in 1972, for example, forced universities and research centres to pay women the same rates as men. Anti-nepotism rules were gradually declared unconstitutional. Feminist philosophers of science critiqued the ideology of science itself, and programs were slowly put into place to encourage women to pursue scientific careers. By the end of the twentieth century women began to achieve parity with men in some sciences (most particularly the biological and medical sciences), while physics and computer science continued as male-dominated fields into the twenty-first century.

Science Produces Consumer Goods

In the postwar years, as women struggled to be taken seriously as scientists and as the status of home economics, the scientific study of the home, lost ground to other science disciplines, the relationship between the laboratory and domestic and work worlds continued to change. Science research was expected to produce useful products not just for the high-technology industries but for the average person. These products were often spinoffs of other research, such as the development of the microwave oven. The use of microwaves to heat things was a by-product of research on magnetron tubes, which were at the heart of radar-detection equipment developed during World War II. In 1946, Percy Spencer (1894–1970) was working for the Raytheon Corporation on ways to improve the manufacturing of magnetron tubes when he noticed that they heated things. In tests, he popped popcorn and exploded an egg by placing them close to a working magnetron tube. He applied for a patent (1 of 120 he received) for a microwave cooker, and in 1947 the first "Radarange" was produced. It was about the size and weight of a refrigerator and required plumbing for the cooling system. However, by 1965 domestic countertop models were introduced and are now a standard kitchen appliance in the industrial world.

Another example of how developments in research were folded back into scientific work to produce consumer goods occurred in the electronics industry.

Around the beginning of the twentieth century, a host of scientists and inventors such as Thomas Edison (1847–1931) and Nikola Tesla (1856–1943) had bridged the gap between theoretical and experimental science and application. Their inventions, the teams they created, and the companies that developed their work brought electrification to industry and the home. The electrical industries, including the telecommunications sector, required workers who were part engineer, part scientist, and part technician. These people could work with the basic principles of electricity, but they were generally asking the question "what can I make this do?" rather than the question "how does the universe function so this is possible?" This is sometimes called "applied science," but there is no clear line between applied and pure science, either historically or institutionally, particularly when the products of applied science were then used in fundamental research in the form of new instruments.

Two discoveries in particular combined to change the practice of science. They were the development of computing and the introduction of the solid state transistor. Independently, each was an important innovation, but together they transformed industrial society. As Big Science coordinated the work of larger and larger teams, computers offered the technology to manage both the equipment of Big Science and the oceans of information such projects generated.

Computing

The origins of computing were analogous with the development of mathematics. The use of tally bones dates from around 100,000 BCE, while the first successful mechanical calculating device was the abacus. Some historians identify a kind of counting table that worked on the same principle as the more familiar frame abacus as far back as 3000 BCE in Babylonia. The Babylonians, who were excellent mathematicians, made clay tablets listing tables of important numerical information as early as 2300 BCE. By about 400 BCE the abacus was used around the Mediterranean. Whether through independent development or trade contact, the abacus was developed in China in its modern form of beads on rods or wires sometime after 600 CE.

The abacus was useful for rapid calculations involving basic operations, and it remained the primary calculating device for most of the world until the twentieth century. However, it was less useful for the kinds of questions examined by natural philosophers and mathematicians by the seventeenth century. Problems in trigonometry involving sines and tangents and greater facility with complex calculations such as square roots were either very time-consuming or impossible using an abacus. John Napier (1550–1617) turned his mind to the theory of calculation in 1594. His work was on algorithms, or methods of calculation. In 1614 he published *Mirifici logarithmorum canons descriptio* (*Descriptions of the Marvelous Rule of Logarithms*), which contained his tables, the rules for their use especially for trigonometry, and the principles of logarithmic construction. The tables had taken him 20 years to complete, but they were so well done that no serious revision was considered necessary for over 100 years.

Napier's work was expanded by other mathematicians who sought to transform logarithmic tables into mechanical form. Both Edmund Gunter (1581–1626) and William Oughtred (1574–1660) produced types of slide rules that used logarithms; Victor Amédée Mannheim (1831–1906) developed the modern slide rule about 1850.

Mechanical calculators, usually based on some system of gears, also have a long history. Descriptions run back into antiquity, with a technological overlap between mechanical calculating devices and other machines such as clocks and astrolabes. In the early modern period Wilhelm Schickard (1592–1632), an astronomer and cartographer, described a mechanical calculator to Kepler in a letter in 1623, but it was not until 1642 that Blaise Pascal presented a working mechanical calculator.

One of the constraints on calculators was the precision necessary to manufacture the component parts. In the nineteenth century precision fabrication reached new levels of quality, so when Charles Babbage (1792–1871) conceived of his Difference Engine in 1821, the technology was available to construct his device. The Difference Engine was designed to calculate and print mathematical tables using polynomial functions, all without the necessity of human intervention. (See figure 12.2.) It was financed in large part by the British government, which paid Babbage £17,500, a huge sum of money. The government hoped that the Difference Engine would be useful in such areas as navigation, astronomy, and calendars. Early on it appeared that the investment might offer some return. Babbage completed enough of the Engine so that in 1827 it was used to calculate tables of logarithms, but instead of finishing the work, he continued adding to its design, requiring it to be partially dismantled and thus delaying its completion. Then he abruptly abandoned the enterprise altogether in 1834 to begin work on a new type of calculating device he called the Analytical Engine. This new device could be programmed by using punched cards so that it was a general purpose machine, rather than simply a kind of mechanized adding machine. The British government, having received nothing for their earlier investment, refused to fund the new machine regardless of Babbage's claims for its amazing potential. Without funding, the Analytical Engine was never built.

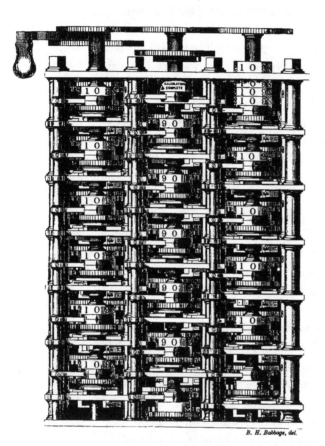

B. H. Babbage, del.

12.2 BABBAGE'S DIFFERENCE ENGINE

A small portion of Babbage's Difference Engine No. 1 from his autobiography, *Passages from the Life of a Philosopher* (1864).

The idea of a calculator whose functions could be changed, or programmed, to suit the circumstances was a crucial step toward the creation of computers. Babbage had conceived of the Analytical Engine storing results (the Store) separately from the part of the machine that did the calculating (the Mill). The idea of punched cards to control the engine had been borrowed from the textile industry, where Joseph-Marie Jacquard (1752–1834) had introduced a system of punched cards to control the patterns woven by looms. The best explanation of the engine's operations came not from Babbage but from Augusta Ada Byron, Countess of Lovelace (1815–52). The daughter of the poet Lord Byron, Ada met Babbage in 1833 and asked him to teach her mathematics. In 1842 Babbage asked her to translate a French account of the technical aspects of the Analytical Engine, but her commentary was longer than the original document. In 1843 she wrote about a special class of calculus functions—this was one of the first descriptions of programming.

Historians disagree about Ada's mathematical abilities and her role in Babbage's work. While she is sometimes called the first "computer programmer," this is somewhat ahistorical. Babbage and his son, who assisted him on the development of the Analytical Engine, were both conversant with the theory and practice of the engine's operations before she was involved in the project. Nonetheless, Ada's contribution in an age when women were actively discouraged from such intellectual activity was remarkable. In honour of her work, in 1979 a programming language developed for the American Department of Defense was named Ada.

Although Babbage never completed his Analytical Engine, the abandoned Difference Engine No. 2 was built at the Science Museum of London in 1991. Weighing two-and-a-half tons and containing more than 4,000 moving parts, it worked just as Babbage had said it would.

Although a small number of difference engines were built, including a motorized version exhibited by George Grant in 1876 in Philadelphia, they were not widely used because there was little market for them. They were generally too expensive for business applications and not versatile enough for mathematical or scientific work. It was not until the introduction of vacuum tubes that offered high-speed switching and the technical demands of World War II that the development of electronic calculating devices moved beyond the theoretical complexity of Babbage's mechanical designs. The Ultra program allowed the British to decrypt German codes, but it took too long. In 1943 Tommy Flowers (1905–98) was sent to Bletchley Park, home of Britain's secret team of code breakers, to build an electronic code breaker. The result was Colossus, which gave the Allies a huge advantage over German forces during the invasion of Europe. Despite its importance, little remains of the Colossus project. The original machines were destroyed in 1946, and even their existence was kept secret until 1970.

On the other side of the Atlantic John Mauchly (1907–80) and Presper Eckert (1919–95) built ENIAC (Electronic Numerical Integrator and Computer). The suggestion that a fully electronic calculator (except for input and output elements) could be developed was made around 1942. One of the main motivations for such a device was the increasing complexity of long-range artillery, which could fire at targets far out of sight. Ballistics tables and firing solutions were so complex and time-consuming that the most complex might take hours to compute by hand. So the American Army Ordnance Corps funded the development of ENIAC, spending almost $500,000 on the project. Mauchly and Eckert worked on and oversaw the construction of the massive machine. It weighed over 30 tons, contained 19,000 vacuum tubes, and consumed almost 200 kilowatts per hour of electrical power to run both the electronics and the cooling system that kept it from burning out.

ENIAC was completed in 1946 at the Moore School of Electrical Engineering at the University of Pennsylvania. After it was tested, it was taken apart and transferred to the Aberdeen Proving Grounds, where it went into general operation in 1947. Although ENIAC was not operational during the war, it was used for a number of other projects, including weather forecasting, satellite trajectories and orbits, and nuclear weapons programs. Colossus had been the first electronic computer, but it was from ENIAC that the computer industry sprang.

Computing Theory: Turing and von Neumann

While computers were at some level an engineering problem, the theory of computers was greatly affected by two men: John von Neumann (1903–57) and Alan Turing (1912–54). Von Neumann was a mathematical prodigy who turned his hand to a wide range of problems from game theory to subatomic physics. He arrived in the United States from Hungary in 1930 and was one of the first people to be appointed to the Institute for Advanced Studies when it was founded in 1933. During the war, von Neumann was a consultant to the Manhattan Project and to the Ordnance Corps, which brought him into contact with the ENIAC group. Turing was also a prodigy, with interests in science and mathematics. His wartime experience was at Bletchley Park, where his work on Project Ultra broke the German Enigma code. Like von Neumann, his work gave him access to actual computing devices.

In 1935 Turing began exploring the theoretical possibilities of computation. In his 1937 paper "On Computable Numbers with an Application to the Entscheidungs-problem," he speculated about a machine that could carry out calculations based on a finite table of operations and reading or deleting a series of instructions on a paper tape. This theoretical device, later called a "Turing machine," offered a number of theoretical ideas that underpin modern computers. After the war Turing was invited to join the National Physics Laboratory and asked to build a computer. His plans for the Automatic Computing Engine were never carried out, however, and he moved to the University of Manchester to work on the Manchester Automatic Digital Machine. In 1950 he published "Computing Machinery and Intelligence" in the journal *Mind*. It outlined many important issues in computing, speculated on the relationship between computers and human thought, and outlined the "Turing test," which argued that machine intelligence could be evaluated by observation of the interaction between a computer and a human. If a human asking questions of a hidden respondent could not discern whether the respondent was a computer or another human, then the computer had likely achieved a level of intelligence equivalent to human intelligence.

Von Neumann's experiences with computers likewise led him to see them as offering a powerful tool. In his "First Draft of a Report on the EDVAC" (1945), he described a general purpose, stored program computer. Many of his ideas were translated into actual computer equipment, and, as a consultant to IBM he contributed to the development of commercial computers in the 1950s. In 1953 the first production model, the IBM 701, was marketed. Nineteen were sold, mostly to aerospace contractors and the government. One of its main uses was, in fact, the development of nuclear weapons. In these two realms—computing and nuclear weapons—von Neumann had a profound influence, which he used to try to persuade the American government to take military action against the Soviet Union. Partly based on his anti-communist political views and partly on ideas that he had developed when he created the economic/mathematical field of game theory, von Neumann advocated using the atomic bomb on the Soviet Union before it could develop its own nuclear weapons.

Both von Neumann and Turing died young. When Turing was arrested for having a homosexual relationship (illegal at the time), the British government removed his security clearance, effectively driving him out of government work amid Cold War paranoia. He committed suicide in 1954. Von Neumann died of cancer only three years later. While many people worked on computers, these two were leaders in the theoretical and conceptual aspects that made the electronic computer the most versatile device ever created.

The Solid State Transistor

Turning computing machines from gigantic monsters full of temperamental vacuum tubes that had to be tended by a phalanx of technicians into something more manageable and affordable required the introduction of a new technology. That technology, which had started at almost the same moment as ENIAC began to calculate ballistic tables for the American Army, was the solid state transistor.

The solid state transistor originated as a solution to a fundamental problem in electrical communication. As the distance an electrical impulse (the signal, whether dots and dashes, voice, or later other kinds of information) had to travel increased, it progressively weakened. This was a serious problem for long-distance telephones, so many people worked to figure out how to boost the signal strength without distorting the signal. It was partly solved by the introduction in 1912 of the Audion tube and its many descendants. For telephone companies with growing numbers of subscribers, the tube solution created its own technical problems, since the amplifiers were bulky, used a great deal of power, and could knock out service for long periods by burning out.

In 1936 the physicist William Shockley (1910–89) was hired by Bell Laboratories and, with his knowledge of quantum mechanics, was asked to look for ways to create a reliable solid state device to replace the old system. His preliminary efforts were unsuccessful, and then the war interrupted work. When he returned to Bell Laboratories in 1945, he returned to the solid state question, but another design failed. The project was then turned over to John Bardeen (1908–91) and Walter Brattain (1902–87), two scientists working in Shockley's lab.

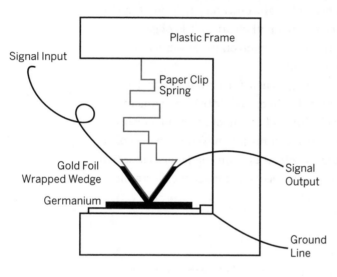

Signal Input

Plastic Frame

Paper Clip Spring

Gold Foil Wrapped Wedge

Germanium

Signal Output

Ground Line

12.3 SOLID STATE AMPLIFIER

They came up with a new approach. Taking a small piece of germanium, some gold foil, and a paper clip as a spring, they were able to amplify an electrical signal almost 100 times its original strength. (See figure 12.3.)

On December 23, 1947, Bardeen and Brattain demonstrated their invention to Bell executives. It was a major breakthrough, although beyond its application to telephone amplification, no one really knew what it would mean for the electronics industry. Further, Shockley was both pleased with the success of his research group and angry that he had not been the inventor after so many years of effort.

He went on to develop a different form of transistor that replaced the fine wires used to channel the signal through the semiconducting material of the Bardeen/Brattain design (called "point contacts") with a more robust system based on "junctions" or boundaries in the semiconducting material itself.

In the beginning the transistor found limited use. In 1952 the first commercial product using the new technology was a hearing aid. Large electronics companies such as General Electric, Philco, and RCA had too much invested in tube technology to simply change to the new transistors. The first mass market effort was the production of junction transistors by Texas Instruments for a portable radio in 1954. In 1958 Jack Kilby (1923–2005) at Texas Instruments put different electronic components on the same piece of semiconductor, while in 1959 Robert Noyce (1927–90) at Fairchild Semiconductors worked out a way to link such components. Between them, the two men created the integrated circuit. Texas Instruments got out of the radio market, and a small Japanese company, Sony, jumped in. Transistor radios, relatively inexpensive and small enough to fit in a shirt pocket, hit the market just as postwar prosperity and the Baby Boom made electronic entertainment big business. By the middle of the 1960s Sony was manufacturing television sets using transistor technology, and Japan was starting to dominate the consumer electronics industry.

The application of the transistor affected various aspects of computer technology, and in 1964 the Comcor Company produced the CI 5000, the first fully transistorized, general purpose computer. The biggest breakthrough after the integrated circuit was the 1971 development by Ted Hoff (1937–) at Intel Corporation of the microprocessor, which placed all the logical elements necessary for computing on a single chip. The microprocessor controlled the states (the on/off conditions) that made programming and calculation possible, as well as handling the input and output of signals. After that, electronic technology exploded. Microprocessors were made not just for computers but were put into almost anything that used electricity, from children's toys to elevators, automobiles, and pacemakers. What had begun as a question of signal strength became the backbone of modern industrial society, controlling the flow of information, monitoring the environment, and containing the massive amounts of information necessary to keep all the computer-controlled systems operating.

The microchip revolution was predicted in 1965 by Gordon E. Moore (1929–), the head of research at Fairchild Semiconductors. He observed that the complexity of integrated circuits had roughly doubled each year since 1959; using that basic rate, he predicted that by 1975 there would be 65,000 transistor chips on the market. Although that took a bit longer—it wasn't until 1981 that chips hit the 65,000 mark—his observation about the rate of increase was labelled "Moore's Law" and has been a kind of rule of thumb for microprocessors ever since. While superchips, at the very limits of the materials and manufacturing systems, are already in the planning and development stage, it is not the hardware alone but the software that ultimately makes the computer useful.

The introduction of computers, both large and small, transformed scientific practice in almost all disciplines. Anything requiring many mathematical operations or large data sets, such as astronomy and subatomic physics, could be handled more quickly and accurately by computers. However, computers could also be used to control equipment, so everything from microscopes to electrophoresis became mechanized. The utility of computing power was so significant that scientists around the world also worked on systems to connect computers and to gain access to computer power at distant locations, giving birth to computer networks and ultimately to the Internet.

The Pill: Science and Gender Relations

While radios, televisions, electronic game consoles, and home computers made inexpensive by transistor technology were entertaining the population and providing a means of stimulating demand for more consumer products, another branch of science was about to change gender relations. This change took place in the bedrooms of the industrial world. In the struggle for gender equality, science had been used as a justification both for and against women's rights. Much of the debate hinged on reproduction. For generations, many people (both men and women) had argued that, since childbearing was the highest purpose for women, anything that took women away from that goal (such as education, science research, or work outside the home) was not just a bad idea but contravened nature. A number of women's rights advocates had worked on birth control as a key both to women's independence and to a better life for everyone. To make matters more complicated, some birth control advocates were connected to the eugenics movement in the early twentieth century, which sought a differential birth-rate for the "fit" and "unfit." Originally, birth control depended on barrier methods such as condoms and cervical coverings or on less reliable ovulation timing methods such as the rhythm method. This all changed in 1951 when Carl Djerassi (1923–2015), while conducting research on the hormonal control of reproduction with a team at the University of Mexico, showed that ovulation could be controlled by administering the hormone progestin orally. An oral contraceptive became a possibility.

The search for a reliable form of hormonal contraceptive was continued by Gregory Pincus (1903–67) and Min Chueh Chang (1908–91) at the Worcester Foundation for Experimental Biology. Margaret Sanger (1879–1966), an early birth control advocate, introduced Katherine Dexter McCormick (1875–1967), heir to the International Harvester fortune, to the researchers, and McCormick agreed to help fund their work. In all, she contributed some $3 million to research on oral birth control. In 1956 large-scale human trials were conducted by John Rock

(1890–1984), and in 1960 the American Food and Drug Administration granted approval for the sale of oral contraceptives. Although there were some problems with the initial drug (it was found to contain about ten times more hormone than was needed for contraception) and some jurisdictions were slow to legalize this new contraceptive, it had a profound effect on Western society. Because of this invention, relationships between the sexes changed radically and made the sexual revolution of the 1960s possible.

Following another line of inquiry, in 1961 Jack Lippes (1924–) introduced the Lippesloop, an IUD (intrauterine device). The IUD was a small inert object (usually made of plastic) inserted in the uterus to prevent fertilization, although the exact reason why it works is still not completely understood. Various forms of IUD may have existed from ancient times, but their modern development began in 1909 with work by Ernst Grafenberg (1881–1957). It was initially seen as a safe, low-cost method of birth control, which was also under the control of women. IUDs were popular, but their use in North America plunged after the A.H. Robins Company was sued over problems with the Dalkon Shield. The Dalkon Shield had been developed by Hugh J. Davis, a physician at Johns Hopkins Medical School during the late 1960s. Design problems were linked to pelvic inflammatory disease, uterine damage, sterility, and a number of deaths. Almost 400,000 claims were brought against the manufacturer, which stopped sales in the United States in 1974 but not in other parts of the world until several years later. Since IUDs were not drugs, there was little control over their development, testing, or marketing. The legal case highlighted the difficulty of using scientific evidence in legal matters, as both sides brought in scientific experts to argue their case. According to some, the Dalkon Shield was safe, while others presented evidence suggesting it was dangerous. The problems with the Dalkon Shield affected the perception of IUDs in general and, especially in the industrialized world, their use dropped dramatically.

Exploring Space

The Age of Aquarius, the Woodstock generation, and the sexual revolution of the late 1960s and 1970s were the creation of modern contraceptives. Science shaped this new age in a very real way, but it also raised fears and disenchantment. Concerns about nuclear Armageddon, the war in Vietnam, and the close ties between scientists and the military contributed to a sense of social unease with science and technology, especially in the United States. Even the flagship of American science and technology, the Apollo program, ended in 1972 when Apollo 17 became the last manned mission to the moon. In 1973, Skylab was launched, but it largely failed to capture the imagination of the public. Many scientists also questioned the utility of the Skylab missions, although they did help provide information about the long-term effects of zero-gravity environments on the human body.

The exploration of space was given over to probes, which made more sense technologically, financially, and scientifically but lacked the romance of a human explorer, a Columbus or Neil Armstrong. Mariner 2 travelled to Venus in 1962, Venera 4 made it into Venus's atmosphere in 1967, and Venera 9 broadcast television pictures from the planet's surface in 1975. Venus, once thought to be a water world, turned out to be a hell of high temperatures, carbon dioxide atmosphere, and acid rain.

Mars, Earth's other close neighbour, was also the target of many probes. In 1971 two Soviet and one American spacecraft went into orbit around the red planet. At first the images they sent back were disappointing, revealing little about the surface of the planet, but that turned out to be due to a planet-wide dust storm. Over time, the Mars satellites sent back large amounts of information. Several attempts were made to land probes on the surface, but it was not until 1974 that Viking 1 and 2 landed and remained operational. While the old science fiction idea of a Martian civilization had long disappeared, there were hopes that life in the form of microbes might be found, but the probes found nothing. To date, no trace of life on Mars has been found, but there remain theories about traces of life in the polar regions where there is more moisture, and recently running water seems to have been detected, with the Mars Exploration Rovers sending back evidence of water in 2015.

Also in 1974, Mariner 10 reached the vicinity of Mercury, showing it to be a lump of baked rock with a lunar-like surface of craters. Probes to the outer planets started with Pioneer 10, which flew by Jupiter in 1973, sending back information about the giant planet's magnetic fields and moons, before heading out into deep space beyond the orbit of the furthermost planet in the solar system. Pioneer 11 reached Saturn in 1979, followed by Voyager 1 and 2 in 1980 and 1981. They sent back spectacular images of Saturn's rings and revealed the presence of previously undetected moons. While Voyager 1 headed into deep space, Voyager 2 continued on to Uranus, flying past it in 1986.

Physics: Particles within Particles

As astronomers were rapidly learning a tremendous amount about outer space, physicists investigated the minuscule. Since the era of Rutherford, physicists had been looking at the structure of atomic particles such as photons, electrons, and neutrons. They had been able to study the behaviour of particles in a number of ways, but the 1912 invention of the "cloud chamber" by Charles Wilson (1869–1959), which used water vapour (and later alcohol) in a small enclosed chamber to track particles, made the process much more versatile. Particles entering the chamber caused a small amount of condensation along the path they followed. Because of the behaviour of the particles, these paths curved or made coils, allowing scientists to calculate energy, charge, and life span.

In 1932, Carl D. Anderson (1905–91) used a cloud chamber to study cosmic rays, attempting to determine if they were particles or waves. He found that the rays could be deflected by a strong magnetic field, indicating a particle nature. At the same time he noticed a track for another particle that was the same as that for an electron, but it curved in the opposite direction, indicating that it had the opposite charge. This particle was an "anti-electron" or positron (from "positive electron"). Its existence had been predicted several years earlier by the theoretical physicist Paul A.M. Dirac (1902–84), who suggested that every subatomic particle would have a corresponding antiparticle.

Other kinds of particles were discovered from cosmic ray studies, such as two "mu-mesons" (shortened to "muon") found by Anderson in 1936. These particles had opposite charges and were much more massive than electrons but lighter than protons. This was followed by the discovery of three forms (positive, negative, and neutral) of "pimeson" (or "pion") in 1947 and in 1952 of the "lambda" particle, which had no charge of its own but was found to break down or decay into a proton and a negative pion.

The particles that made up atoms were literally and theoretically coming apart, but studying the bits was a technical challenge. To study the particles it was necessary to accelerate them, smack them into a target (other particles), and see what came apart. This was accomplished with machines known (not surprisingly) as accelerators, invented in 1928 by John Cockcroft (1897–1967) and Ernest Walton (1903–95). These were linear accelerators that projected the particles down long straight tubes or tunnels. While useful, they had limitations, and in 1931 Ernest Lawrence (1901–58) developed the first working cyclotron, which used a ring of magnets to accelerate particles. Cyclotrons confirmed Einstein's theory of relativity since, as the electrons were accelerated toward the speed of light, they gained mass. This became apparent because the magnetic field required to keep the electrons moving around the cyclotron had to be greatly increased as they sped up. But to break apart the atomic particles in order to try and find the range of subatomic particles predicted by theorists required huge amounts of energy, and eventually scientists built a new class of accelerators that synchronized the increase in magnetic field needed as the particle circled through it. These machines were called "synchrotrons"; the first one was created in 1946. Further refinements also allowed experimenters to link accelerators, so that in 1983 Fermilab created an accelerator that used a linear accelerator, a 500-foot diameter synchrotron, and a four-mile diameter synchrotron.

In addition to the new accelerators a more sensitive method of recording the particle tracks was developed. In the early 1950s Donald A. Glaser (1926–2013) created the "bubble chamber," which worked on the same principle as the cloud chamber but used liquid hydrogen heated to just below its boiling point rather than water or alcohol vapour. When charged particles passed through the hydrogen, they added just enough energy to cause it to boil and release bubbles. By measuring the path and the number of bubbles produced by the passage of a particle, information about its mass and velocity could be determined.

With all this new equipment, a host of subatomic particles were found. In addition to the old proton and electron were the muon, kaon, gluon, sigma, xi, lambda, eta, and a host of neutrinos. In 1961 Murray Gell-Mann (1929–) and Yuval Ne'eman (1925–2006) independently created a theoretical model to classify the many particles into families, somewhat like Mendeleev's periodic table of elements. Yet there was a mystery about the apparent creation of large numbers of subatomic particles. A group of particles known as "hadrons" seemed to be almost limitless in number, an odd finding for something that was supposed to be more fundamental or elementary than the large bits of the atom. In 1964 Gell-Mann and George Zweig (1937–) theorized that there was a parallel between hadrons and molecules. There were, of course, a huge number of different molecules, but only a small and limited number of elemental atoms. The hadron particles, Gell-Mann and Zweig argued, were to the subatomic world what molecules were to the atomic world. They predicted the existence of truly elementary particles that Gell-Mann named "quarks," borrowed from a line in James Joyce's novel *Finnegans Wake*. In this model, each proton or neutron was made up of three quarks plus "gluons" that held the quarks together.

Physics: The Very Big

The very big and the very small had to intersect, because the creation of all these particles had to come from some source. According to the Big Bang theory, before the beginning of the universe all energy and matter were confined to a point with no dimensions. It was very hot, and the energy of the universe was so high that there were no particles and all forces were essentially unified or completely symmetrical. When the universe came into being, it started to cool and parts started to differentiate. This differentiation, or asymmetry, led to the creation of matter, building up from the subatomic realm, which clumped together in various ways to give us the universe as we now see it. This was the Big Bang theory as outlined by Friedmann and expanded in 1970 by Roger Penrose (1931–) and Stephen Hawking (1942–).

Although there are still many points of debate about the details of the origin of the universe, subatomic physics and astrophysics have tended to confirm the Big Bang model. The study of the very big (astronomy) and the very small (subatomic physics) also come together in theoretical efforts to create a "grand unified theory," or gut, that will provide the mathematical model for all four forces (electromagnetic, gravitational, weak nuclear, and strong nuclear), in effect rolling particles back to the original symmetrical condition at the Big Bang. Experimental results have shown that the distinction between the electromagnetic force and the weak nuclear force (which is responsible for nuclear decay, as opposed to the strong force that holds the atomic nucleus together) disappears at very high energy levels, so the two forces look like just one force.

In the 1970s a new tool was conceived to help look for evidence of the early state and evolution of the universe as well as to broaden optical astronomy. A space-based telescope was first proposed as a joint project between the European Space Agency and NASA. The Hubble Space Telescope, named after Edwin Hubble who had first observed the expansion of the universe, was launched into orbit aboard the space shuttle Discovery on April 25, 1990. It provides images free of the distortion of the atmosphere, although a flaw in the optical system had to be corrected in space in 1993. The Hubble Telescope provides huge amounts of data, about 14,000 megabytes daily. (See plate 8. The Eagle Nebula is one of the most famous pictures taken by the Hubble Telescope.)

Ecology and the Environment

Although subatomic particles and the origin of the universe were brought to the attention of the public by people such as Carl Sagan and his television program *Cosmos*, and Stephen Hawking's best-selling book *A Brief History of Time* (1988), public interest definitely shifted back to the home planet during the 1970s and 1980s. One of the key reasons for this was the rise of the ecological movement. What we think of today as environmentalism is not a new development but can be traced back to efforts in the nineteenth century to conserve wilderness areas and to limit water and air pollution. Even earlier, societies from the ancient Egyptians on had laws about water use, garbage collection, and human waste disposal, and these laws existed in various forms wherever urban centres appeared. Yet the scientific study of ecology came relatively late. Biology tended to look first at individual organisms and later at genetics, rather than the organism in the environment. Similarly, chemists were focused on the controlled use of chemicals, not on the effects of uncontrolled chemicals, while physicians were trained to look for particular pathogens and were less aware of the interaction of people with the environment.

In 1962 the inseparable relationship between people and their environment was brought forcibly before the public when Rachel Carson (1907–64) published *Silent Spring*. Aimed at a popular audience, her book introduced many to the problem of environmental pollution, in particular, the problem of pesticide use. Carson pictured nature as an interlocked system rather than a series of independent components that could be treated individually. A farmer spraying DDT on crops to control insects was not just exposing the crops to the chemical but also leaving the chemicals in the ground where they could be absorbed into the tissues of the plants they were meant to protect and by other crops long after the spraying ended. These levels could build up, remaining in the soil for many years.

But the persistence of pesticides was not the worst part. The chemicals worked their way into the food chain, accumulating in higher and higher levels in the tissues of those animals at the top of the predator cycle. Plants contaminated insects, which in turn were eaten by song birds. The concentration of toxic chemicals could grow so high in the song birds that they would die or be unable to reproduce. Their death was the inspiration for the title of Carson's book. If action was not taken, argued Carson, there could truly be a silent spring. The death of song birds was one of a litany of dangerous effects caused by pesticides, and Carson presented numerous stories about the outright abuse of chemical agents, a lack of scientific study of the long-term effects of pesticides by either government or manufacturers, and inadvertent or deliberate cover-ups of mismanagement of chemical agents. For instance, the pesticide Aldrin was sprayed over parts of Detroit by Michigan state officials to control Japanese beetle. When the spray planes began to fly with no prior warning, concerned citizens flooded city officials and the Federal Aviation Agency with calls. Citizens were told that the spraying was completely safe, even though the American Public Health Service and the Fish and Wildlife Service both had published reports of the toxicity of Aldrin. Large numbers of birds died, while animals and humans exposed to the chemical got sick.

Carson's research struck a responsive chord with a public increasingly disenchanted with the changes brought about by science and fearful of its power. By 1969 there was enough popular support for government action on environmental concerns for the United States to pass the National Environmental Policy Act; in 1970 the Environmental Protection Agency (EPA) was founded to enforce that legislation as well as other laws such as the Clean Air Act. In the same year the EPA started operating, the first Earth Day was celebrated. Based on the teach-ins of the movement protesting the war in Vietnam, it was designed to bring the issues of pollution to the public. Such public interest also helped legislators ban the pesticide DDT in 1972.

The dangers of chemical contamination were further thrust into public view when the Love Canal story broke in 1978. Love Canal was a residential development in Niagara Falls, New York, that had been built on or beside a chemical waste dump that the Hooker Chemical Company operated from 1920 to 1953. In 1976 residents began to complain about bad odours and oozing black or brown sludge. By 1978 a major battle over the safety of the neighbourhood was underway. Toxic chemicals including dioxin, known to be a powerful carcinogen, were found on the site, but the degree of danger was very much at issue, with experts lining up on both sides of the case. The federal government stepped in when President Jimmy Carter declared the site a Federal Emergency zone, and the residents were relocated. Claims and counter-claims about the health risks continue to be made to this day, with some environmentalists and scientists claiming that the danger was very great, linking such environmental damage to cancer, miscarriage, and declines in sperm counts not just in Love Canal but worldwide. Other scientists have argued that there was no sound evidence of increased health problems, again showing that scientists could be found to support opposing positions. Such controversies tended to confirm the public's lack of faith in an "objective" scientific answer about dangers that appeared to be created by scientists in the first place.

CONNECTIONS

Earth Day and the Rise of Environmentalism

While scientists and politicians had been concerned about clean air and water for generations, and while various pieces of anti-nuisance and anti-pollution legislation had been passed in Europe beginning in the nineteenth century, it was not until the 1960s that popular action forced governments and businesses to act decisively on these issues. One of the popular calls to action was the creation of Earth Day, a day for citizens to advocate for a cleaner and more sustainable environment. The original idea for Earth Day came from two sources: peace and environmental activist John McConnell (1915–2012) proposed a day to celebrate peace and ecology at a UNESCO conference in San Francisco in 1969; and a month later US Senator Gaylord Nelson (1916–2005) proposed an Earth Day at an environmental teach-in after witnessing the damage caused by an oil spill on the coast of Santa Barbara. Much of the early organizing for Earth Day was done by environmental activist Denis Hayes (1944–), and the first Earth Day was celebrated on April 22, 1970, with thousands of participants at 2,000 colleges and universities and 10,000 primary and secondary schools across the United States.

Earth Day was the result of the growing global concern about human degradation of the environment. One of the biggest concerns for environmental activists in this period was smog. Combining the words "smoky" and "fog," smog was seen to be obscuring the sky over every major city in the world. In Britain "killer fogs" or "pea soupers"

The environmental movement represents an intersection of scientific exploration and public participation. At one end of the spectrum are the eco-warriors for whom any scientific information is suspect because science is unnatural and scientists have helped make the destruction of nature possible. Therefore, only direct action to stop environmental destruction will do any good. At the opposite end are the pro-science groups and individuals who range from scientific optimists, who believe that science will solve many environmental problems, to the debunkers, who use their scientific skills to counter any claims of ecological damage they see as being based on bad science. In between are both non-governmental organizations, such as the Sierra Club, Pollution Probe, and the Club of Rome, and governmental organizations, such as the EPA and the World Health Organization, that attempt to use science to indicate where problems exist and to support actions to solve them.

were created by natural fog combining with toxic sulphur dioxide and soot from coal fires to produce smog so thick it was almost opaque and could actually kill people, especially the elderly or those with breathing problems. The Great Smog of 1952 in London made more than 100,000 people sick and perhaps caused as many as 12,000 deaths. This crisis led directly to the British Clean Air Act of 1956, which limited the domestic use of coal in British cities.

Another aspect of environmental degradation was brought to public attention by Rachel Carson. When she published *Silent Spring* (1962) to make people aware of the environmental damage being caused by pesticides, particularly DDT, American citizens cried out for government legislation to protect people and nature. US public concern about air pollution, water pollution, and pesticides led to the passing of the National Environmental Policy Act of 1969 and the creation of the Environmental Protection Agency (EPA) in 1970. The EPA was given the mandate to enforce both new rules and a range of environmental laws that were already in place.

The concern that the environment was not just a national issue but a global one was heightened by Paul R. Ehrlich's book *The Population Bomb* (1968) and the Club of Rome report *The Limits to Growth* (1972), both of which made dire predictions about the potential for disaster caused by the growth of global populations and environmental collapse. In 1971 Greenpeace was founded, originally to protest against nuclear tests and radioactive fallout, but quickly becoming concerned with wider environmental issues. In 1972 the UN Environment Program began in an attempt to coordinate international efforts. There have been some successes from these environmental efforts: DDT was banned, most industrial countries began to produce less smog, and whaling was outlawed, as were ozone-depleting chemicals.

Scientists have been important players in these environmental movements. But they could not have made these changes without the support and participation of citizens around the world. The challenge was to engage the public and give them the tools to fight for legislative and regulatory change.

DNA and the Human Genome Project

Human control of the environment became even more direct during the postwar period because of genetic research. The discovery of the structure of DNA offered the possibility of directly manipulating the control mechanism of cell activity, thereby giving us the ability to change and ultimately create new kinds of life forms. It has commercial possibilities in genetically altered crops and food, as well as invoking the recurring spectre of modern eugenics made possible by the direct manipulation of humans.

Since Watson and Crick uncovered the structure of DNA, thousands of research hours have been poured into studying its function and working out methods to manipulate it to give an organism new characteristics. In many ways uncovering the structure of DNA was like being given the owner's manual to an expensive car and finding that it was written in code. Certain things were obvious from observation, but the details of how DNA worked still needed to be figured out. At the heart of the decoding effort were the base-pairs, the a-t, c-g combinations that made up the rungs of the ladder. Somehow, the sequence of bases controlled cell function. In 1961 Sydney Brenner (1927–), Francis Crick, and their team argued that the bases could be read in strings of three (such as t-t-g and the complementary a-a-c), which they call "codons." Codons allowed RNA (ribonucleic acid, a kind of molecular robot) to control the actual production of proteins. They called the molecules "transfer RNA" or tRNA.

In the same year, Brenner, François Jacob (1920–2013), and Matthew Meselson (1930–) discovered messenger RNA (mRNA), which carries the pattern of part of the DNA to the ribosomes, the site of protein synthesis in the cell. With the system to transfer information from DNA to protein production, the method of cellular control was revealed. (See figure 12.4.)

This discovery opened up the possibility of interacting with DNA, first by identifying what parts of the huge molecule were responsible for what enzyme or protein chains, and then by getting DNA to do what was wanted. The isolation of restriction enzymes that could cut DNA at specific sites meant that parts of DNA could be looked at separately. The first of these molecular scalpels was identified by Hamilton Smith (1931–) and Kent Wilcox in 1970, while in 1971 the first attempt to write out or sequence base pairs started with the lambda virus. Since viruses need few instructions to live, they were the logical choice for early sequencing work.

By 1973 the techniques of cellular control were well enough understood that scientists could begin manipulating them. Stanley Cohen (1922–), Herbert Boyer (1936–), and Robert Helling (1936–2006) placed foreign DNA in a host organism (in this case the bacterium E. coli), which then replicated itself. This was the technique of recombinant DNA, and it was the foundation of DNA cloning. In 1978 somatostatin became the first human hormone produced by recombinant DNA technology.

When Fred Sanger (1918–2013) introduced his chain termination method for DNA sequencing technique in 1977, essentially all the theoretical elements were in place for the big project: mapping the human genome. This idea had been around since the discovery of the structure of DNA, but the sheer size of the undertaking made it prohibitively difficult, and the tools to record or map the bases accurately were not available. As interest in decoding grew, new tools and methods were

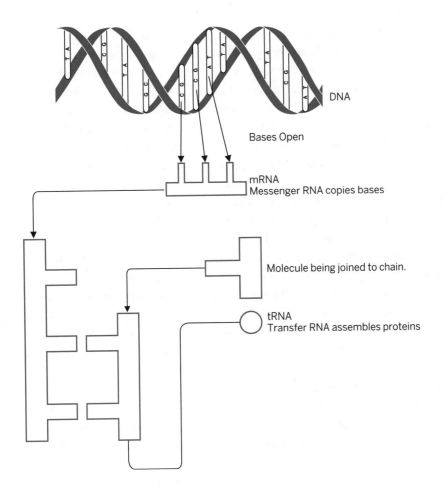

DNA

Bases Open

mRNA
Messenger RNA copies bases

Molecule being joined to chain.

tRNA
Transfer RNA assembles proteins

12.4 PROTEIN ASSEMBLY

developed, and by 1983 a number of labs began to decode chromosomes in a variety of organisms. At about this time the American Department of Energy (DOE) was considering what to do with its various biological researchers. These scientists, many of whom were in the Office of Health and Environmental Research (OHER), had worked on those aspects of the nuclear weapons and nuclear energy programs that looked at the biological effects of radiation as well as doing basic research on cellular biology. These were not as pressing concerns by 1983 as they had been earlier, and so the DOE was looking for ways to employ them on other projects.

In 1985 Chancellor Robert Sinsheimer held a meeting at the University of California, Santa Cruz, to discuss the possibility of mapping the whole DNA list of bases in human chromosomes. The sequencing of the human genome would be a massive project, since there were an estimated 3 billion bases, and the average cost to sequence a base was about $10. Yet the potential benefits were enormous, from cures for disease to extended life. With the firm belief that techniques and equipment would improve, thus dropping the cost of sequencing, participants were enthusiastic about the future of the project. That same year OHER held a meeting at Santa Fe, New Mexico, on the feasibility of the Human Genome Initiative, while James Watson held a similar meeting at the Cold Harbor research centre. Their conclusion was that the mapping of the entire human genome was both possible and desirable. It would be the Manhattan Project of biology, with big teams, big money, leading scientists, and a potentially world-changing result. At last, biologists could compete with physicists for money and prestige.

The following year, the DOE allocated $5.3 million for a pilot project. However, the National Institutes of Health (NIH) also began to fund genetics research. Given the medical and biological aspects of the research, it seemed strange to some officials that the DOE was running the show, but it had several advantages over the NIH. It had operated many large multi-laboratory research projects so had in place a management infrastructure for big projects. It also had a great deal of money and political clout. Most of the conflicting issues were resolved when the DOE and the NIH signed an agreement in 1988 to work together on the project, which became international when HUGO (the Human Genome Organization) was formed to coordinate international research.

The year 1988 was the real start of the Human Genome Project (HGP), and it began with one of the discoverers of the structure of DNA at its head: James Watson, who held the position until 1992. Under his leadership, the HGP made great strides. Better computer technology, both hardware and software, was developed, and computer-automated equipment was integrated into research. The project was not only concerned with finding the code. It also addressed ethical and legal issues by allocating 3 per cent of grant money to fund research on the social implications of genetic research. In 1990, DOE submitted its proposal to Congress. Entitled *Understanding Our Genetic Inheritance: The US Human Genome Project*, it laid out a five-year budget of $200 million a year as the first phase of a 15-year project.

Because a significant portion of the budget was dedicated to research and the development of technology, the speed at which the sequencing took place was remarkable. The cost of analyzing bases dropped from $10 each to 10¢ each. The HGP met its five-year goals in four years and by 1995 had high resolution maps of chromosomes 16 and 19, as well as large sections of 3, 11, 12, and 22. The following year a large international conference on the HGP, sponsored by the Wellcome Trust in Britain, one of the world's largest private funders of medical research, brought even more attention to the project. Global concern about the potential ethical issues associated with the HGP were serious enough that in 1997 UNESCO released the *Universal Declaration on the Human Genome and Human Rights*, which attempted to provide international agreement on the ethical use of genetic information.

In 1999 the HGP passed the 1 billion base mark, and with new advances in robotic sequencing equipment, on February 12, 2001, the HGP announced that it had completed what it called the "working draft" of the entire human genome. The genome sequence was completed in 2003, effectively ending the mapping part of the project, but many years of analysis and research remain to be done. This is only a general map based on a single individual, not a genetic blue print for all humanity. While the map is hugely important, the lasting legacy of the HGP will be the techniques and technology that were created to do this international project.

There were fewer genes, the functional parts of DNA that actually controlled cellular activity, than some scientists had anticipated, but they had decoded the owner's manual to the human organism. With recombinant technology and cloning, scientists could also turn bacteria and higher organisms into biological factories to produce drugs or other useful products. However, to understand why the development of recombinant genetics was more than an important scientific development, the laboratory work must be read along with two significant legal decisions: *Diamond v. Chakrabarty* and *John Moore v. The Regents of the University of California*. These two cases are not the only legal events concerning genetic material, but they illustrate why a kind of genetics gold rush developed.

On June 16, 1980, the Supreme Court of the United States decided in favour of the respondent in the case of *Diamond v. Chakrabarty*. It ruled that the Patents and Trademarks Office (USPTO), represented by Commissioner Diamond, was incorrect in its decision to deny a patent to Ananda Chakrabarty, a microbiologist working for General Electric. What made the case so important was that Chakrabarty's patent was for a microbe that he had genetically modified so that it "ate" parts of crude oil, converting them into harmless by-products in order to clean up oil spills. In effect, the Supreme Court ruled that the microbe was not naturally occurring and was a new and useful "composition of matter." The decision was a close one, going five to four in favour of the patent. Although the Court offered a fairly narrow window for the patenting of living organisms, it nonetheless granted the legality of modern genetic patents, and the race was on for genetic commerce. The decision overturned the long-standing policy of the USPTO of denying patents on living things.[1] Since the USPTO had become the *de facto* venue of record for patents, a policy change in the American patent rules had far-reaching implications for everyone. International patent agreements recognized the protection of patents granted in one country as being valid in all signatory countries.

The USPTO continued to deny patents on living things until 1987 but then issued a formal policy statement allowing the patenting of nonhuman multicellular living organisms. This was partly in response to the Chakrabarty decision and other cases, but it was also a change to promote genetic research. Many researchers, universities, and private companies had argued that without patent protection there would be less incentive to do genetic research and other countries might gain an advantage in the area.

Among the first "higher" organisms to be patented was the Harvard Mouse, also known as oncomouse. This mouse was genetically modified to be more susceptible to cancer, and it was sold as a tool for cancer research. Its creators were granted an American patent in 1988.[2]

The second important case was *John Moore v. The Regents of the University of California.* John Moore, who suffered from a particular type of leukemia, was treated at the UCLA Medical Center in 1980 and had his spleen removed. He was also asked to return several times for tests and to give other kinds of tissue samples. Moore's doctors felt that his tissues or cell lines might be extremely useful for research, so they developed research material from his spleen and the other samples. In 1981 they applied for and were granted a patent on his cell line, which was then sold commercially. Three years later Moore discovered that his cells were patented and being sold without his knowledge, consent, or financial return, so he sued. In 1990 the Supreme Court of California ruled on the case and effectively denied the suit.

This ruling declared that in California, once cells were out of a person's body, they no longer belonged to the person they came from and that anyone

1. The USPTO policy was not without exception, since a patent had been granted to Louis Pasteur for a type of purified yeast in 1873.
2. In 2002 the Supreme Court of Canada denied an application for a patent on the "Harvard mouse," arguing that higher life forms could not be considered new inventions.

who got them could claim them as their own property. To many, the ruling seemed absurd on several levels. First, if a person lost a hand (a collection of cells) in some terrible accident and someone came along and took the hand away, that would clearly be theft; yet, in *Moore v. Regents*, the court ruled that researchers taking cells were not doing anything illegal. Second, the UCLA lawyers argued that it was in the public interest to allow genetic material to be collected and used by researchers but that it would be against the public interest if those researchers had to keep track of and compensate the original source of the material. The added cost and work would inhibit research. Thus, their argument seemed to be that it was good for the public interest that universities and private companies profit from genetic research, but it was bad for the public interest that the public (the source of the genetic material) profit from genetic research.

A flash-point for many of the concerns about genetics was the decision in 1998 by the Icelandic government to sell, or perhaps more correctly lease, the entire genetic heritage of the Icelandic people to a private company. Kari Stefansson, a genetic researcher, established the company deCODE and, in part funded by $200 million from the pharmaceutical company Roche, proposed the creation of the Iceland Health Sector Database (IHD). This database would include genealogical records going back hundreds of years and public health records starting in 1915, as well as genetic information on almost all Icelanders. Iceland made a good target of study for several reasons. The population is relatively small, around 275,000 people, and relatively homogeneous. Although a tiny country, it is a First World nation with universal literacy and a high level of general education and public health care, as well as the longest parliamentary tradition in the world. The extensive genealogical records that reach back to the Norse era allow researchers to identify genetic groups historically and to correlate them with genetic information from the current population.

Supporters of the project argued that Iceland benefits from deCODE's activities by getting money, high-technology research facilities, genetic information, and free access to any drugs or therapies that might arise from the information in the IHD. Detractors argue that the government has overstepped its authority and thus violated human rights by collecting the information (and in fact outlawing the private sale of genetic information by individual citizens) and that there are problems with the privacy system used to protect patient confidentiality, since DNA is by definition the perfect identifier. The government has allocated to itself the right to gather and use genetic information as if it were no different than information for a driver's licence, census data, or tax returns.

As ambitious as Stefansson's project was, it lacked a clear business model. In 2012 deCODE was in financial trouble having never turned a profit and was purchased by Amgen, a biopharmaceutical company based in California. The genetics system and database part of deCODE was then sold to WuXi PharmaTech in 2013. deCODE Genetics of Iceland continues as a subsidiary company of Amgen, continuing the search for genetic markers for disease.

In a wider sense, some commentators have pointed out that what the IHD provides is not just a tool for the detection of disease but a baseline for what is a "normal" human. The healthy, blond-haired, blue-eyed Icelanders strike some critics as a potential model for human appearance. This fear, played out in any number of science fiction horror stories such as the film *Gattaca* (1997), has in turn been dismissed by supporters of genetic research as both unlikely from a technical point of view and socially undesirable. The social implications of genetics, whether modified canola or human beings, offer a frightening vision of the future for some and a potentially utopian future through the mastery of nature to others.

Conclusion

Whether it is the power of computers or the potential of the genetic revolution, the utility of science had been brought home in the most direct way by the end of the twentieth century. Science and society are now inexorably intertwined. The public has come to expect that science will produce the consumer goods being sought by an increasingly affluent society, and in the creation of many of those goods, such as the transistor and the computer, new tools for research and avenues of inquiry have been opened. Scientific investigation of the natural world helped environmentalists think about the planet as a closed ecosystem and argue for better stewardship of the environment. At the same time, science has developed new ideas and techniques for manipulating that very nature we are supposed to be preserving. For good or ill, we now live in a world where the scientific point of view takes precedence over most other ways of knowing the world. Science has, in a very real way, made the world we live in and changed it forever.

Essay Questions

1. Was Ada Byron, Countess of Lovelace, the first computer programmer?

2. How did solid state physics transform modern consumer electronics?

3. Why did some biologists think that the Human Genome Project would be like the Manhattan Project for biology?

4. Why is Rachel Carson's book *Silent Spring* seen as one of the founding documents of modern ecology?

CHAPTER TIMELINE

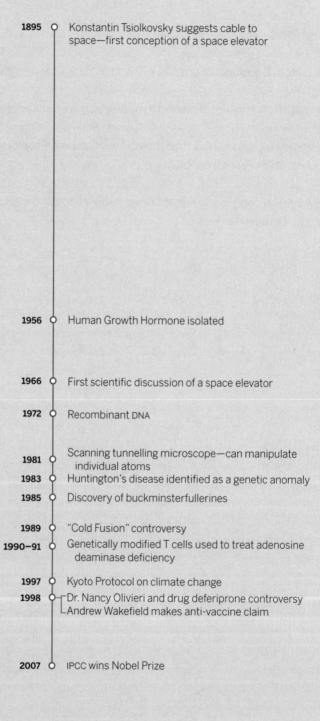

1895 — Konstantin Tsiolkovsky suggests cable to space—first conception of a space elevator

1956 — Human Growth Hormone isolated

1966 — First scientific discussion of a space elevator

1972 — Recombinant DNA

1981 — Scanning tunnelling microscope—can manipulate individual atoms

1983 — Huntington's disease identified as a genetic anomaly

1985 — Discovery of buckminsterfullerines

1989 — "Cold Fusion" controversy

1990–91 — Genetically modified T cells used to treat adenosine deaminase deficiency

1997 — Kyoto Protocol on climate change

1998 — Dr. Nancy Olivieri and drug deferiprone controversy
Andrew Wakefield makes anti-vaccine claim

2007 — IPCC wins Nobel Prize

SCIENCE AND NEW FRONTIERS: POTENTIAL AND PERIL IN THE NEW MILLENNIUM

13

I n 1995 the Natural Law Party ran candidate John Hagelin, a Transcendental Meditation disciple with a doctorate in physics from Harvard University, in the American presidential election. During the campaign the party offered a scientific demonstration of the power of Transcendental Meditation to bring peace to a troubled world. More than 5,000 disciples gathered in Washington, DC, to meditate, with the purpose of bringing peace to a notoriously violent city. A year later the party released a study that "scientifically" proved that Washington had indeed been more peaceful because of the waves of love and harmony projected by the meditators. When it was pointed out that the crime rate in the city during the demonstration was actually much higher than normal, the party defended its study on the basis that, without the meditators, the crime rate would have been even higher.

The case of the yogic peace waves highlights the power and problem of science at the beginning of the twenty-first century. Because we have come to accept and even expect wondrous things to come from science, it has become easier to make wild or even fraudulent claims of discovery so long as the claim is draped in scientific terms or supported by someone who claims to have scientific credentials. If a person with a PhD in physics says something is scientifically proven, shouldn't the general public accept the reality of the claim? After all, what higher standard of scientific expertise is there than a doctoral degree in physics?

Our understanding of what science is and how it interacts with our society has been fundamentally altered. For the Greeks, the study of nature was an elite, highly controlled activity, undertaken by a tiny intellectual cadre with philosophical and religious objectives. For people living in the twenty-first century, science is an incredibly powerful tool for political, economic, and social change. Everyone now claims some knowledge of science, and many, such as the Natural Law Party, claim the mantle of science as a way of demonstrating the importance, objectivity, and ultimate truth of their ideas. At the same time, scientific investigations, still performed by elite, highly trained experts, continue to make great forays into new understandings and manipulations of nature.

How did this change take place? Gradually, state leaders—princes and governments—began to see the utility of science and scientists, either to elevate their status as cultural and intellectual leaders or to boost their military and economic power and well being. Sixteenth-century natural philosophers, providing status and spectacle for European princely courts, are the linear ancestors of the Manhattan Project and Big Science funded by the government and military today. Scientists themselves contributed to this transformation, arguing for the usefulness of their investigations even when this was not the case. Here, the parallel between the rhetoric of early Royal Society statements and that of modern grant applications is instructive. Granting agencies want to know what they will get out of the relationship in exchange for the money and connections they offer. As the power of scientists has increased, so has the value of association with science.

The result of this transformation of science is paradoxical. On the one hand, twenty-first-century science has the power to transform our lives and our understanding of the universe in ways we can barely imagine. The huge resources now available to scientists, the tools at their disposal, and society's belief in the beneficence of science in general all contribute to the potential flowering of fascinating and significant scientific results. On the other hand, this widespread exposure to science has led to both a fear of its power and a credulity that allows scientific charlatans to flog their wares with impunity. There is reason in both these positions. Science showed its terrifyingly brutal face in the twentieth century, and for many interested in peace, equity, or the environment, the way of science does not seem to offer answers to the massive problems facing us. Equally, science has become so complex that lay people cannot understand it, and so those claiming scientific discoveries and breakthroughs receive favourable reporting in the press and widespread acceptance with little critical examination.

The power of science, the legitimate fear of its misuse, and a gullible misunderstanding of scientific principles are all the result of the triumph of science since the scientific revolution. The challenge now will be to expand our understanding of the idea of science and to use our informed judgment to improve the world and keep it safe.

What Is Science in the Twenty-first Century?

One of the ways we can see the transformation of science in our society is in the act of talking about science, which has become more complicated. The term "science" has been remade. While it has never had a universally accepted definition, it once was commonly regarded as referring only to the study of the physical world and the tools and methods of that study. The term has now become a general indicator of a claim for any profound or specialized knowledge. Adding "science" to a phrase is often an attempt to make whatever it is attached to seem more certain, insightful, true, or useful. The phrases "science of hair care," "the science of business management," "political science," or "created by a leading scientist" attempt to tie a product to the idea of science, even if under minimal scrutiny there is no practical or historical link to science whatsoever.

Even before the members of the Royal Society and the Académie des Sciences began to promote the concept of the utility of science as a justification for the existence of their organizations, the utility of science had been integral to the ethos of research. Patrons were looking for more than philosophical insight when they employed natural philosophers, and, like Grand Duke Cosimo, we have come to expect science to provide more than esoteric knowledge. The successful exploitation of science has been demonstrated so powerfully over the last 400 years that nations now neglect it at their peril. It has changed the course of wars, helped raise up countries economically, and changed gender relations. Science is now so closely linked to the success of nations that it is a mandatory subject of education in all industrialized nations and many others. Every child must learn science to gain productive employment and to be a good citizen. The degree of integration of science education is one indicator of what separates the developed from the developing world.

Science has been so broadly injected into industrial society that it is difficult to distinguish what science is and even who is a "scientist." While we readily acknowledge the science Nobel Prize winners as scientists, does the category include people with advanced degrees who do no original research, such as purity control chemists or physicists who work on Wall Street? Physicians receive a great deal of scientific training, so are general practitioners scientists or does the term apply only to medical researchers? Psychoanalysts, homeopaths, computer programmers, and sociologists have at one time or another claimed scientist status. A hominid paleobiologist and a cultural anthropologist may work in the same department at a university, but are they both scientists? Clearly, there is a spectrum of careers from "very scientific" to "no science required," but the range and number of people claiming science status has expanded enormously.

The functional definition that "Science is what scientists do" has started to break down in the modern world and will break down even more in the future as the trend to claim scientific status by a wider and wider range of people continues. This will make the problem of informed choices about scientific issues even harder. In the era of the greatest number of working scientists and the widest teaching of science in human history, the very concept of scientific knowledge can become muddled. The number of people claiming to offer scientific insight has become legion, so there is frequently conflicting "expert" opinion on a variety of socially important issues. Whether it is in a murder trial or a debate about global warming, scientific experts with equivalent credentials may offer diametrically opposed opinions. As science has had a greater direct effect on society and more people can claim to be doing science, we have seen a rise in the misunderstanding, or misrepresentation, of scientific ideas. Further, outright fraud cloaked as science has become easier.

Science Must Perform Miracles: Genetic Testing and Nanotechnology

Governments, industries, and ordinary people all believe that science should do wondrous things. Some of the miracles are big, such as the discovery of insulin to treat diabetes, saving millions of lives, while other breakthroughs are small and often go unnoticed by the general public, such as the discovery of a new species of ichthyosaur that resembles a cross between a crocodile and a dolphin. Two areas that promise major discoveries that have a direct effect on human life are genetic testing and therapy, and the materials revolution in nanotechnology. While each offers the potential for great benefits, each also carries with it concerns about application, especially ethical issues about transforming the human condition.

Genetic Testing and Genetic Therapies

Genetic testing is, in some ways, a continuation of other forms of testing for diseases. Whether it is culturing blood to discover the presence of a disease organism or looking at tissue samples for signs of cancer, the first stage of genetic testing was based on looking for an indicator of some problem. For example, in

1983 James Gusella (1952–) and his team identified Huntington's disease as a genetic anomaly on chromosome 4. As mapping has improved, the number of problem sites that can be identified has increased dramatically. Hundreds of tests now exist for diseases such as cystic fibrosis, Tay-Sachs, Down syndrome, and many others. Testing is also starting to reveal not only existing conditions but also the potential for disease. This means that a person might, for example, have an elevated risk of getting a particular kind of cancer, although it is not a certainty. This raises ethical questions, since genetic testing could be used to formulate public health policy or on a more individual level be used by health insurance companies to determine the cost of coverage based on the likelihood of a person being affected by genetically linked diseases.

There is also an ethical issue regarding the ownership of tests. Since genes can be patented, tests based on certain gene lines have become private property. Patent holders have forced laboratories, both medical testing labs and research labs, to pay fees for certain tests or stop doing them, not because equipment or even techniques are being used without compensation, but because of ownership of specific genetic material. Who owns the genetic material of the globe is being hotly debated, especially in cases where the genetic material has been gathered from indigenous people, who may or may not have been informed of its potential use.

Genetic testing of a wide population has also been made possible, raising issues of privacy versus social benefit and of who has the right to information. Health care providers might be able to plan more rationally for services if they had genetic information about the general population, but such broad testing might also be used to deny insurance to people likely to develop costly health problems. Should employers be allowed to test employees to see if they are likely to get sick? Should potential parents be screened before they have children? Equally, a number of researchers have claimed that there are genetic links to behaviour, so genetic testing might be used to identify people who are more susceptible to addiction or criminal acts. While it may sound like a bad science fiction story, we have the technology to create mass genetic databases on whole national populations, as was done in Iceland (see Chapter 12) and which is limited only by computer storage capacities.

While testing can indicate problems, it is the ability to use genetics to repair problems that now attracts most interest. The first use of recombinant DNA in therapy was undertaken by a team working for the American National Institutes of Health (NIH). In 1990 and 1991 they used retroviruses to modify T cells (part of the immune system) taken from two girls with a rare genetic disorder called adenosine deaminase deficiency. When the T cells were returned to the patients, a certain portion of the T cells being produced by the patients continued to be free of the genetic problem. Their health improved to the point where they were able to reduce their drug intake to half the amount used in conventional treatment of the disease. While this was not a complete cure, it was a remarkable change for the patients.

By 2014 more than 2,000 protocols for gene therapy had been conducted by the NIH. Many of the ethical concerns about genetic therapy have to do with limits to its use. While treating some genetically based diseases seems unproblematic, at what point is a therapy no longer about defence against disease but about changing a person in order to reach some desired condition? Should gene therapy be developed for baldness or to make people taller?

The case of height is an interesting one, since it bridges both the pre-genetic and genetic treatment eras. The hormone HGH or Human Growth Hormone was isolated by Choh Hao Li (1913–87) and his team in 1956. While HGH is not the only control mechanism that affects human height, it was developed as a treatment for dwarfism in children in the 1960s. Over time doctors started getting requests for HGH treatment for children who did not suffer from dwarfism but were just shorter than average height or even at average height. This raised ethical questions about the use of therapy to improve the human body rather than ameliorate a debilitating condition.

The uncertainty associated with genetic therapy and problems such as the non-clinical use of HGH led the NIH to establish a panel to examine the issue. As the history of HGH shows, the ethical questions are not theoretical. Although the panel's 1995 report argued that genetic therapy had many potential benefits, and future work looked very promising, it also warned people about too much enthusiasm:

> Overselling of the results of laboratory and clinical studies by investigators and their sponsors—be they academic, federal, or industrial—has led to the mistaken and widespread perception that gene therapy is further developed and more successful than it actually is. Such inaccurate portrayals threaten confidence in the integrity of the field and may ultimately hinder progress toward successful application of gene therapy to human disease.[1]

In 1985 the Food and Drug Administration approved the marketing of HGH produced by genetically modified bacteria, the second genetically engineered drug

---

1. Stuart H. Orkin and Arno G. Motulsky, *Report and Recommendations of the Panel to Assess the NIH Investment in Research on Gene Therapy* (Bethesda, MD: NIH, December 7, 1995).

after insulin to be introduced. While geneticists and physicians worked on these products with therapy in mind, promoters of the use of HGH touted it as a wonder drug that would increase muscle mass, decrease body fat, make a person look and feel younger, and even increase libido. While these claims are largely untrue, or at best unsubstantiated, promoters have used the scientific foundation of HGH work to justify wild claims and to send e-mail spam advertising their product to millions of people. Marketing and a fear of failing to keep up genetically with the next-door neighbours may drive the demand for therapy and provide ample opportunities for charlatans offering a kind of Elixir of Life.

As our knowledge of cell function increases, it is likely that genetic therapies will improve. One of the principal targets will be cancer cells, where the ultimate goal will be to get the cancerous cells to return in effect to their normal state. There is, however, another path to dealing with the problems of cells, and that is to create them to order rather than trying to fix them up later. Genetic modification, which has already been introduced to a number of food crops and the production of the Harvard mouse, as well as drug-producing bacteria, will come to humans. Some observers, such as Jeremy Rifkin (1945–), have already speculated on the potential smorgasbord of genetic choices that may be available to parents in the future. Everything from eye and hair colour, disease resistance, breast size, height, length of natural life, intelligence, and even musical ability may be modifiable. While the movies have tended to portray the genetic modification of people as an evil plot to create a super race (often in the form of relentless killing machine-soldiers), in reality the choice to modify human fetuses will be based largely on parental concern about providing the best life possible for their children. What will the world look like when the rich can modify themselves and the poor cannot?

The Materials Revolution

In 1959, the physicist Richard Feynman (1918–88) gave a lecture entitled "There's Plenty of Room at the Bottom," about the possibility of the direct manipulation of atoms as a way of doing synthetic chemistry. This has been seen as the start of nanotechnology both theoretically and in practical terms because Feynman concluded his talk with two challenges: 1.) Make a working electrical motor that would fit in a cube with sides $\frac{1}{64}$ of an inch (0.39 mm) and 2.) Reduce a page of text to be 25,000 times smaller than normal print. The first challenge was finished in 1960 by William McLellan (1924–2011), but it took until 1985 for Tom Newman to inscribe the first page of A Tale of Two Cities on the head of a pin using an electron beam. At that size, the whole of the *Encyclopedia Britannica* could be written on the pin.

The main material that has become the focus of nanotechnology has been carbon. Carbon as a structural material really started with the introduction of carbonized polyacrylonitrile by Akio Shindo (1926–) in the early 1960s. By bonding carbon fibres together and making thread and cloth from carbon, it could be used to replace heavier materials. With a tensile strength equal to steel, but only a fraction of the weight, carbon fibres have increasingly become the material of choice for designers and builders. Although there were some problems with delamination and brittleness in the early carbon fibre components, by the 1990s better polymers were being used and carbon fibre was being used in everything from tennis racquets to airplane wings.

While carbon fibres are part of nanotechnology, it was another form of carbon that started a new carbon revolution and was first called nanotechnology. In 1985 Harold Kroto (1939–) at the University of Sussex was wondering about carbon chains in space. Evidence suggested that certain stars, red dwarfs, produce a kind of soot. If this were the case, these carbon chains would be one of the oldest possible molecules, perhaps forming the foundation for a number of celestial objects and providing the materials to make up organic matter in the universe. To test this hypothesis, Kroto asked Richard Smalley (1943–2005) and his team at Rice University in Houston to recreate some of the conditions that exist at the surface of a red dwarf star. By shooting a laser at a block of carbon and collecting the vapourized clusters of molecules, they discovered that some of the molecules contained a fixed number of carbon atoms, either 60 or 70. The resulting molecules looked like geodesic domes as designed by R. Buckminster Fuller (1895–1983) and were formally named *buckminsterfullerene*. (See figure 13.1.) Fullerenes or more simply, bucky balls, have several properties that make them interesting. They can conduct electricity, are very hard, and because of their shape, can capture other atoms.

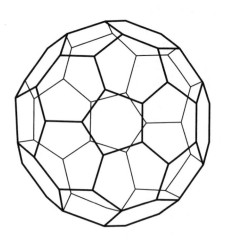

13.1 BUCKMINSTERFULLERENE—
"BUCKY BALL"

Perhaps the most audacious use suggested for such carbon fibres was to create a cable 99,820 kilometres long rising up from the equator to a space platform. The idea of such a system was first visualized by Konstantin Tsiolkovsky (of Russian rocketry fame) in 1895 when he looked at the Eiffel Tower and imagined a cable rising up from it into space. One of the first examinations of the possibility of a space elevator came from the team of John D. Isaacs (1913–80), Allyn C. Vine (1914–94), Hugh Bradner (1915–2008), and George E. Bachus (1930–), who published "Satellite Elongation into a True 'Sky-Hook,'" in the journal *Science* in 1966. (See figure 13.2.) In addition to moving materials cheaply, the space elevator would also serve as a launch platform, flinging things into space at a velocity of more than 25,000 kilometres per hour, like a giant slingshot using the rotation of the Earth.

Although nano fibres are important, the ultimate goal is to create nano devices. A start toward this submicroscopic engineering was made in 1981 when Gerd Binning (1947–) and Heinrich Rohrer (1933–2013), working at the IBM Research Lab in Zurich, created the first scanning tunneling microscope (STM). Although it is called a microscope, it examines materials well below the range of any optical system. Rather than looking at small objects, it feels them in a way analogous to a phonograph needle

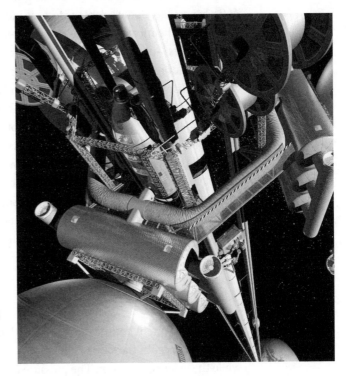

13.2 SPACE ELEVATOR

feeling the contours in the groove of a vinyl record. With a current running through a very sharp needle, the STM can trace the outline of objects down to a hundredth of a nanometre (a millionth of a millimetre). High-powered computers turn the data into a visual image.

This microscope has a second trick up its sleeve, however. In addition to sensing small objects, it can pick them up and move them around. In 1989 another IBM team at Almaden, California, used an STM to arrange 35 xenon atoms to spell "IBM." This feat was topped in 1996 back in Zurich when scientists made a miniature abacus consisting of 11 rows of ten C_{60} fullerene molecules, which the tip of the STM pushed backwards and forwards to count. While an abacus that requires an STM to count may seem like a bit of a science stunt, it actually has a serious purpose. If nanotechnology is to be created, there has to be some way to make the original nanomachines. In a sense, the STM offers the possibility of creating the machine shop that will produce the machine-making machines. Other researchers have developed microscopic pumps and motors, so the production of such micromachines seems entirely possible. Some scientists foresee nanofactories creating useful materials from piles of raw chemicals, from the molecular-level assembly of carbon-fibre materials to whole objects—chairs, computers, and airplanes—out of vats of basic elements.

A kind of macro version of this has already appeared with the invention of additive manufacturing or 3D printing. In 1981 Hideo Kodama (1944–) of the Nagoya Municipal Industrial Research Institute introduced a method of making three-dimensional objects using a liquid plastic that would harden on exposure to ultraviolet light. Since then a variety of printing methods have developed including sintering (using heat to turn a powder to a solid, often using metal), liquid films, and thermoplastics. While the majority of additive manufacturing is still directed at making prototypes or specialized components, consumer machines are becoming available. This technology has the potential to be used for nanoscale production, replacing the extremely expensive STM approach with more commercially affordable production methods.

Good Science Goes Bad: Cold Fusion

There is a dictum in both science and law that extraordinary claims require extraordinary evidence. In the case of cold fusion, the claims were indeed extraordinary while the evidence was not. The pressure to get ideas out of the lab and into the market has opened the door to scientists taking shortcuts in order to preserve priority and commercialize discoveries. One of the most notorious cases highlighting the dangers of circumventing established scientific processes is the story of cold fusion.

Given our general expectation that science can produce wonders, it did not seem completely unrealistic in 1989 that a new energy source had been discovered. When Chase Peterson, president of the University of Utah, announced that two scientists, Stanley Pons (1943–) and Martin Fleischmann (1927–2012), had discovered fusion at room temperature, it caused a sensation. They had created an electrolysis "cell" consisting of a glass flask containing heavy water (deuterium oxide) and run a current through it. At some point, they noted a sharp rise in temperature even though the amount of power being used had not changed. If the scientists had generated even a tiny bit more energy in their fusion cells than they had put in, their accomplishment could revolutionize energy production, rewrite physics, and make large sums of money. They called their discovery "cold fusion" as opposed to the "hot fusion" seen in stars or hydrogen bombs.

In the days immediately after the announcement, it was not clear what Pons and Fleischmann had actually done or even the exact experimental setup they had used, and so the idea that some new aspect of physics had been uncovered seemed possible. Some hastily constructed experiments at other labs appeared to confirm, or at least not clearly disprove, the claim. If cold fusion was real, some parts of physics would have to be revised, but such events had happened in the past. The unexpected results of a simple experiment heating up a black cube of carbon had, for example, helped initiate quantum physics. While most of these kinds of scientific controversies take place within the scientific community, Pons and Fleischmann played out their story in the full glare of media attention and with the backing of tremendous amounts of research money. Part of the controversy came from other scientists, especially physicists, who said that cold fusion could not possibly work, setting up a fight between the chemists (Pons and Fleischmann) and the physicists (such as John R. Huizenga [1921–2014]), which provided an easy conflict to cover in the media.

The stakes of scientific discovery are high: not only money, equipment, and research time are being used, but people's reputations are at stake. It was thus important for Pons and Fleischmann to demonstrate publicly their precedence in this revolutionary scientific discovery, without having gone through the time-consuming process of checks and balances required by peer review. In addition, patrons, in the form of governments, philanthropic organizations or private companies, must make decisions about what types of research they will fund and what people within those areas should be supported. Those patrons are keen to have publicity, demonstrating that their funding has been valuable in the creation of new and potentially useful knowledge. Although funding is not quite a zero-sum game in which the total amount of money available is fixed, it is close to that, so backing the wrong research project or the wrong people can damage the potential for future funding. While it is difficult to calculate how much money and time has been spent on the chimera of cold fusion, it is in the tens of millions of dollars and the loss of thousands of laboratory hours that could have supported other research efforts.

Claims for cold fusion turned out to be completely unfounded, and there was no challenge to our understanding of physics, but the event highlights some of the problems of science in the new millennium. The major characters in the story, Pons and Fleischmann, were reputable scientists with degrees, memberships in professional associations, and employment at major universities. They had all the credentials to make them reliable sources for scientific discovery, so it was not unreasonable to presume that their work merited serious consideration. That other scientists objected to their conclusions did not, in itself, make the work wrong. History is littered with stories of established scientists opposing new discoveries, from Priestley's objection to Lavoisier's oxygen theory to the controversy over Einstein's theory of relativity, or the geological community's condemnation of Wegener's idea that continents move. In their eagerness to establish precedence for a potentially revolutionary idea, however, Pons and Fleischman believed that they needed to present their ideas as proven before they had been subjected to rigorous investigation by others. This demonstrates the danger of a system that has increasingly commercialized research in science by offering great rewards to those who can produce extraordinary results quickly.

As a postscript to the cold fusion story, a small number of private and public groups are still funding cold fusion research despite more than 25 years of failure. This is often held up by supporters as proof that there is a scientific foundation to the original idea, but in fact such funding is a form of marginal investment. In other words, patrons fund a few unlikely projects as a bet on a long shot, particularly in tax jurisdictions that allow them to write off research costs.

Science in the Corporate World

Most commentators have attributed the "discovery" of cold fusion to wishful thinking and poor experimental procedures rather than malfeasance. Science does have a method of self-regulation, designed to squeeze unsound science out of the range of topics considered by scientists as legitimate or worthwhile. Scientists point to the peer review system in journals (not used by Pons and Fleischmann, who turned instead to the mass media) and to the replication of experiments as the internal means of weeding out bad science. This is, however, more problematic than you might think, since it turns out to be difficult to repeat big experiments, and few experiments are ever repeated in practice. Since scientists' livelihood is based on discovery, there is not much support for repeating work that has already been done. As well, since the cost of large-scale experiments, such as supercollider tests, can run into the millions of dollars, it may not be possible economically to repeat certain experiments even if there were some interest in doing so. Scientists, therefore, often rely on the consilience of experimental results, rather than on repeated experiments. In other words, experimental results are accepted as valid, even if not independently tested, as long as the results fall within expected norms, corresponding to already established theories, and the experiment has been conducted following accepted procedures.

A second and growing problem for self-regulation is the use of secrecy in research. While aspects of research of military importance have long been kept secret, more non-military research is being kept confidential, based on the idea of proprietary information not only in corporate-funded research but also increasingly in public research, as universities and governments look to spin off research into profit-making businesses. If there is no evidence to evaluate, claims are hard to test, and timely decisions about research or products become difficult. Secrecy can affect both the discoveries, which are often shielded from public scrutiny, and the application of research, also kept secret. Legal settlements over product liability that have used scientific research often have included non-disclosure agreements or "gag orders" to prevent parties from revealing to others what problems were uncovered. Such restrictions have been applied to a range of products, from cigarettes to cosmetics. If scientific work cannot be examined by other scientists, then the self-regulation system breaks down.

Added to this problem are efforts by interested parties to protect their investments by interfering with scientific inquiry. This can take the form of biased research or efforts to block research that might indicate problems with a product or procedure. The most publicized cases have been in the pharmaceutical business, in which a number of researchers have falsified results in favour of a drug. Conversely, other researchers have been fired, threatened with legal action, sued, or had funding withdrawn for publishing negative results or suggesting problems with drugs. Such was the case for Dr. Nancy Olivieri, who was threatened with legal action and removed from her position at Toronto's Hospital for Sick Children after publishing a negative report on the drug deferiprone in the *New England Journal of Medicine* in 1998. Although she was later reinstated, her case was far from isolated.

In the anti-vaccine scandal starting in 1998, surgeon and medical researcher Andrew Wakefield (1957–) claimed at a press conference to have found a link between the measles, mumps, and rubella vaccine (MMR) and autism. His research was published in the journal *Lancet*, but was later discovered to have been poor quality and misreported to support his position, leading to the withdrawal of the paper by the journal. It was further discovered that he had undisclosed financial interests, receiving money from lawyers in a lawsuit against the MMR vaccine and connections with two pharmaceutical companies, one of which was planning to develop an alternative vaccine.

The problem of biased research has become so urgent that in 2001 the International Committee of Medical Journal Editors issued a warning that they would no longer publish drug trial reports from researchers who were bound by agreements that limited academic freedom. In other words, it was the whole story or no story.

Since new drugs can cost millions of dollars to produce and can generate billions of dollars in revenues, the pressure to publish positive research is very great. An increasing number of scientific journals now require a disclosure of financial interest, such as funding sources or corporate remuneration, from scientists submitting papers.

While interested parties may try to circumvent the publication of negative results, science journals are not always neutral players, so the self-regulatory mechanism of peer review is not completely reliable either. In recent years, even the most prestigious journals such as *Science* and *Nature* have been accused of rushing results into print in order to be the first to publish cutting-edge work. Because of the importance of publication, a kind of feedback loop exists between the scientists and the journals. Journals gain prestige by publishing exciting and groundbreaking results. Scientists who want to make a name for themselves aim to have their work published in prestige journals. This convergence of interests is not necessarily a problem, although it can encourage both players to take shortcuts. As well, the reliance on "blind" or anonymous reviewing for publication is not always impartial. In fields where everyone's work is known to everyone else, the control mechanism of peer review may break down. Since only scientists in the same field can understand and evaluate work, when they act as reviewers they may be reluctant to criticize other members of the same small community. The editors of scientific journals have little defence against bad work if the peer reviews on which they rely are biased or incomplete. The stakes are high. Millions of dollars in research money, leading research posts, international prestige, and the ultimate prize—a Nobel Prize—may ride on the status of publications.

The various problems of verification are a natural by-product of a complex and self-regulating profession, but it can make the interaction between scientists and the general public difficult. Lay people are left troubled by conflicting and often contradictory scientific claims, so it is no wonder that some have rejected the whole enterprise of science, greeting all new discoveries with skepticism. Ignoring science is putting one's head in the sand as it continues to change our lives, whether we like it or not. In the coming years, we as individuals and as a society will be faced with an increasing number of choices presented to us by science.

Where should research money go? How will we assess the differential importance of demands for cash for a giant synchrotron to look into the interior of subatomic particles against space stations or the search for the cure for cancer? How do we balance the dangers against the potential benefits of creating genetically modified food in a world where starvation and pestilence are common? How do we assess arguments about global warming? What about deeply personal choices, such as the potential to genetically modify our children or even ourselves?

Denialism

One of the results of the power of science and of concerns about the interference of funding agencies and industries has been a growing movement to deny the results of scientific research. Denialism is based on the idea that only 100 per cent agreement of scientists on a scientific topic is good enough to justify social action. This conflates two ideas that are commonly held by the general public. The first is that in an argument, both sides should be heard and given equal coverage. The second is that in the past minority views in science have been shown to be correct. Deniers often portray themselves as being like Galileo standing against the Roman Catholic Church or Wegener challenging the authority of geologists. The problem is that these people are usually confusing scientific debate over details with lack of consensus on the overarching theory. Additionally, deniers use the idea that science is part of a larger social context to suggest that any particular idea is contingent and therefore not to be trusted.

There are three classic cases of denialism: Smoking causing cancer, particularly lung cancer; acid rain; and climate change. In each case, large commercial interests were under attack by the mainstream scientific community and in response funded or hired scientists to counter or simply deny the scientific evidence. When that failed, they sought to obscure the public reception of the evidence by arguing that more study was needed or to delay action by keeping the controversy going for as long as possible. These minority scientists have often portrayed themselves as crusaders for truth, but unlike Galileo, Einstein, or Wegener, they were not making any new discoveries, only attacking the discoveries of others. It is perfectly reasonable to test the ideas of those who said a large portion of lung cancers were caused by smoking or that acid rain was caused by sulphur dioxide and nitrogen oxides in air pollution. However, failing to generate actual scientific evidence is often a signal that contrary opinions are sliding toward denialism rather than true scientific debate.

Climate Change

In 2005, Hurricane Katrina crashed into the coast of the United States leaving behind more than 1,800 dead and doing billions of dollars of damage. Was the hurricane just the storm of the century, or was it part of global climate change? Record temperatures, receding glaciers, and concerns about changing ocean currents have all been pointed to as part of a much larger change in global climate. The science of climate change is complex, and over the years it has become a battle-ground for various interest groups, with one side arguing that it is human created and potentially catastrophic, especially if we do not take significant measures to limit greenhouse gases as soon as possible. The other side argues that the climate is not really being affected by human activity, which is minor compared to natural forces, and that radical activity will stifle the economy and do nothing but waste time and money. Each side has claimed that science supports its position. By 2015, the scientific community has reached a general consensus, with over 99 per cent of the climate researchers agreeing that the major factors in climate change are anthropogenic (human made). While there are still strong advocates against this position, it has become more difficult to find credible scientists willing to take the side of the climate change deniers. On the other hand, what to do about climate change is proving just as difficult as persuading governments and industry that it exists.

The prevailing scientific position has been articulated by the Intergovernmental Panel on Climate Change (IPCC). This panel, created by the World Meteorological Organization and the UN Environment Program in 1988, produced a series of assessment reports from 1990 to 2007. These assessments became the basis for the UN Framework Convention on Climate Change, which in 1997 produced the Kyoto Protocol, a plan to reduce greenhouse gases. The IPCC received the Nobel Peace Prize, jointly with Al Gore, in 2007. What is interesting about the IPCC is that it is not a body that conducts scientific research. Rather, it gathers all the scientific research previously conducted and decides, through democratic processes, what the current state of scientific knowledge is. Given that the membership of the IPCC is based on fair representation of all UN member countries, rather than on who are the leading world scientists in the area, the result is science by consensus.

One hundred and ninety-nine states agreed to follow the Kyoto Protocol by 2010, although Japan and Russia have said they would not set new targets and Canada formally withdrew from the Protocol in 2012. The United Nations Climate Change Conference of 2015 (COP 21) in Paris resulted in a new agreement to take stronger measures to reduce greenhouse gas emissions, in order to keep global warming below 2°C. Despite these actions, the average temperature of the Earth continues to rise, and the ability of states and industry to turn this around will tax both science and governments. The history of this conflict reveals the difficulty of applying science to problems with political and economic implications. Developing countries don't want to stop increasing their economic activity and point out that on a per-person basis they pollute far less than people in developed countries. Industrial nations have resisted regulations that apply only to industry in developed countries, pointing out that industries in developing countries are often poorly regulated and use old technology. From a scientific point of view it may not matter who produces the pollution, but from a political point of view it certainly does. This demonstrates the need, as we move forward, for scientists and social scientists to work together with government support to change social attitudes as well as industry expectations.

Pseudoscience

While the case of cold fusion demonstrates the dangers of commercial pressures and the drug scandals represent the problem of the misuse of science, our scientific age has also produced a rise in efforts to deceive people by using ideas that sound like science, but are in fact wrong or even fraudulent. The term "pseudoscience" refers to claims that something is scientific but lacks actual scientific evidence. The major problem with pseudoscience is that the use or application of such ideas or products uses up resources and can even endanger people. In some cases, the historical foundation of pseudoscientific beliefs had some justification such as vernalization used by Lysenko or the idea that personality could be determined by the shape of the skull in phrenology. In these cases what started as a reasonable hypothesis was demonstrated to be false, a process that happens all the time in science. They became pseudoscience because significant numbers of people continued to believe in them despite actual evidence to the contrary.

The most notorious pseudoscience in Western history has been homeopathy. The field of homeopathy was created in 1796 by Samuel Hahnemann (1755–1843) in an era before the germ theory of disease was discovered and bleeding was still a common practice among physicians. Hahnemann's impulse was a good one—do no harm—but he created a system based on pure fantasy. At its heart, homeopathic theory is based on the idea that "like cures like," so that if, for example, you have a fever, you should take something that would make you feel hot such as chili peppers.

Added to the "like cures like" dictum was the even more fanciful idea that the lower the dose of the curative, the greater its curative strength. This is achieved by serial dilution of one part to one hundred parts of solvent (1C), and then repeated. Hahnemann often used 30C dilutions (1 molecule of active ingredient to 10^{60} molecules of solute). In molecular terms, a 13C homeopathic remedy does not contain a single molecule of the original material. Dilutions up to 200C are used in modern homeopathic materials. To understand the scale of this, 200C is $1:10^{400}$ while the estimated number of atoms in the observable universe is only 10^{80}. In other words, most homeopathic remedies contain nothing but a small amount of solvent, usually distilled water, and the original material has no medical utility.

The immunologist Jacques Benveniste (1935–2004) tried to rescue homeo-pathy with claims that water had memory, so that dilution to no original material was irrelevant. Because Benveniste, like Pons and Fleischmann, was a respected scientist, his ideas were scientifically tested even though they defied basic prin-ciples of science and even logic (why, for example, would the water remember the homeopathic material and not everything else it had been in contact with?). Benveniste went even further in 1997 when he argued that the memory could be transmitted over telephone lines and later over the Internet. A small number of people claimed to have replicated Benveniste's work, but in every case they failed to replicate the work when observed by third parties.

Homeopathy has depended for its continued existence on the support of celebrities and the fear of many people about the problems with the medical system. While it was perfectly reasonable for scientists to investigate homeopathic remedies, it is also perfectly reasonable to say that such remedies have no utility. They do, however, pose a threat to people who may not seek real medical help and they may also put others at risk by the use of homeopathic "vaccinations" that open a path for infectious diseases that are easily controlled to re-occur.

Conclusion

As science has had a greater and greater role in shaping society, there has always been resistance to the changes it offers. At one level, it is wise to be cautious about the introduction of new products in a complex system. As the critic of technology Neil Postman (1931–2003) pointed out, the introduction of a new "thing"—be it a device, practice, or ideology—changes society. It is not "society plus the computer," but a new society.[2] The conundrum of science in society is that the producers of

----

2. Neil Postman, *Technopoly* (New York: Vintage, 1993).

science may not be the best people to judge what the impact of their products will be; however, because of the highly technical nature of the work, those who lack training may not understand the work well enough to make informed choices. Errors, flawed work, and other problems are inevitable, and the examples of DDT, thalidomide, and eugenics should stand as a warning that scientific mistakes can have dangerous consequences. The fact that there will be problems does not, however, mean that science should be rejected. Rather, it means that we must work to understand the potential benefits and problems of using scientific developments.

Scientific research represents a complex interplay of social demands, technical constraints, and personal interests and abilities. It is not driven solely by ideas, but neither can it be produced to order. While science has provided some profound insights into the structure of nature, it has also presented us with some difficult questions about how to use that knowledge. Ironically, knowing more has made our choices more difficult rather than less. Understanding the history of science offers another venue for approaching these difficult questions, since it can show us the power and the danger of past choices and explain how we have arrived at the world we live in. Science has, for example, been claimed as the basis for both Marxism and modern democracy.

The history of science can also be useful because it reveals the broader context of science rather than looking only at its products. No one owns science. If we wish to make informed choices, we must never forget that science exists because people created it, and it cannot exist separate from the community. Behind all the patents, prizes, and professional degrees, the idea of science—our long effort to understand nature—and the knowledge that radiates from that search are part of our shared human heritage.

Essay Questions

1. What issues about science does the case of cold fusion expose?

2. What does the debate about GMO foods tell us about the place of science in the modern world?

3. Why did climate change deniers work against the scientific consensus?

4. How does corporate scientific research threaten the ideology of public knowledge?

FURTHER
READING

ONE   *THE ORIGINS OF NATURAL PHILOSOPHY*

Aristotle. *Meteorologica*. Trans. H.D.P. Lee. Cambridge, MA: Harvard University Press, 1952.

Aristotle. *Physics*. Trans. Hippocrates G. Apostle. Grinnell, IA: Peripatetic Press, 1980.

Aristotle. *Posterior Analytics*. Trans. Jonathan Barnes. Oxford: Clarendon Press, 1975.

Bernal, Martin. *Black Athena: The Afroasiatic Roots of Classical Civilization*. New Brunswick, NJ: Rutgers University Press, 1987.

Byrne, Patrick Hugh. *Analysis and Science in Aristotle*. Albany, NY: State University of New York Press, 1997.

Clagett, Marshall. *Greek Science in Antiquity*. Freeport, NY: Books for Libraries Press, 1971.

Irby-Massie, Georgia L., and Paul T. Keyser, eds. *Greek Science of the Hellenistic Era: A Sourcebook*. London: Routledge, 2002.

Lloyd, G.E.R. *Early Greek Science: Thales to Aristotle*. London: Chatto and Windus, 1970.

Lloyd, G.E.R. *Greek Science after Aristotle*. London: Chatto and Windus, 1973.

Lloyd, G.E.R. *Magic, Reason and Experience: Studies in the Origin and Development of Greek Science*. Cambridge: Cambridge University Press, 1979.

Lloyd, G.E.R., and Nathan Sivin. *The Way and the Word: Science and Medicine in Early China and Greece*. New Haven, CT: Yale University Press, 2002.

Plato. *The Republic*. Trans. G.M.A. Grube. Indianapolis, IN: Hackett Publishing, 1992.

Plato. *Timaeus*. Trans. Francis M. Cornford. Indianapolis, IN: Bobbs-Merrill, 1959.

Rihll, T.E. *Greek Science*. Oxford: Oxford University Press, 1999.

Tuplin, C.J., and T.E. Rihll, eds. *Science and Mathematics in Ancient Greek Culture*. Oxford: Oxford University Press, 2002.

Zhmud, Leonid. *Pythagoras and the Early Pythagoreans*. Trans. Kevin Windle and Rosh Ireland. Oxford: Oxford University Press, 2012.

TWO   *THE ROMAN ERA AND THE RISE OF ISLAM*

Baker, Osman. *The History and Philosophy of Islamic Science*. Cambridge: Islamic Texts Society, 1999.

Beagon, Mary. *Roman Nature: The Thought of Pliny the Elder*. Oxford: Oxford University Press, 1992.

Bricker, Harvey M., and Victoria R. Bricker. *Astronomy in the Maya Codices*. Philadelphia: American Philosophical Society, 2011.

Dallal, Ahmad S. *Islam, Science, and the Challenge of History*. New Haven, CT: Yale University Press, 2010.

French, Roger, and Frank Greenaway, eds. *Science in the Early Roman Empire: Pliny the Elder, His Sources and His Influence*. London: Croom Helm, 1986.

Glasner, Ruth. *Averroes' Physics: A Turning Point in Medieval Natural Philosophy*. Oxford: Oxford University Press, 2009.

Harley, J.B., and David Woodward. *History of Cartography, Volume II, Book I. Cartography in the Traditional Islamic and South Asian Societies*. Chicago: University of Chicago Press, 1992.

Hogendijk, J.P. *The Enterprise of Science in Islam: New Perspectives*. Cambridge, MA; London: MIT Press, 2003.

Huff, Toby E. *The Rise of Early Modern Science: Islam, China, and the West*. Cambridge: Cambridge University Press, 1993.

Lehoux, Daryn. *What Did the Romans Know? An Inquiry into Science and Worldmaking*. Chicago: University of Chicago Press, 2012.

Masood, Ehsan. *Science & Islam: A History*. London: Icon, 2009.

Principe, Lawrence M. *The Secrets of Alchemy*. Chicago: University of Chicago Press, 2013.

Qadir, C.A. *Philosophy and Science in the Islamic World*. London: Routledge, 1990.

THREE   *THE REVIVAL OF NATURAL PHILOSOPHY IN WESTERN EUROPE*

Brotton, Jerry. *The Renaissance Bazaar: From the Silk Road to Michelangelo*. Oxford: Oxford University Press, 2002.

Grant, Edward. *The Foundation of Modern Science in the Middle Ages, Their Religious, Institutional and Intellectual Contexts*. Cambridge: Cambridge University Press, 1996.

Grant, Edward. *Planets, Stars and Orbs: The Medieval Cosmos 1200–1687*. Cambridge: Cambridge University Press, 1994.

Grant, Edward, ed. *A Source Book in Medieval Science*. Cambridge, MA: Harvard University Press, 1974.

Kibre, Pearl. *Studies in Medieval Science: Alchemy, Astrology, Mathematics and Medicine*. London: Hambledon, 1984.

Lindberg, David C. *The Beginnings of Western Science: The European Scientific Tradition in Philosophical, Religious, and Institutional Context, 600 B.C. to A.D. 1450*. Chicago: University of Chicago Press, 1992.

FOUR *SCIENCE IN THE RENAISSANCE: THE COURTLY PHILOSOPHERS*

Biagioli, Mario. *Galileo, Courtier: The Practice of Science in the Culture of Absolutism*. Chicago: University of Chicago Press, 1994.

Biagioli, Mario. *Galileo's Instruments of Credit: Telescopes, Images, Secrecy*. Chicago: University of Chicago Press, 2006.

Blair, Ann. *The Theater of Nature: Jean Bodin and Renaissance Science*. Princeton, NJ: Princeton University Press, 1997.

Bono, James J. *The Word of God and the Languages of Man: Interpreting Nature in Early Modern Science and Medicine*. Madison: University of Wisconsin Press, 1995.

Cormack, Lesley B. *Charting an Empire: Geography at the English Universities, 1580–1620*. Chicago: University of Chicago Press, 1997.

Daston, Lorraine. *Wonders and the Order of Nature, 1150–1750*. New York: Zone Books, 1998.

Drake, Stillman. *Galileo: Pioneer Scientist*. Toronto: University of Toronto Press, 1990.

Finocchiaro, Maurice A. *Defending Copernicus and Galileo: Critical Reasoning in the Two Affairs*. New York: Springer, 2010.

Galilei, Galileo. *Dialogue Concerning the Two Chief World Systems—Ptolemaic and Copernican*. Trans. Stillman Drake. Foreword by Albert Einstein. Berkeley: University of California Press, 1967.

Galilei, Galileo. *Two New Sciences, Including Centers of Gravity and Force of Percussion*. Trans. Stillman Drake. Madison: University of Wisconsin Press, 1974.

Gingerich, Owen. *The Book Nobody Read: Chasing the Revolutions of Nicolaus Copernicus*. New York: Penguin Books, 2004.

Grafton, Anthony, with April Shelfor and Nancy Siraisi. *New World, Ancient Texts: The Power of Tradition and the Shock of Discovery*. Cambridge, MA: Belknap Press of Harvard University Press, 1992.

Magnus, Albertus. *The Book of Secrets of Albertus Magnus of the Virtues of Herbs, Stones and Certain Beasts, also A Book of the Marvels of the World*. Ed. Michael R. Best and Frank H. Brightman. Oxford: Clarendon Press, 1973.

Moran, Bruce T., ed. *Patronage and Institutions: Science, Technology, and Medicine at the European Court, 1500–1750*. Rochester, NY: Boydell Press, 1991.

Newman, William Royall, and Anthony Grafton, eds. *Secrets of Nature: Astrology and Alchemy in Early Modern Europe*. Cambridge, MA: MIT Press, 2001.

Saliba, George. *Islamic Science and the Making of the European Renaissance*. Cambridge, MA: MIT Press, 2007.

Swerdlow, Noel, and Otto Neugebauer. *Mathematical Astronomy in Copernicus's* De Revolutionibus Part 1–2. Studies in the History of Mathematics and Physical Sciences 10. New York: Springer-Verlag, 1984.

Westman, Robert. *The Copernican Question: Prognostication, Skepticism, and the Celestial Order*. Berkeley: University of California Press, 2011.

Vollmann, William T. *Uncentering the Earth: Copernicus and* On the Revolutions of the Heavenly Spheres. New York: Norton, 2006.

FIVE   *THE SCIENTIFIC REVOLUTION: CONTESTED TERRITORY*

Bala, Arun, ed. *Asia, Europe and the Emergence of Modern Science: Knowledge Crossing Boundaries*. New York: Palgrave, 2012.

Dear, Peter Robert. *Revolutionizing the Sciences: European Knowledge and Its Ambitions, 1500–1700*. Princeton, NJ: Princeton University Press, 2001.

Harkness, Deborah. *The Jewel House: Elizabethan London and the Scientific Revolution*. New Haven, CT: Yale University Press, 2007.

Hunter, Lynette, and Sarah Hutton, eds. *Women, Science and Medicine 1500–1700: Mothers and Sisters of the Royal Society*. Stroud, UK: Sutton, 1997.

Jardine, Lisa. *Ingenious Pursuits: Building the Scientific Revolution*. New York: Nan A. Talese, 1999.

Lindberg, David C., and Robert S. Westman, eds. *Reappraisals of the Scientific Revolution*. Cambridge: Cambridge University Press, 1990.

Long, Pamela O. *Artisans/Practitioners and the Rise of the New Sciences, 1400–1600*. Corvallis: Oregon State University Press, 2011.

Newman, William Royall. *Atoms and Alchemy: Chymistry and the Experimental Origins of the Scientific Revolution*. Chicago: University of Chicago Press, 2006.

Newton, Isaac. *Opticks*. New York: Prometheus, 2003.

Newton, Isaac. *The Principia*. Trans. Andrew Motte. New York: Prometheus, 1995.

Osler, Margaret J. *Reconfiguring the World: Nature, God, and Human Understanding from the Middle Ages to Early Modern Europe*. Baltimore, MD: Johns Hopkins University Press, 2010.

Osler, Margaret J. *Rethinking the Scientific Revolution*. Cambridge: Cambridge University Press, 2000.

Park, Katherine, and Lorraine Daston, eds. *Early Modern Science. The Cambridge History of Science*, vol. 3. Cambridge: Cambridge University Press, 2003.

Schiebinger, Londa L. *The Mind Has No Sex?: Women in the Origins of Modern Science*. Cambridge, MA: Harvard University Press, 1989.

Shapin, Steven. *The Scientific Revolution*. Chicago: University of Chicago Press, 1996.

Smith, Pamela H. *The Body of the Artisan: Art and Experience in the Scientific Revolution*. Chicago: University of Chicago Press, 2004.

Westfall, Richard S. *Never at Rest: A Biography of Isaac Newton*. Cambridge: Cambridge University Press, 1980.

SIX   *THE ENLIGHTENMENT AND ENTERPRISE*

Bell, Madison Smartt. *Lavoisier in the Year One: The Birth of a New Science in an Age of Revolution*. New York: W.W. Norton, 2006.

Binnema, Ted. *Enlightened Zeal: The Hudson's Bay Company and Scientific Networks, 1670–1870*. Toronto: University of Toronto Press, 2014.

Diderot, Denis. *Encyclopédie ou dictionnaire raisonné des sciences des arts et des métiers*. Stuttgart-Bad Cannstatt: F. Frommann Verlag (G. Holzboog), 1966.

Fox, Christopher, Roy Porter, and Robert Wokler, eds. *Inventing Human Science:*

Eighteenth-Century Domains. Berkeley: University of California Press, 1995.

Gascoigne, John. *Encountering the Pacific in the Age of Enlightenment*. Cambridge: Cambridge University Press, 2014.

Holmes, Frederic Lawrence. *Lavoisier and the Chemistry of Life*. Madison: University of Wisconsin Press, 1985.

Lavoisier, Antoine. *Elements of Chemistry*. Trans. Robert Kerr. New York: Dover, 1965.

Lynn, Michael R. *Popular Science and Public Opinion in Eighteenth-Century France*. Manchester: Manchester University Press, 2006.

Poirier, Jean-Pierre. *Lavoisier, Chemist, Biologist, Economist*. Trans. Rebecca Balinski. Philadelphia: University of Pennsylvania Press, 1996.

Porter, Roy, ed. *Eighteenth-Century Science. The Cambridge History of Science*, vol. 4. Cambridge: Cambridge University Press, 2003.

Schiebinger, Londa L. *Colonial Botany: Science, Commerce, and Politics in the Early Modern World*. Philadelphia: University of Pennsylvania Press, 2005.

SEVEN   *SCIENCE AND EMPIRE*

Bartholomew, James. *The Formation of Science in Japan: Building a Research Tradition*. New Haven, CT: Yale University Press, 1989.

Bowler, Peter J. *Evolution: The History of an Idea*. Berkeley: University of California Press, 1984.

Bowler, Peter J. *Life's Splendid Drama: Evolutionary Biology and the Reconstruction of Life's Ancestry, 1860–1940*. Chicago: University of Chicago Press, 1996.

Darwin, Charles. *The Origin of Species by Means of Natural Selection, or, The Preservation of Favoured Races in the Struggle for Life*. London: J. Murray, 1860.

Garfield, Simon. *Mauve: How One Man Invented a Color That Changed the World*. New York: Norton, 2001.

Geison, Gerald L. *The Private Science of Louis Pasteur*. Princeton, NJ: Princeton University Press, 1995.

Greene, Mott T. *Geology in the Nineteenth Century: Changing Views of a Changing World*. Ithaca, NY: Cornell University Press, 1982.

Lyell, Charles. *Principles of Geology*. Chicago: University of Chicago Press, 1990.

Nye, Mary Jo. *Before Big Science: The Pursuit of Modern Chemistry and Physics, 1800–1940*. New York: Twayne Publishers, 1996.

Paul, Harry W. *From Knowledge to Power: The Rise of the Science Empire in France, 1860–1939*. Cambridge: Cambridge University Press, 1985.

Strathern, Paul. *Mendeleyev's Dream: The Quest for the Elements*. New York: St. Martin's Press, 2000.

EIGHT   *ENTERING THE ATOMIC AGE*

Dry, Sarah. *Curie*. London: Haus Publishers, 2003.

Ede, Andrew. *The Rise and Decline of Colloid Science in North America, 1900–1935: The Neglected Dimension*. Aldershot, UK: Ashgate, 2007.

Hunt, Bruce J. *Pursuing Power and Light: Technology and Physics from James Watt to Albert Einstein*. Baltimore, MD: Johns Hopkins University Press, 2010.

Knight, David. *The Making of Modern Science: Science, Technology, Medicine and Modernity, 1789–1914*. Cambridge: Polity Press, 2009.

Levinovitz, Agneta Wallin. *The Nobel Prize: The First 100 Years*. London: Imperial College Press, 2001.

Navarro, James. *A History of the Electron: J.J. and G.P. Thomson*. Cambridge: Cambridge University Press, 2012.

Nye, Mary Jo, ed. *The Modern Physical and Mathematical Sciences. The Cambridge History of Science,* vol. 5. Cambridge: Cambridge University Press, 2003.

Pasachoff, Naomi E. *Marie Curie and the Science of Radioactivity*. New York: Oxford University Press, 1997.

Reeves, Richard. *A Force of Nature: The Frontier Genius of Ernest Rutherford*. New York: W.W. Norton, 2008.

Wilson, David. *Rutherford, Simple Genius*. Cambridge, MA: MIT Press, 1983.

NINE   *SCIENCE AND WAR*

van Dongen, Jeroen. *Einstein's Unification*. Cambridge: Cambridge University Press, 2010.

Freemantle, Michael. *The Chemists' War, 1914–1918*. Cambridge: Royal Society of Chemistry, 2015.

Henig, Robin Marantz. *The Monk in the Garden: The Lost and Found Genius of Gregor Mendel, the Father of Genetics*. Boston: Mariner Books, 2001.

Isaacson, Walter. *Einstein: His Life and Universe*. New York: Simon & Schuster, 2008.

Mawer, Simon. *Gregor Mendel: Planting the Seeds of Genetics*. New York: Abrams, in association with the Field Museum, Chicago, 2006.

Russell, Edmund. *War and Nature: Fighting Humans and Insects with Chemicals from World War I to* Silent Spring. Cambridge: Cambridge University Press, 2001.

Shapiro, Adam R. *Trying Biology: The Scopes Trial, Textbooks, and the Antievolution Movement in American Schools*. London: Pickering and Chatto, 2013.

Stachel, John, ed. *Einstein's Miraculous Year: Five Papers That Changed the Face of Physics*. Princeton, NJ: Princeton University Press, 1998.

TEN   *THE DEATH OF CERTAINTY*

Brennan, Richard P. *Heisenberg Probably Slept Here: The Lives, Times, and Ideas of the Great Physicists of the 20th Century*. New York: Wiley, 1997.

Cassidy, David Charles. *Uncertainty: The Life and Science of Werner Heisenberg*. New York: W.H. Freeman, 1992.

Galison, Peter. *How Experiments End*. Chicago: University of Chicago Press, 1987.

Galison, Peter, and Bruce Hevly, eds. *Big Science: The Growth of Large-Scale Research*. Stanford, CA: Stanford University Press, 1992.

Gimbel, Steven. *Einstein's Jewish Science: Physics at the Intersection of Politics and Religion*. Baltimore, MD: Johns Hopkins University Press, 2012.

Hughes, Jeff. *The Manhattan Project: Big Science and the Atom Bomb*. Cambridge: Icon Books, 2002.

Kohler, Robert E. *Partners in Science: Foundations and Natural Scientists*. Chicago: University of Chicago Press, 1991.

Lindee, M. Susan. *Suffering Made Real: American Science and the Survivors at Hiroshima*. Chicago: University of Chicago Press, 1994.

Maddox, Robert James. *Weapons for Victory: The Hiroshima Decision Fifty Years Later*. Columbia, MO: University of Missouri Press, 1995.

Olby, Robert C. *The Path to the Double Helix: The Discovery of DNA*. Foreword by Francis Crick. New York: Dover Publications, 1994.

Sime, Ruth Lewin. *Lise Meitner: A Life in Physics*. Berkeley: University of California Press, 1996.

Watson, James D. *The Double Helix: A Personal Account of the Discovery of the Structure of DNA*. London: Weidenfeld and Nicolson, 1997.

ELEVEN *1957: THE YEAR THE WORLD BECAME A PLANET*

Anderson, Frank Walter. *Orders of Magnitude: A History of NACA and NASA, 1915–1980*. Washington, DC: National Aeronautics and Space Administration, Scientific and Technical Information Office, 1981.

Brzezinski, Matthew. *Red Moon Rising: Sputnik and the Hidden Rivalries That Ignited the Space Age*. New York: Times Books, 2007.

Dickson, Paul. *Sputnik: The Launch of the Space Race*. Toronto: Macfarlane, Walter and Ross, 2001.

Killian, James Rhyne. *Sputnik, Scientists, and Eisenhower: A Memoir of the First Special Assistant to the President for Science and Technology*. Cambridge, MA: MIT Press, 1977.

Marvin, Ursula B. *Continental Drift: The Evolution of a Concept*. Washington, DC: Smithsonian Institution Press, 1973.

NASA. *Space Flight: The First 30 Years*. Washington, DC: National Aeronautics and Space Administration, Office of Space Flight, 1991.

Oreskes, Naomi. *The Rejection of Continental Drift: Theory and Method in American Earth Science*. New York: Oxford University Press, 1999.

Stine, G. Harry. *ICBM: The Making of the Weapon That Changed the World*. New York: Orion Books, 1991.

Verschuur, Gerrit L. *The Invisible Universe Revealed: The Story of Radio Astronomy*. New York: Springer-Verlag, 1987.

Wang, Zuoyue. *In Sputnik's Shadow: The President's Science Advisory Committee and Cold War America*. New Brunswick, NJ: Rutgers University Press, 2008.

Wegener, Alfred. *The Origin of Continents and Oceans*. Trans. John Biram. London: Methuen, 1966.

Wilson, J. Tuzo. *IGY: The Year of the New Moons*. London: Michael Joseph, 1961.

TWELVE *MAN ON THE MOON, MICROWAVE IN THE KITCHEN*

Agar, Jon. *Turing and the Universal Machine: The Making of the Modern Computer*. Cambridge: Icon, 2001.

Aspray, William. *John von Neumann and the Origins of Modern Computing*. Cambridge, MA: MIT Press, 1990.

Campbell-Kelly, Martin, William Aspray, Nathan Ensmenger, and Jeffery R. Yost. *Computer: A History of the Information Machine*. 3rd ed. Boulder, CO: Westview Press, 2014.

Carson, Rachel. *Silent Spring*. Introduction by Al Gore. Boston: Houghton Mifflin, 1994.

Clancey, William J. *Working on Mars: Voyages of Scientific Discovery with the Mars Exploration Rovers*. Cambridge, MA: MIT Press, 2012.

Coles, Peter. *Cosmology: The Origin and Evolution of Cosmic Structure*. Chichester, UK: John Wiley, 2002.

Hodges, Andrew. *Turing: A Natural Philosopher*. London: Phoenix, 1997.

McElheny, Victor K. *Drawing the Map of Life: Inside the Human Genome Project*. New York: Basic Books, 2010.

McLaren, Angus. *A History of Contraception: From Antiquity to the Present Day*. Oxford: Basil Blackwell, 1990.

Riordan, Michael. *The Hunting of the Quark: A True Story of Modern Physics*. New York: Simon & Schuster, 1987.

Rossiter, Margaret. *Women Scientists in America: Before Affirmative Action, 1940–1972*. Baltimore, MD: Johns Hopkins University Press, 1995.

Rossiter, Margaret. *Women Scientists in America: Struggles and Strategies to 1940*. Baltimore, MD: Johns Hopkins University Press, 1982.

Sideris, Lisa H. *Rachel Carson: Legacy and Challenge*. Albany, NY: State University of New York Press, 2008.

Smith, Robert W. *The Space Telescope: A Study of NASA, Science, Technology and Politics*. 2nd ed. Cambridge: Cambridge University Press, 1993.

Stein, Dorothy. *Ada, A Life and a Legacy*. Cambridge, MA: MIT Press, 1985.

Swade, Doron. *The Difference Engine: Charles Babbage and the Quest to Build the First Computer*. New York: Viking, 2001.

THIRTEEN *SCIENCE AND NEW FRONTIERS: POTENTIAL AND PERIL IN THE NEW MILLENNIUM*

Aldersey-Williams, Hugh. *The Most Beautiful Molecule: The Discovery of the Buckyball*. New York: Wiley, 1995.

Baldi, Pierre. *The Shattered Self: The End of Natural Evolution*. Cambridge, MA: MIT Press, 2001.

Clarke, Arthur Charles. *Ascent to Orbit: A Scientific Autobiography: The Technical Writings of Arthur C. Clarke*. New York: Wiley, 1984.

Cowan, Ruth Schwartz. *Heredity and Hope: The Case for Genetic Screening*. Cambridge, MA: Harvard University Press, 2008.

Drexler, K. Eric. *Engines of Creation: The Coming Era of Nanotechnology*. London: Fourth Estate, 1996.

Mulhall, Douglas. *Our Molecular Future: How Nanotechnology, Robotics, Genetics, and Artificial Intelligence Will Transform Our World*. Amherst, NY: Prometheus Books, 2002.

Oreskes, Naomi, and Eric M. Conway. *Merchants of Doubt: How a Handful of Scientists Obscured the Truth on Issues from Tobacco Smoke to Global Warming*. New York: Bloomsbury Press, 2010.

Rifkin, Jeremy. *The Biotech Century: Harnessing the Gene and Remaking the World*. New York: Jeremy P. Tarcher/Putnam, 1998.

Taubes, Gary. *Bad Science: The Short Life and Weird Times of Cold Fusion*. New York: Random House, 1993.

GENERAL READINGS IN THE HISTORY OF SCIENCE

Asimov, Isaac. *The History of Physics*. New York: Walker, 1984.

Bowler, Peter J., and Iwan Rhys Morus. *Making Modern Science: A Historical Survey*. Chicago: University of Chicago Press, 2005.

Brock, William H. *The Norton History of Chemistry*. New York: Norton, 1992.

Bronowski, Jacob. *The Ascent of Man*. Boston: Little, Brown, 1973.

Bryson, Bill. *A Short History of Nearly Everything*. Toronto: Anchor Canada, 2004.

Diamond, Jared. *Guns, Germs, and Steel: The Fates of Human Societies*. New York: W.W. Norton, 2005.

Golinski, Jan. *Making Natural Knowledge: Constructivism and the History of Science*. Cambridge: Cambridge University Press, 1998.

North, John David. *The Norton History of Astronomy and Cosmology*. New York: Norton, 1995.

Olby, Robert C. *Fontana History of Biology*. New York: Fontana, 2002.

INDEX

Illustrations indicated by page numbers in italics